Ion Transport Through Membranes

Academic Press Rapid Manuscript Reproduction

Ion Transport Through Membranes

Edited by

Kunio Yagi

Institutc of Applied Biochemistry
Gifu, Japan

Bernard Pullman

Institute de Biologie Physico-Chimique
Paris, France

1987

ACADEMIC PRESS, INC.

Harcourt Brace Jovanovich, Publishers

Tokyo Orlando San Diego New York
Austin Boston London Sydney Toronto

ACADEMIC PRESS JAPAN, INC.
Hokoku Bldg. 3-11-13, Iidabashi, Chiyoda-ku, Tokyo 102

United States Edition published by
ACADEMIC PRESS, INC.
Orlando, Florida 32887

United Kingdom Edition published by
ACADEMIC PRESS, INC. (LONDON) LTD.
24/28 Oval Road, London NW1 7DX

Library of Congress Cataloging-in-Publication Data

Ion transport through membranes.

Based on an international symposium held Nov. 3-5, 1985
in Nagoya, Japan.
Includes index.
1. Biological transport, Active--Congresses.
2. Ion-permeable membranes--Congresses. I. Yagi, Kunio,
Date. II. Pullman, Bernard, Date. [DNLM:
1. Cell Membrane--metabolism--congresses. 2. Cell
membrane Permeability--congresses. 3. Ion Channels--
metabolism--congresses. QH 601 I6473 1985]
QH509.I66 1986 574.87'5 86-32233
ISBN 0-12-768052-7 (U.S.)

Printed in Japan

87 88 89 9 8 7 6 5 4 3 2 1

Contents

Preface

Concentration gradients of ions between cell compartments are signals which provoke successive reactions for cell functions such as energy transduction, muscle contraction, and nerve excitation. Such a gradient can only be brought about by ion transport through the biomembranes separating cell compartments. Accordingly, ion transport is an essential process for many normal physiological functions. Understanding the importance of this problem, we decided to organize an international symposium, which we entitled "Ion Transport through Membranes."

The symposium, sponsored by the Japanese Society of Biochemistry, International Union of Biochemistry, Suzuken Memorial Foundation, and Yagi Memorial Foundation, was held from November 3 to 5, 1985, at the Nagoya Kanko Hotel in Nagoya, Japan. Leading specialists in this field from around the world were invited.

On that occasion, each speaker was asked to present a mini-review, including his or her own findings, of a given area of this broad topic. These manuscripts were then edited to provide a monograph which surveys the most up-to-date knowledge on this exciting topic.

The editors wish to express thanks to the above-mentioned organizations for their financial support, to the authors for their cooperation, and to Madame Alberte Pullman for her help during the planning of the symposium.

Kunio Yagi
Bernard Pullman

DESIGN OF IONOPHORES FOR A CARRIER-INDUCED ION TRANSPORT THROUGH BULK MEMBRANES

W. Simon
D. Ammann
E. Pretsch
W. E. Morf
U. Oesch
M. Huser
P. Schulthess

Department of Organic Chemistry
Swiss Federal Institute of Technology
Zurich, Switzerland

I. INTRODUCTION

Ion carriers (a class of ionophores) are lipophilic complexing agents having the capability to bind ions reversibly and to transport them across organic membranes by carrier translocation (1,2). Ideally, selective ion carriers render a membrane permeable for one given sort of ion I only. If such ionophore-based membranes are used in cell assemblies of the type

External reference electrode	Sample solution (solution 1)	Membrane	Internal filling solution (solution 2)	Internal reference electrode

and if an electric potential gradient is applied between solutions 1 and 2, an exclusive transport of the ions I across the membrane should result, the transport number

This work was partly supported by the Swiss National Science Foundation, Orion Research Inc. and Corning Ltd.

being then $t_I = 1$. For the same membrane electrode cell the electric potential difference at zero current (e.m.f.) between the external and the internal reference electrodes depends then on the ratio of the activities of the ion I in the solutions 1 and 2. If a constant composition of solution 2 is used, the activities of the selected ion in solution 1 may therefore be measured potentiometrically according to the Nernst equation (3). In particular, neutral carriers, defined as ionophores that carry no charge when not complexed by the transported ion, have led to such cell assemblies (ion-selective electrodes) with a wide range of different selectivities (2). Figure 1 shows a selection of ion carriers which we designed in view of an analytically relevant application in bulk membranes or which we found to exhibit the required properties for such a use. Ionophores 1 - 19 in Fig.1 are neutral carriers for cations, 21 (4) is a charged carrier for anions and 20 (5,6) is a neutral carrier for anions. Some of them have found widespread application in cation sensors for clinical use (for a review see (2)). The ionophores 2 (7), 3 (7), 4 (8), 7 (9), 11 (10), 12 (10), 14 (11), 15 (12) and 18 (13) have been described more recently. Among the different bulk membranes, solvent polymeric membranes based on poly(vinyl chloride) (14) have been studied in especially great detail. Membranes with ∿32 wt.-% poly(vinyl chloride), ∿65 wt.-% plasticizer and ∿1 - 3 wt.-% ionophore exhibit diffusion coefficients of free carriers and carrier/ion complexes of about 10^{-7} to 10^{-8} cm^2 sec^{-1} (15) comparable to values for carriers in black lipid membranes (16).

II. REQUIREMENTS FOR THE DESIGN OF IONOPHORES

Ionophores for an analytically relevant application in solvent polymeric membranes have to meet at least the following three requirements (see also (17)):

A. Ion-Permeability Selectivity

The ionophore has to induce ion-permeability selectivity in membranes. As the ion selectivity of a membrane is related to the free energies of transfer of the ions from the aqueous phase (sample) to the membrane phase, the selectivity of neutral carrier-based membranes obviously depends on various factors. Such factors are mainly (2): (a) the selectivity behaviour of the carrier used, which can be completely characterized by complex stability constants, (b) the extraction properties of the membrane solvent (plasticizer), (c) the

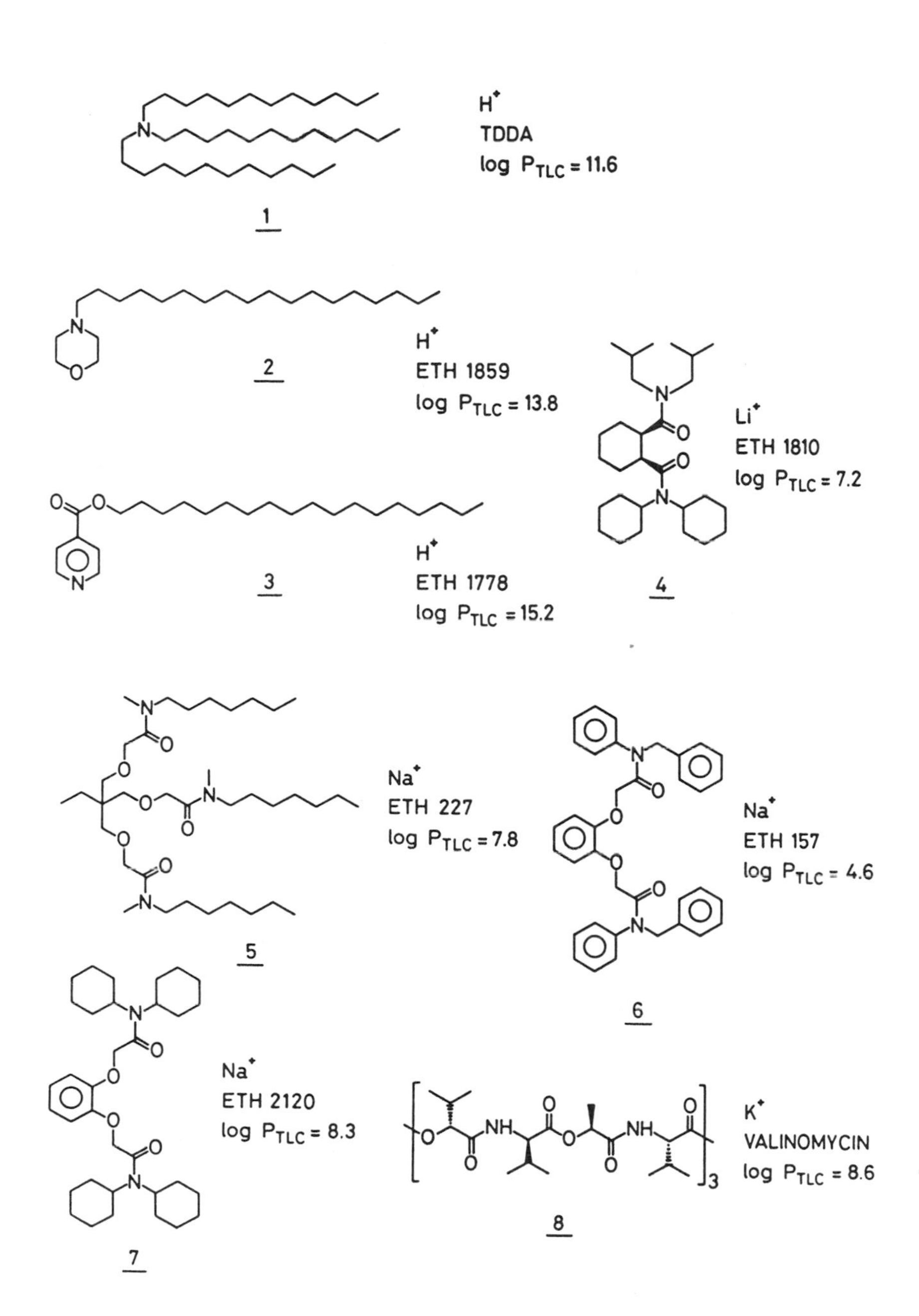

FIGURE 1. Constitutions of the ionophores discussed.

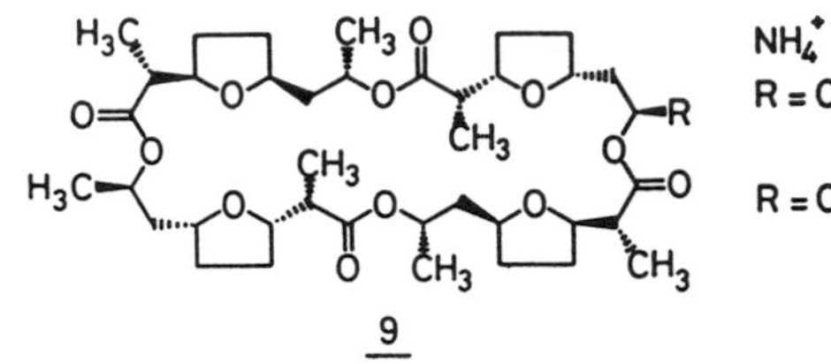
NH4+
R = CH3 : NONACTIN
log PTLC = 5.8
R = C2H5 : MONACTIN
log PTLC = 6.5
9

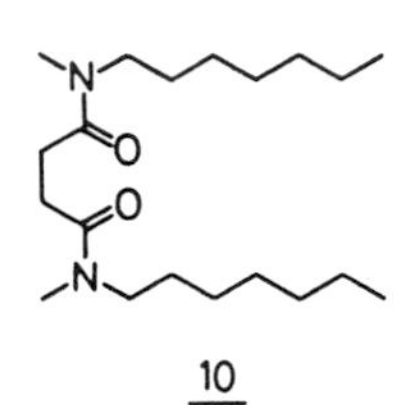
10
Mg2+
ETH 1117
log PTLC = 5.8

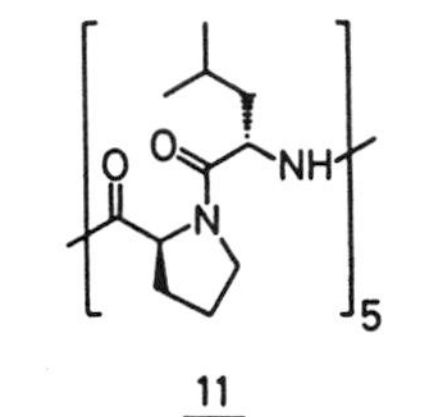
11
Mg2+
CYCLO(L-PRO-L-LEU)5
log PTLC = 3.6

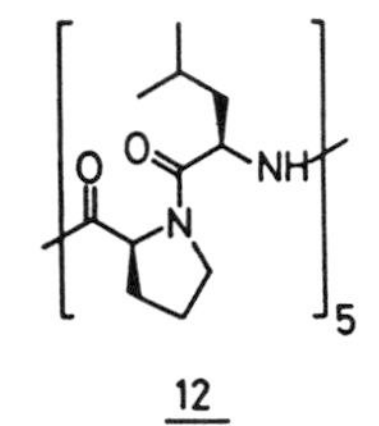
12
Mg2+
CYCLO (L-PRO-D-LEU)5

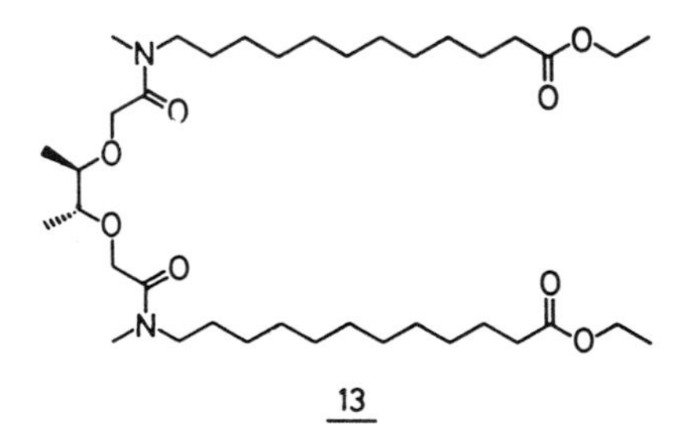
13
Ca2+
ETH 1001
log PTLC = 7.5

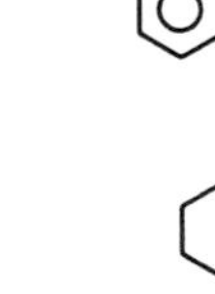
14
Ca2+
ETH 129
log PTLC = 7.2

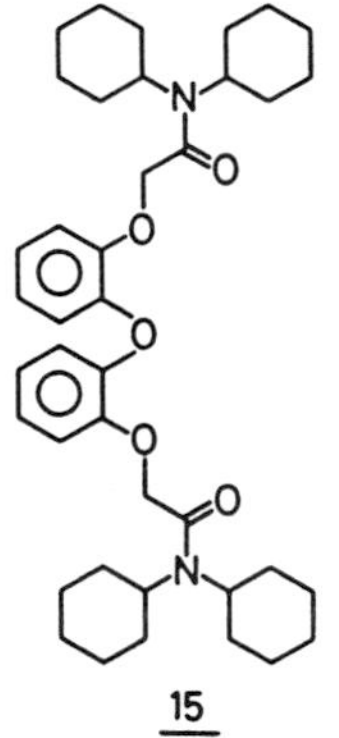
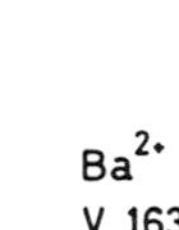
15
Ba2+
V 163
log PTLC = 8.2

Figure 1. (Continued).

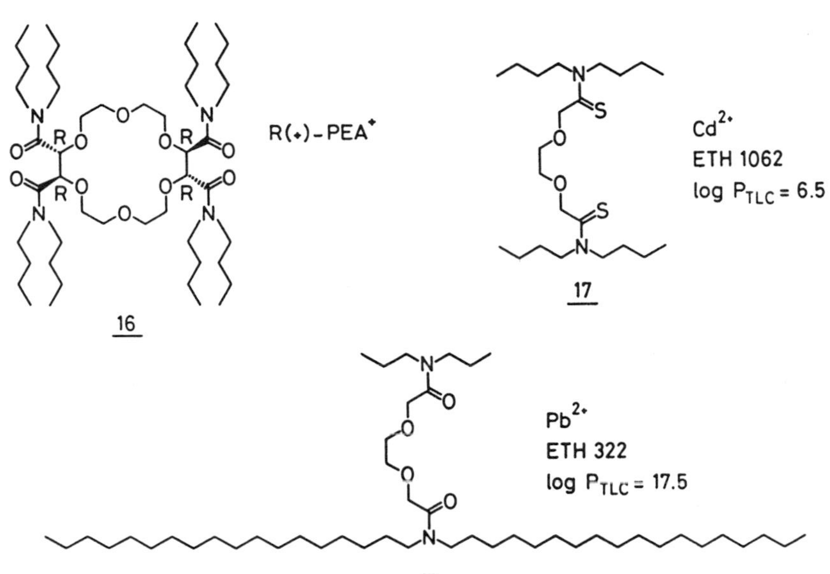

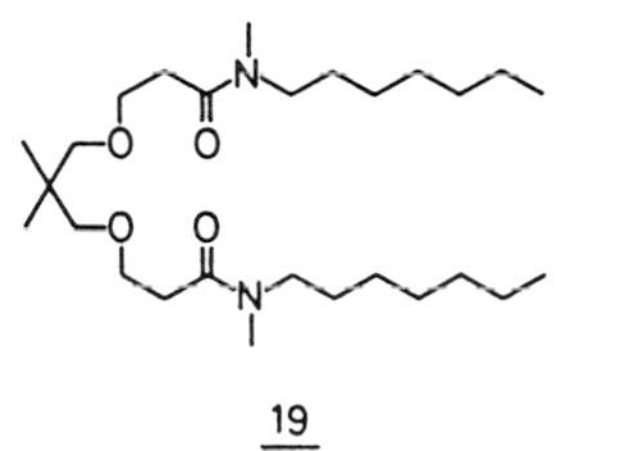

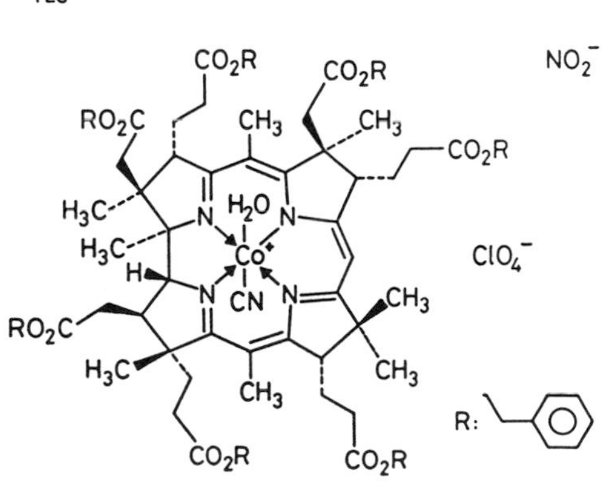

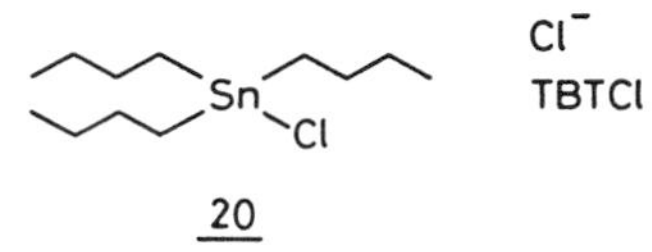

FIGURE 1. (continued).

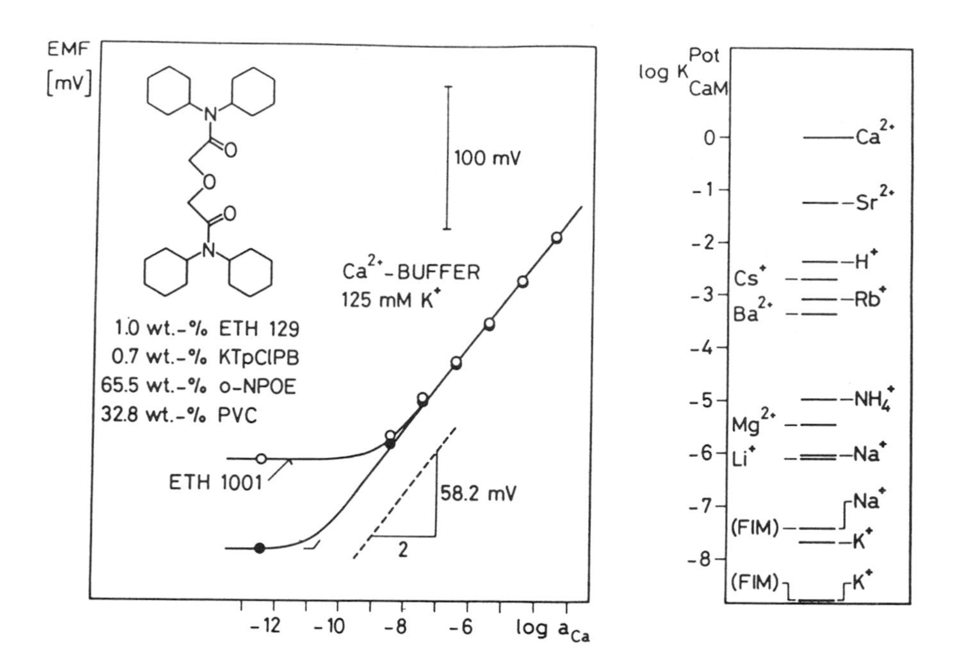

FIGURE 2. Electrode function (left, dots) and potentiometric selectivities (right) of a PVC membrane containing ionophore 14 (ETH 129). The electrode function is determined in calcium-buffered solutions containing a constant background of 125 mM K^+. For comparison the electrode function of a membrane containing ionophore 13 (ETH 1001) is given (circles). The selectivity coefficients ($\log K_{CaM}^{Pot}$) are determined by the separate solution method (SSM) (20). For Na^+ and K^+ they are also given as determined in calcium-buffered solutions by the fixed interference method (FIM) (20).

concentration of the free ligand in the membrane phase and (d) the concentration of ionic sites in the membrane. The effects of (b), (c) and (d) can be described to a large extent by model calculations and may be controlled by adequate membrane technology (2,18,19). Figure 2 corroborates that neutral carriers can induce outstanding selectivities in membranes if adequate membrane technology is applied (11). Reciprocal selectivities $1/K_{CaK}^{Pot}$ (preference of Ca^{2+} over K^+ by the membrane as measured potentiometrically (20)) of up to 10^9 have been obtained (11).

As expected theoretically, the selectivity of complex formation of charged ion carriers in membranes containing predominantly neutral associates (ion / charged ionophore complexes) is only partially exhibited in potentiometric

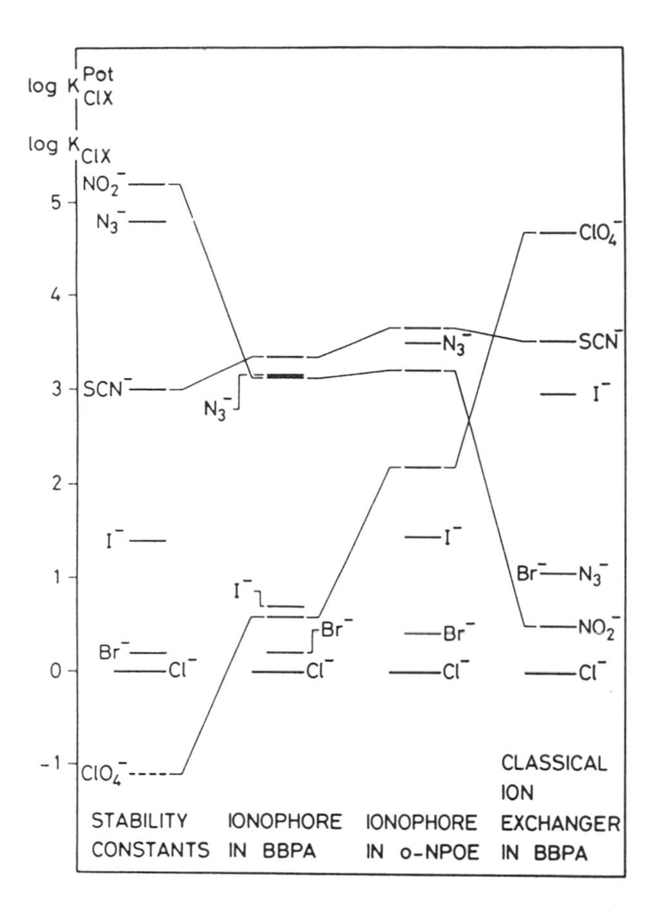

FIGURE 3. Comparison of the relative stability constants ($\log K_{ClX}$) of anion complexes of vitamin B_{12} (column 1) with potentiometric selectivities ($\log K^{Pot}_{ClX}$) of PVC membranes with an analog of ionophore <u>21</u> (a,b,d,e,f,g-hexamethyl-c-octadecyl Co-aquo-Co-cyanocobyrinate perchlorate) (columns 2 and 3) and of a PVC membrane based on a classical anion exchanger (BBPA: bis(1-butylpentyl) adipate, o-NPOE: o-nitrophenyl octyl ether).

studies on cell assemblies of the type described above (19). Thus the potentiometric selectivity is found to lie within the limits given by the ratio of the complex formation constants and the selectivity of liquid membranes based on dissociated ion exchangers (21,22). An example is given in Fig. 3 for an anion carrier of the type <u>21</u> (for details see (23)).

The design of ionophores with a given selectivity is more problematic (a). Although an overwhelming activity in designing host molecules (ionophores) for selected guest species (ions) is in evidence (24-31), only modest use has been made of model calculations describing the host-guest interactions (25,32-34).

Classical electrostatic models are useful for calculating the ion-molecule interactions near the energy minima for Group IA/IIA cations (35). However, they require a knowledge of molecular parameters normally not available. It has been shown (35) that semi-empirical quantum chemical treatments of ion-ligand interactions often lead to unrealistic results. In contrast, ab initio computations give reliable results even if small but well balanced basis sets are chosen (36-39). The application to realistic ionophores is usually prohibited by the extent of the computation even if small basis sets are used. Through ab initio calculations on model complexes and a representation of the interaction energies by pair potentials (40-44), such large systems are easily made amenable to analysis. Another approach for the description of large systems on the basis of ab initio calculations on small systems is described in (39). For a discussion of the stability of the complexes the sum of the interaction energies and the conformational energy relative to the most stable conformation of the free ligand is relevant. Unfortunately, the computation of such conformational energy changes is still too uncertain (45) (see also (46)). Recently a model based on ab initio calculations has been proposed for the evaluation of conformational energies (47).

B. Lipophilicity of Ionophores

The ionophore must exhibit a high lipophilicity to be confined to the membrane phase for an appropriate period of time. To guarantee a continuous use lifetime of at least one month of a typical solvent polymeric membrane sensor in contact with whole blood or undiluted blood serum, a lipophilicity P (partition coefficient between water and octane-1-ol) of the carrier larger than $10^{8.4}$ is necessary (49). Incorporating adequate structural elements (e.g. alkyl groups) into ionophores, this required lipophilicity may easily be obtained. A reliable estimate for P may be obtained by thin-layer chromatography (17). Such estimated values log P_{TLC} are given in Figure 1 for the neutral ionophores studied. Obviously, most of the ionophores have been designed to exhibit a sufficient lipophilicity for such a desired membrane lifetime.

C. Ligand Exchange Kinetics

The ionophores should form relatively stable complexes with the primary ion I (high selectivity) but, on the other hand, the exchange reaction of the selected ions at the membrane / solution interface must be sufficiently reversible. Therefore, the free energy of activation of the ligand exchange reaction

$$IS + S' \rightleftharpoons IS' + S$$

where S and S' are ionophores, has to be relatively low.

For Zn^{2+} or Cd^{2+} complexes of ligand 17 (Fig. 1), free energies of activation of the ligand exchange reaction of < 45 kJ mol^{-1} (in acetonitrile) have been measured (17,50). Cation permselectivity is indeed observed with $CdCl_2$ in the sample solution (17). As $CdCl^+$ is probably the permeating species, a slope of the electrode response of approximately 60 mV (25°C) is obtained (17). However, in systems with a free energy of activation of the ligand exchange reaction of > 65 kJ mol^{-1} (in acetonitrile) the cationic complexes of the ionophore act as anion exchangers (e.g., complexes with Pt^{2+} or Pd^{2+} (50)). An electrode containing the $PdCl_2$ complex of ligand 17 in the membrane phase therefore responds to the chloride anions in a sample solution of $CdCl_2$ (17). Theoretically related is the requirement that a sufficiently high and constant concentration of ionophore should be present in the membrane phase in the unloaded form (19). If a cation ionophore is predominantly present within the membrane phase in the loaded form, anion-permselectivity is induced (see Fig. 4) (19,48). These findings are in agreement with the experimental evidence that the transport rate of ionophores passes a maximum when increasing the stability constant of the ionophore / ion interaction (51).

In order to keep the free energy of activation of the ligand exchange reaction sufficiently small, the design of ionophores has been focused on non-macrocyclic structures (see Fig. 1).

Using model calculations (section A), CPK model building and adequate membrane technology, it has been possible to design neutral carrier-based membranes that show analytically relevant ion selectivities for a wide range of ions (Fig. 1). Although the synthetic ionophores shown in Fig. 1 are predominantly non-macrocyclic (see section C), there are several reports on the successful application of macrocyclic ionophores (e.g., crown compounds) in ion-selective electrodes (52-54) (for a review, see (2)).

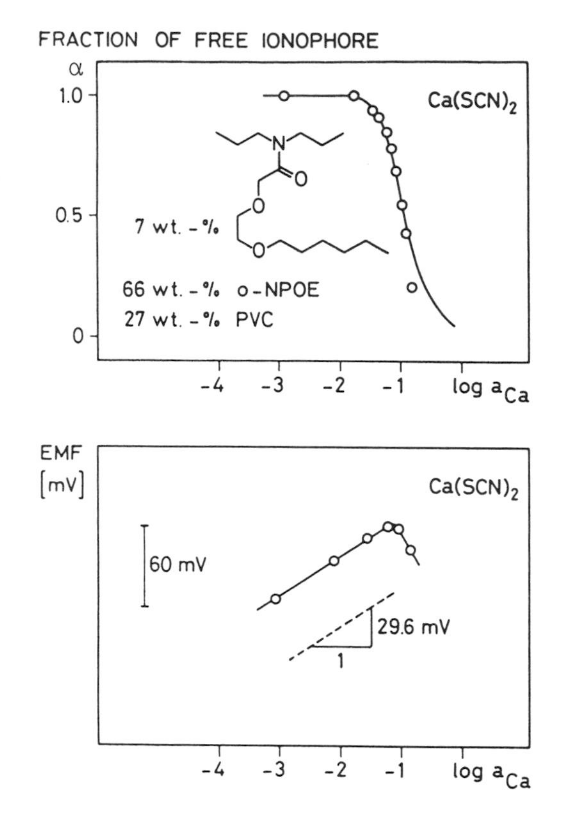

FIGURE 4. Extraction behaviour of a PVC membrane containing a calcium ionophore as determined by ^{13}C-NMR (top) and electrode function of a membrane of the same composition (bottom) in contact with calcium thiocyanate solutions (19,48).

III. ION TRANSPORT

In a series of earlier contributions (31,55-61) we have shown that there exists a close relationship between the potentiometric ion selectivity of liquid membrane electrodes and the selectivity exhibited by the same membranes in ion transport experiments. Two different types of transport studies can be performed. In electrodialysis experiments (56-59) a voltage is applied across the bulk membrane and the transport number of the permeating ion is determined (i.e. the fraction of electric current carried by this

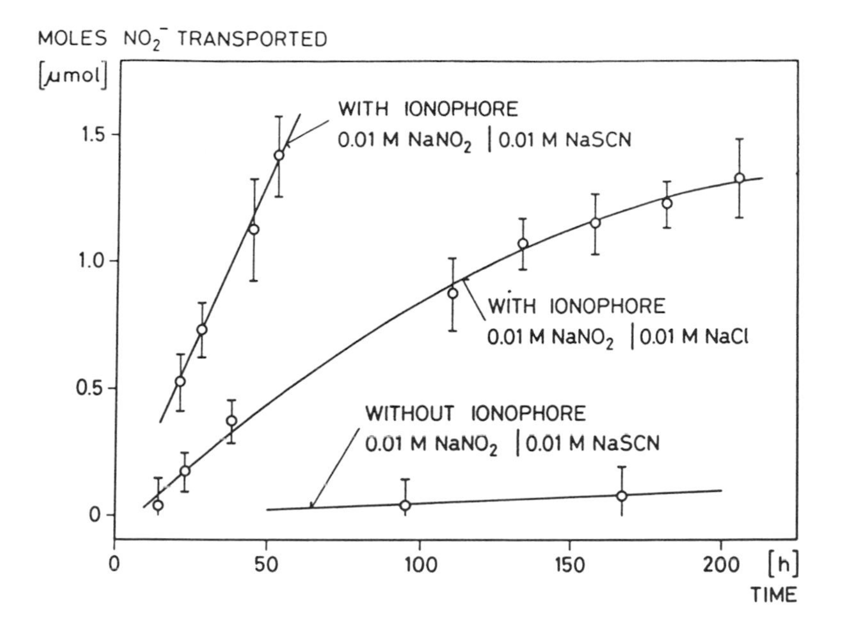

FIGURE 5. Zero-current transport of NO_2^- through bulk membranes containing the NO_2^--selective ionophore <u>21</u> ($d \approx 40$ µm, $A = 12.6$ cm^2, composition: 33.2 wt.-% PVC, 65.8 wt.-% bis(1-butylpentyl)decane-1,10-diyl diglutarate (ETH 469), 1.0 % ionophore). The driving forces are anion concentration gradients as indicated between solution 1 (NO_2^-) and solution 2 (Cl^- or SCN^-). The transport of NO_2^- into solution 2 is measured by differential pulse polarography.

species). On the other hand, in zero-current countertransport systems (60-63) a selective transport of cations across the membrane is induced and compensated by an oppositely directed flow of a second sort of cations. The striking ion selectivity of neutral carrier-based membranes can be understood as a permeability selectivity. Thus, the same membranes as used in potentiometric sensors for specific ions should be able to pump these ions selectively across when a transmembrane potential is applied. In the case of ideally specific systems, the electric current should be carried exclusively by one sort of ions. This corresponds to a transport number of $t_I = 1$ for the permeating species. Such ideal selectivity in ion transport is generally observed for neutral carrier membranes (see (64)).

Cation transport under zero-current conditions was first realized for liquid membranes with negatively charged ligands (60,62,63). The primary energy source for such transport of Na^{+} (62,63) or Mg^{2+} ions (60) came from the simultaneous counterflow of hydrogen ions being driven by a pH gradient. Application of the same ion-pumping principle to neutral carrier-based membranes was problematic, however, since neutral ligands - in contrast to the negatively charged counterparts - hardly interact with protons. To overcome this, a proton carrier (a lipophilic weak acid) was incorporated into the membrane, in addition to the neutral ionophore (61). This finally allowed us to couple the net transport of specific cations to the facilitated diffusion of hydrogen ions. A detailed description of this countertransport system as well as a theoretical treatment are given elsewhere (61).

The electrically charged ionophore 21 exhibits surprisingly high selectivity for anions in potentiometric studies on solvent polymeric membranes (see (4) and Fig. 3). A zero-current anion transport experiment indeed corroborates a high selectivity of the ionophore (Fig. 5).

REFERENCES

1. Burgermeister, W., and Winkler-Oswatitsch, R., Topics in Current Chemistry 69:91 (1979).
2. Ammann, D., Morf, W.E., Anker, P., Meier, P.C., Pretsch, E., and Simon, W., Ion Select. Electrode Rev. 5:3 (1983).
3. Nernst, W., Z. Phys. Chem. 2:613 (1889); 4:129 (1889).
4. Stepánek, R., Kräutler, B., Schulthess, P., Lindemann, B., Ammann, D., and Simon, W., Anal. Chim. Acta, in press.
5. Wuthier, U., Pham, H.V., Pretsch, E., Ammann, D., Beck, A.K., Seebach, D., and Simon, W., Helv. Chim. Acta, in press.
6. Wuthier, U., Pham, H.V., Zünd, R., Welti, D., Funck, R.J.J., Bezegh, A., Ammann, D., Pretsch, E., and Simon, W., Anal. Chem. 56:535 (1984).
7. Oesch, U., Brzózka, Z., and Simon, W., in preparation.
8. Metzger, E., Ammann, D., Schefer, U., Pretsch, E., and Simon, W., Chimia 38:440 (1984).
9. Maruizumi, T., Wegmann, D., Ammann, D., and Simon, W., in preparation.
10. Behm, F., Ammann, D., Simon, W., Brunfeldt, K., and Halstrøm, J., Helv. Chim. Acta 68:110 (1985).
11. Schefer, U., Ammann, D., Pretsch, E., and Simon, W., in preparation.
12. Kleiner, T., Bongardt, F., Vögtle, F., Läubli, M.W., Dinten, O., and Simon, W., Chem. Ber. 118:1071 (1985).

13. Lindner, E., Tóth, K., Pungor, E., Behm, F., Oggenfuss, P., Welti, D.H., Ammann, D., Morf, W.E., Pretsch, E., and Simon, W., Anal. Chem. 56:1127 (1984).
14. Moody, G.J., Oke, R.B., and Thomas, J.D.R., Analyst 95:910 (1970).
15. Oesch, U., and Simon, W., Anal. Chem. 52:692 (1980).
16. Läuger, P., and Stark, G., Biochim. Biophys. Acta 211:458 (1970).
17. Simon, W., Pretsch, E., Morf, W.E., Ammann, D., Oesch, U., and Dinten, O., Analyst 109:207 (1984).
18. Meier, P.C., Morf, W.E., Läubli, M., and Simon, W., Anal. Chim. Acta 156:1 (1984).
19. Morf, W.E., and Simon, W., in "Ion Selective Electrodes in Analytical Chemistry" (H. Freiser, ed.), p. 211. Plenum Press, New York, London, Washington, Boston, 1978.
20. Guilbault, G.G., Durst, R.A., Frant, M.S., Freiser, H., Hansen, E.H., Light, T.S., Pungor, E., Rechnitz, G., Rice, N.M., Rohm, T.J., Simon, W., and Thomas, J.D.R., Pure Appl. Chem. 48:127 (1976).
21. Sandblom, J., Eisenman, G., and Walker, J.L., J. Phys. Chem. 71:3862 (1967).
22. Sandblom, J., and Orme, F., in "Membranes" (G. Eisenman, ed.), vol. 1. Marcel Dekker, New York, 1972.
23. Schulthess, P., Diss. ETH Zürich, 1986.
24. Cram, D.J., and Cram, J.M., Science 83:803 (1974).
25. Lehn, J.-M., Struct. Bonding (Berlin) 16:1 (1973).
26. Weber, E., and Vögtle, F., Chem. Ber. 109:1803 (1976).
27. Pedersen, C.J., J. Am. Chem. Soc. 89:2495 (1967); 89:7017 (1967); 92:386 (1970); 92:391 (1970).
28. Prelog, V., Pure Appl. Chem. 50:893 (1978).
29. Stoddard, J.F., Chem. Soc. Rev. 8:85 (1979).
30. Poonia, N.S., and Bajaj, A.V., Chem. Rev. 79:389 (1979).
31. Morf, W.E., Ammann, D., Bissig, R., Pretsch, E., and Simon, W., in "Progress in Macrocyclic Chemistry" (R.M. Izatt, and J.J. Christensen, eds.), vol. 1, p. 1. Wiley-Interscience, New York, 1979.
32. Morf, W.E., and Simon, W., Helv. Chim. Acta 54:2683 (1971).
33. Simon, W., Morf, W.E., and Meier, P.C., Struct. Bonding (Berlin) 16:113 (1973).
34. Lifson, S., Felder, C.E., and Schanzer, A., J. Am. Chem. Soc. 105:3866 (1983).
35. Schuster, P., Jakubetz, W., and Marius, W., Topics in Current Chemistry 60:1 (1975).
36. Pullman, A., Berthod, H., and Gresh, N., Int. J. Quantum Chem. Symp. 10:59 (1976).
37. Kolos, W., Theor. Chim. Acta 54:187 (1980).
38. Gianolio, L., and Clementi, E., Gazz. Chim. Ital. 110:179 (1980).
39. Gresh, N., and Pullman, A., Int. J. Quantum Chem. 22:709

(1982).
40. Clementi, E., "Lecture Notes in Chemistry", vol. 2. Springer-Verlag, Berlin, 1976; vol. 19. Springer-Verlag, Berlin, 1980.
41. Corongiu, G., Clementi, E., Pretsch, E., and Simon, W., J. Chem. Phys. 70:1266 (1979).
42. Corongiu, G., Clementi, E., Pretsch, E., and Simon, W., J. Chem. Phys. 72:3096 (1980).
43. Pretsch, E., Bendl, J., Portmann, P., and Welti, M., in "Proceedings of the Symposium on Steric Effects in Biomolecules" (G. Naray-Szabo, ed.), p. 85. Akadémiai Kiadó, Budapest, 1982.
44. Pretsch, E., Gratzl, M., Pungor, E., and Simon, W., in "Proceedings of the 3rd Symposium on Ion-Selective Electrodes" (E. Pungor, and I. Buzás, eds.), p. 315. Akadémiai Kiadó, Budapest, 1981.
45. Bendl, J., and Pretsch, E., J. Comput. Chem. 3:580 (1982).
46. Badertscher, M., Welti, M., Portmann, P., and Pretsch, E., Topics in Current Chemistry, in press.
47. Gresh, N., Claverie, P., and Pullman, A., Theor. Chim. Acta 66:1 (1984).
48. Büchi, R., Pretsch, E., Morf, W.E., and Simon, W., Helv. Chim. Acta 59:2047 (1976).
49. Oesch, U., Anker, P., Ammann, D., and Simon, W., in "Ion-Selective Electrodes" (E. Pungor, and I. Buzás, eds.), p. 81. Akadémiai Kiadó, Budapest, 1985.
50. Hofstetter, P., Pretsch, E., and Simon, W., Helv. Chim. Acta 66:2103 (1983).
51. Behr, J.-P., Kirch, M., and Lehn, J.-M., J. Am. Chem. Soc. 107:241 (1985).
52. Kimura, K., and Shono, T., in "Ion-Selective Electrodes" (E. Pungor, and I. Buzás, eds.), p. 155. Akadémiai Kiadó, Budapest, 1985.
53. Toner, J.L., Daniel, D.S., and Geer, S.M., United States Patent 4 476 007, 1984.
54. Lindner, E., Tóth, K., Horváth, M., Pungor, E., Ágai, B., Bitter, I., Töke, L., and Hell, Z., Fresenius Z. Anal. Chem. 322:157 (1985).
55. Morf, W.E., "The Principles of Ion-Selective Electrodes and of Membrane Transport", Akadémiai Kiadó, Budapest; Elsevier, Amsterdam, New York, 1981.
56. Schneider, J.K., Hofstetter, P., Pretsch, E., Ammann, D., and Simon, W., Helv. Chim. Acta 63:217 (1980).
57. Morf, W.E., Wuhrmann, P., and Simon, W., Anal. Chem. 48:1031 (1976).
58. Morf, W.E., and Simon, W., in "Ion-Selective Electrodes" (E. Pungor, and I. Buzás, eds.), p. 25. Akadémiai Kiadó, Budapest, 1977.
59. Thoma, A.P., Viviani-Nauer, A., Arvanitis, S., Morf, W.E.,

and Simon, W., Anal. Chem. 49:1567 (1977).
60. Erne, D., Morf, W.E., Arvanitis, S., Cimerman, Z., Ammann, D., and Simon, W., Helv. Chim. Acta 62:994 (1979).
61. Morf, W.E., Arvanitis, S., and Simon, W., Chimia 33:452 (1979).
62. Cussler, E.L., AIChE J. 17:1300 (1971); Cussler, E.L., Evans, D.F., and Matesich, M.A., Science 172:377 (1971).
63. Choy, E.M., Evans, D.F., and Cussler, E.L., J. Am. Chem. Soc. 96:7085 (1974).
64. Oggenfuss, P., Morf, W.E., Funck, R.J., Pham, H.V., Zünd, R.E., Pretsch, E., and Simon, W., in "Ion-Selective Electrodes" (E. Pungor, and I. Buzás, eds.), p. 73. Akadémiai Kiadó, Budapest, 1981.

STRUCTURAL, KINETIC AND MECHANISTIC ASPECTS OF CARRIER MEDIATED TRANSMEMBRANE ION TRANSPORT

Kalpathy Easwaran

Molecular Biophysics Unit
Indian Institute of Science
Bangalore, India

I. INTRODUCTION

The unique property of exhibiting selective permeabilities to molecules and ions by biological membranes is of vital importance as it regulates and drives several cellular processes. Although some of the biochemical nutrients can permeate unaided, transport of cations is a process mediated by specific membrane associated proteins (1,2). Most of our present knowledge about the mechanisms of ion transport across membranes has been due to the discovery of macrocyclic and linear antibiotics (ionophores) isolated from microorganisms (3-5) which selctively enhance the cation permeability across natural (6-8) and model membranes (9-13).

The evidence for the postulation of several mechanistic pathways for ion transport originated from electrical measurements of the bilayer lipid membranes (BLMs) incorporated with ionophores and studies on other model systems (14-16). These studies have led to the concept of ion movement across membranes by two distinct mechanisms, namely, a diffusive carrier and a channel mechanism (Fig.1). In a diffusive carrier mechanism, the ionophore diffuses through the membrane interior as an ionophore-cation complex and releases the cation at the other interface. On the other hand, in a channel mechanism the molecule spans the entire length of the membrane normally as a hollow cylinder through which the cations pass from one side of the membrane to the other. A relay carrier mechanism in which the cation is handed over

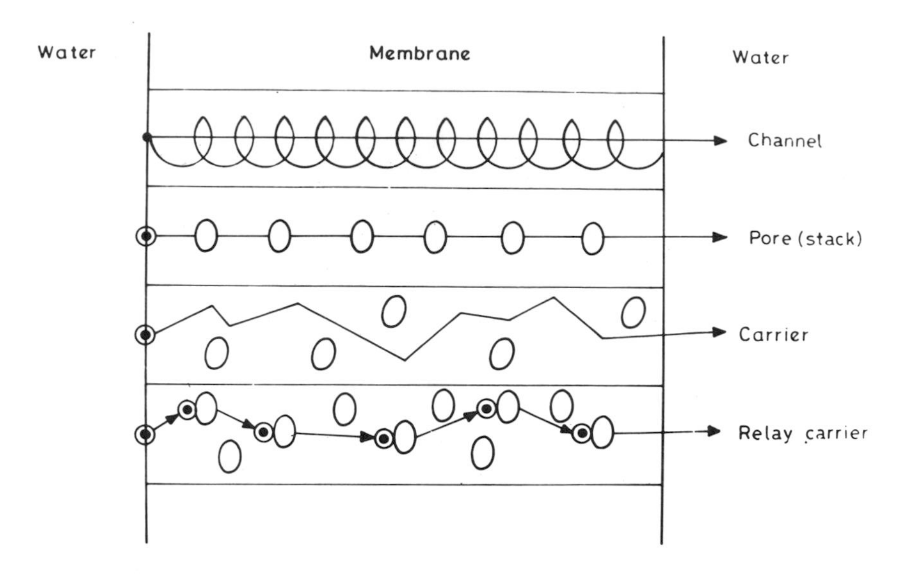

FIGURE 1. Mechanisms of cation transport.

from one ionophore to the other until it reaches the other side has also been proposed (17) although evidence for its operation is not yet available.

The process of ion-transport across membranes at the molecular level need a full understanding of a) the structure and conformation of the various ionophores and their cation complexes under varied conditions, for example, by complexing them with cations of different size and charge and using different solvents and anions, b) ionophore-cation interactions, c) dynamics of cation transport across model and biological membranes.

II. STRUCTURE AND CONFORMATION OF CARRIER IONOPHORES AND THEIR CATION COMPLEXES

During the last decade there has been considerable interest in the structure and conformation of various naturally occurring ionophores and their synthetic analogs as also of synthetic ion binding cyclic peptides (5,17,18-24). The studies in solution have been essentially carried out using a variety of physico-chemical techniques, in particular, nuclear magnetic resonance (NMR) and circular dichroism (CD). Where possible, the solid state conformation has been studied using X-ray crystallography. The studies have been aimed at understanding the molecular basis of carrier mediated transmembrane ion transport. In this article, some of our results obtained on few carrier ionophores and their synthetic analogs are discussed. Relevent literature data of work by other groups are also included. The results discussed clearly show that the ion selectivity of the carrier ionophores depend not only on the nature of the ligands but also on the conformational state of the whole molecule and that the conformation is very much dependent on the size and charge of the cation and the solvent and anion used.

A. Valinomycin-Cation Complexes

Among the carrier ionophores, valinomycin (VM) (Fig.2) has emerged as an extraordinary molecule due to its high specificity and selectivity for K^+ over Na^+ (by a factor of 10^4) in transport across natural (6-8) and artificial (9-13) membranes. The structure and conformation of free VM and its K^+ complex in both solution and solid state have been worked

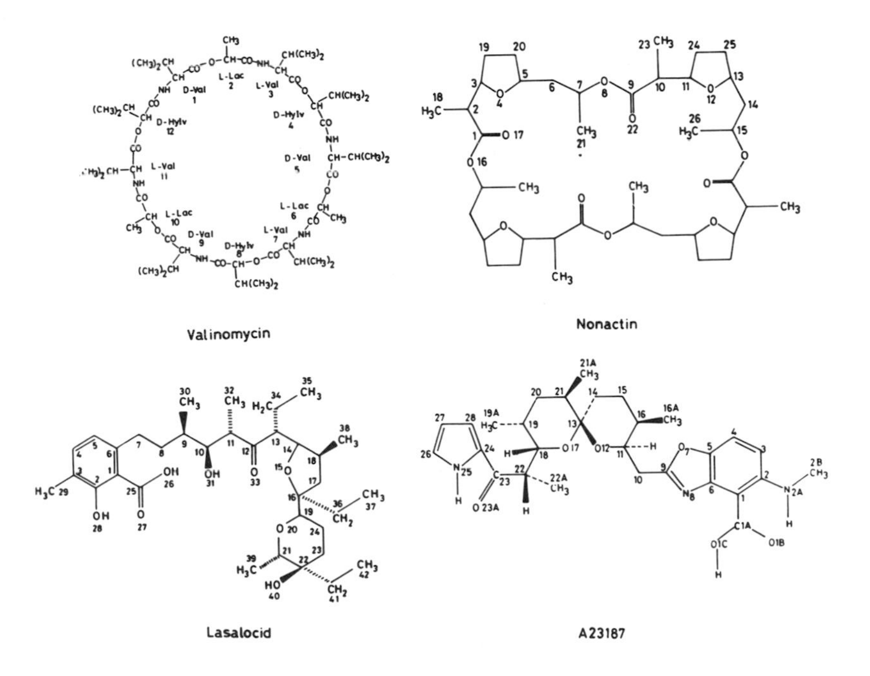

FIGURE 2. Chemical structures of ionophores.

at length (5,25-32). It has been shown that VM adopts different conformations depending on the polarity of the solvent and that the conformation of the VM-K^+ complex is the same as that observed for the free valinomycin in non-polar solvents (bracelet structure) with six ester carbonyls forming ligands to the K^+ ion which is situated at the center of the cavity (5) (Figs. 3 and 4) . However, the solid state structure of free VM is very different from that observed in solution (33). The molecular mechanism for the increase in K^+ transport by valinomycin has been suggested as due to the diffusion of the VM-K^+ complex across the membrane interior after the molecule captures the K^+ ion at one interface and release of the K^+ ion at the other interface. While the complexing ability of VM to K^+ and the transport characteristics of VM-K^+ have been studied at length and understood in some detail, the studies involving other monovalent and divalent cations are only a few. It has been reported that valinomycin binds to divalent cations in methanol forming a 1:1 type complex (34).

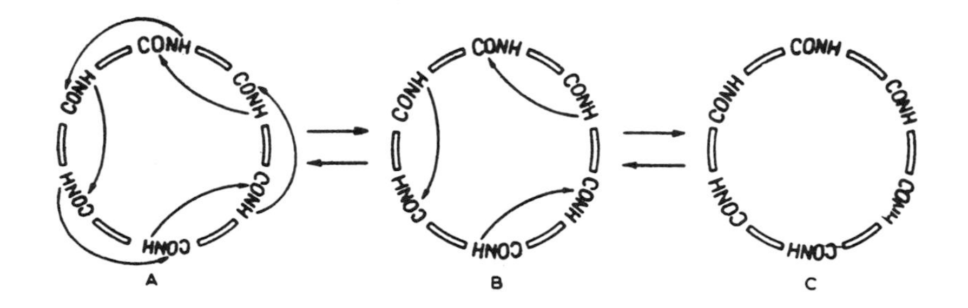

FIGURE 3. A, B and C conformations of valinomycin in different solvents (taken from ref. 5).

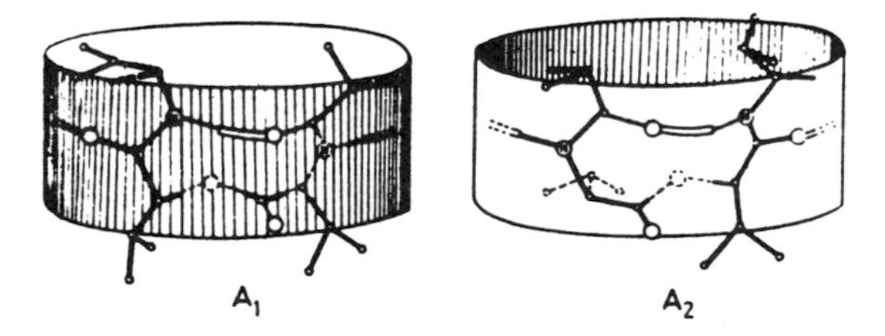

FIGURE 4. Schematic drawing of the bracelet structure of valinomycin (taken from ref. 5).

The studies carried out with other monovalent cations, Rb^+, Cs^+, Na^+, and Li^+ indicated that while K^+, Rb^+ and Cs^+ ions complexes were similar, Na^+ and Li^+ complexes were very different from the well known VM-K^+ conformation. Li^+ and Na^+ ion complex showed a different CD spectra as compared to K^+, Rb^+, and Cs^+ complex (Fig.3 of ref.24). The Na^+ complex of VM differed from the VM-K^+ complex in terms of the observed anion and solvent dependence and the nature of complexation (34,35). It has been shown that Na^+ forms a weaker complex with reduced number of co-ordinating ester carbonyl oxygens and intramolecular hydrogen bonds. Detailed NMR investigations of VM-Li^+ complex (36) showed that Li^+ forms an equimolar complex with the cation complexed to three ester carbonyls on the D-val-L-lac side of the bracelet. It is interesting to note that the single crystal X-ray structure of VM-Na^+ picrate complex showed a similar 1:1 type conformation (different from the VM-K^+ complex) with three ester carbonyls co-ordinating to the cation on the D-val-L-lac side (37). The structure also revealed a water molecule in the center of the cavity of the bracelet occupied by K^+ in the VM-K^+ complex.

The studies carried out on the VM-divalent cations (Ca^{2+}, Ba^{2+}, Mg^{2+} and Sr^{2+}) in acetonitrile showed that the conformation of the VM-divalent complexes were quite different from those of VM-monovalent complexes (36,38-43). The conformation of Ba^{2+} and Sr^{2+} complex were similar and that of VM-Ca^{2+} system similar to the Mg^{2+} complex. The detailed CD and NMR studies carried out on VM-Ba^{2+} system gave a novel conformation for the VM in its barium perchlorate complex (39-42). The CD data on the titration of barium perchlorate with VM in acetonitrile indicated that VM forms complexes with Ba^{2+} ions with stoichiometries of 2:1, 1:1 and 1:2 (VM:barium) , although the conformation stabilized only at high salt concentrations (41). ^1H NMR titration analyisis of the alpha CH and NH proton chemical shifts (Fig. 4 in ref. 41) indicated a stable 1:2 (VM:barium) complex. The final chemical shift positions of the alpha CH and NH protons of the barium complex (Table 5 of ref.24) clearly showed that the conformtaion of the VM-Ba^{2+} complex is very different from that of free VM and VM-K^+ complex. The molecular model consistent with all the available NMR data for the VM-Ba^{2+} complex, in particular, the J_{NH} coupling constant (Table 6 in ref.24) which led to allowed φ angles, the results indicating the absence of any intramolecular hydrogen bonds (24) and the observation that amide carbonyls are involved in metal binding (Fig.5) gave a novel conformation for the VM-barium perchlorate complex in acetonitrile. The conformation can be described as an extended depsipeptide chain with no internal

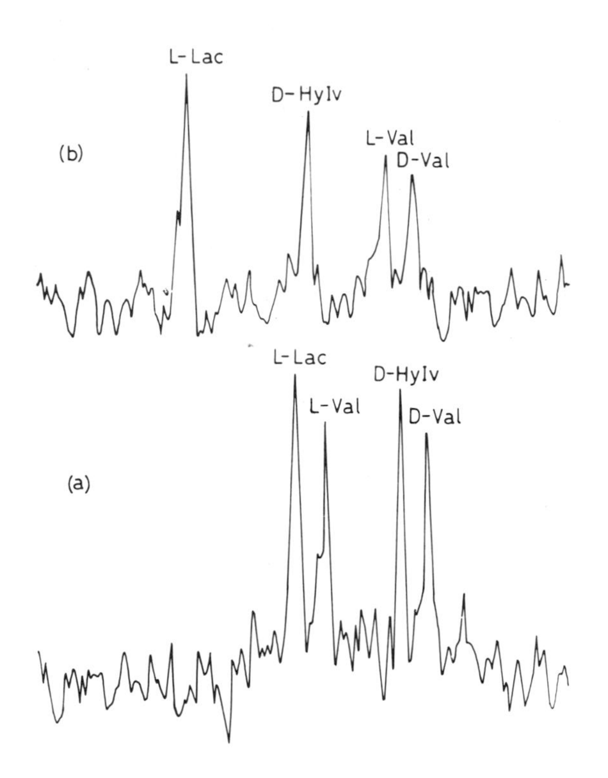

FIGURE 5. The carbonyl region of ^{13}C NMR spectra for (a) valinomycin and (b) its barium perchlorate complex in acetonitrile.

hydrogen bonds and wound in the form of an ellipse with barium ions located at the foci (Fig.4 in ref.24). The possibility of an averaging of two 1:1 type conformations in solution with the Ba^{2+} ion at one focus in one and at the other focus in the other cannot be ruled out. But the support for 1:2 (VM:barium) stoichiometry comes essentially from the analysis of the ^{1}H NMR titration data(41). It has been reported earlier that VM forms a 1:1 type complex with Ba^{2+} ion in methanol (34). The single crystal X-ray structure analysis of VM-barium perchlorate complex gave a flat open conformation for the complex, with two barium ions per molecule similar to that proposed in solution (40) (Fig.6). Each Ba^{2+} ion is liganded to three consecutive amide carbonyls. The two barium ions are separated by 4.57 Å. The crystal structure contains infinite layer of valinomycin molecules inter-

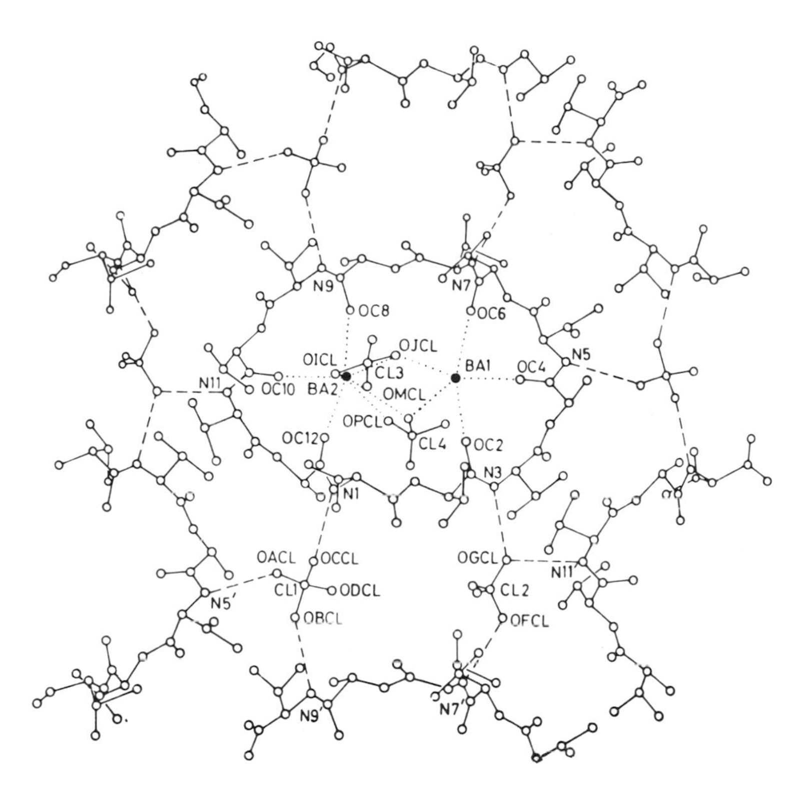

FIGURE 6. A view along the c axis of the molecular layers made up of valinomycin molecules interconnected by perchlorate groups in the VM-barium perchlorate crystal structure.

connected by perchlorate groups. The NMR and X-ray crystallography study of VM-barium thiocyanate indicated that the conformation of this complex is very similar to that of VM-barium perchlorate complex.

It is interesting to note that the barium ion which has a similar ionic radius as K^+ is not able to form the VM-K^+ type bracelet conformation. The reason for this could lie primarily in the difference in the hydrated spheres of the two ions. Ba^{2+} ion with its larger hydration sphere is unable to be close enough to the liganding carbonyl oxygens due to steric hindrance of the valyl side chains, while hydrated K^+ ion has the right size to fit in the cavity and to compete with the liganding ester carbonyls. The conformational studies made with an analog of valinomycin, cyclo(val-gly-gly-pro)$_3$ in which the two D-residues of VM are replaced by glycine and the L-lac with a proline, supported the above

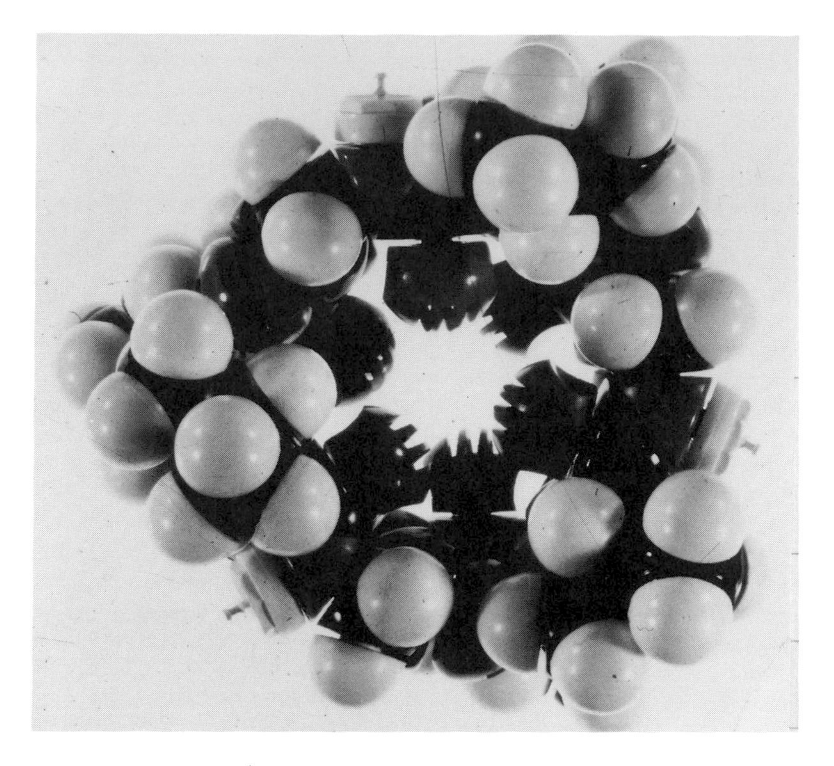

FIGURE 7. Photograph of the CPK model of the proposed conformation of the cyclo(val-gly-gly-pro)$_3$:Ba^{2+} complex.

possibility. The conformation of the K^+ and Ba^{2+} complexes of cyclo(val-gly-gly-pro)$_3$ were identical and has a bracelet structure similar to the VM-K^+ conformation (44) (Fig.7).

Detailed ^1H and ^{13}C NMR spectral analysis carried out on VM-calcium perchlorate complex showed that VM forms a 2:1 (peptide-ion-peptide) sandwich complex with Ca^{2+} at low salt concentrations (38). As the calcium concentration is increased, in addition to the sandwich complex an equimolar (1:1) complex different from that observed for VM-K^+ is also observed. The 1:1 complex is similar to the one proposed for VM-Na^+ or VM-Li^+ complex. The VM-Mg^+ complex has a very similar conformation to that of the calcium complex. The proposed model for the 2:1 ion sandwich VM-Ca^{2+} complex is shown in Fig.8. As can be seen from the figure, the calcium ion is co-ordinated to the three ester carbonyls of the two valinomycin molecules on the D-val-L-lac side of the bracelet structure.

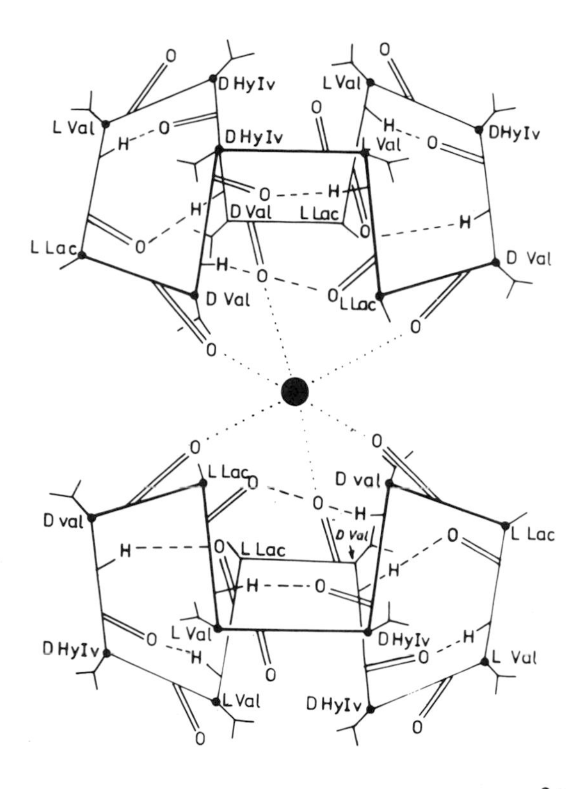

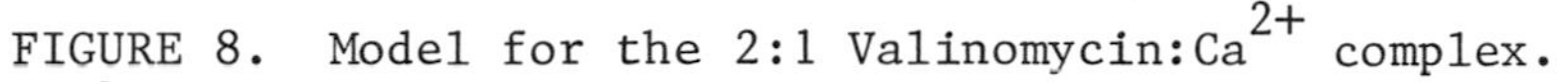
FIGURE 8. Model for the 2:1 Valinomycin:Ca^{2+} complex.

B. Nonactin-Ca^{2+} Complex

Cation complexation studies with nonactin (Fig. 2) showed that this ionophore forms isomorphous 1:1 complex with Na^+, K^+ and Ca^{2+} ions (45,46). The backbone of these complexes resembles the seam of a tennis ball. In the nonactin-Ca^{2+} complex (46,47) the four tetrahydrofuran ring oxygens are at the center while the four carbonyl oxygens, a pair from the top and bottom of the seam, tetrahedrally co-ordinate to the ion. Figure 9 shows the proposed conformation for the nonactin-Ca^{2+} complex from single crystal X-ray structure analysis and from NMR studies. The molecular structure and the crystal packing are similar to those observed in the nonactin complexes of sodium thiocyanate and potassium thiocyanate.

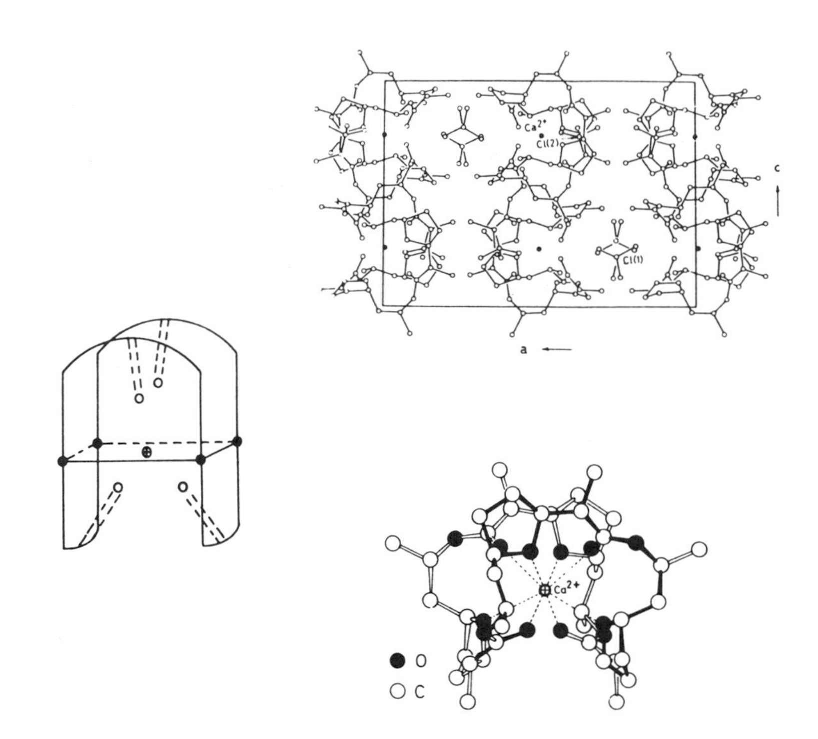

FIGURE 9. Crystal structure of nonaction-Ca^{2+} perchlorate complex as viewed down the b axis (upper-right); perspective view of the nonactin-Ca^{2+} perchlorate complex (lower-right); schematic diagram of nonactin-Ca^{2+} perchlorate complex from NMR studies (lower-left).

C. Lasalocid-Ca^{2+} Complex

Among the carboxylic ionophores, lasalocid(X-537A) and A23187 (Fig.2) form two important members of this class as they are calcium carriers (4,48). Lasalocid exists as monomer in polar solvents (49) and associates into a dimer in non-polar solvents (50). It forms neutral as well as charged complexes both in monomeric and dimeric forms with different conformations depending on the solvent, the cation and deprotonation state at the carboxylic group of the ionophores (51,52). Extensive CD and NMR studies by our group showed that lasalocid forms 2:1 (ion sandwich) and 1:1 (equimolar) type complexes with Ca^{2+} ion (53,54) and Li^{+} ion (55) with the conformation depending on the cation concentration. The conformational model for the complex based on the NMR data

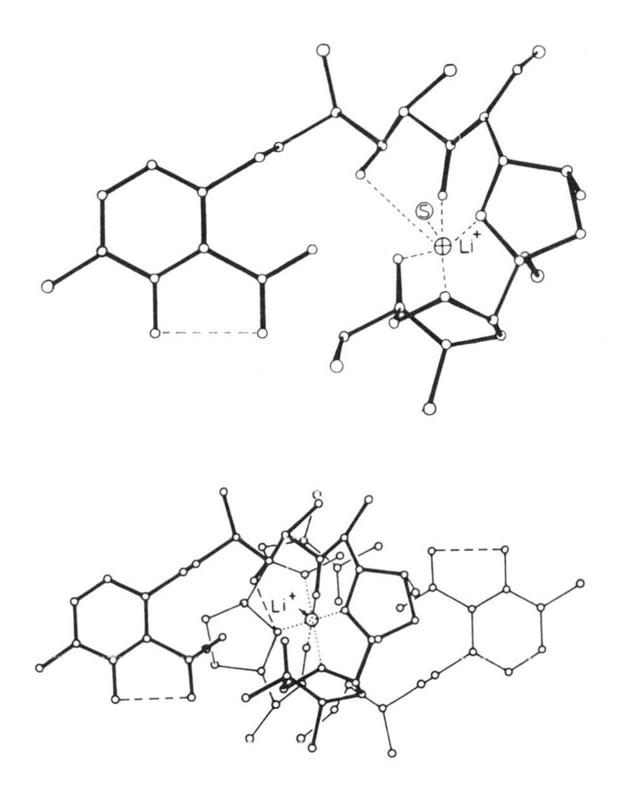

FIGURE 10. Conformational model for 1:1 (top) and 2:1 (bottom) lasalocid :Li^+ complex.

showed that Ca^{2+} or Li^+ ion is preferentially bound to an end of the molecule to three oxygen atoms, the other end (the salicylic acid part) being relatively free (Fig.10). In the 2:1 sandwich complex the cation is sandwiched between two lasalocid molecules with three oxygen atoms binding to it from each molecule.

In conclusion, studies aimed at understanding the ionophoric abilities of carriers revealed a propensity of them to form very stable carrier-cation complex. Irrespective of the stoichiometry of the complex, nature of the ionophore and the cations, several generalizations can be drawn from these observations. All the complexes have non-polar exteriors and effectively shield the cation from the surrounding environment. This is essential for their lipid solubility which reduces the activation energy for transport. In medium polar solvent, the cation is bound by a part of the molecule with the rest of the molecule having a non-polar exterior.
A significant observation resulting from our studies is the

formation of 2:1 ion sandwich complexes. Their formation in non-polar solvents alone suggests that they are formed from pre-formed equimolar complexes in the membrane. Such formation is possible by handing over of the cation from one ionophore to the other, a mode of transport known as relay carrier mechanism. The possibility of such non-equimolar complexes even in the case of VM-K^+ ion complex at high concentration of VM was suggested earlier by Ivanov (17).

III. INTERACTION OF IONOPHORES WITH PHOSPHOLIPID VESICLES

Although the conformation of the ionophore-cation complex under varied conditions is very important for an understanding of the molecular mechanism of tranmembrane cation transport, the solution environment model appears to be an over-simplification as the importance of specific interaction between ionophores and the polar head groups and acyl chains became more evident. Studies on the interaction of some of the ionophores, in particular, valinomycin with model membranes have already indicated the importance of specific interaction of the ionophores with vesicles (22,56-59). While there are difficulties in determining the conformation of ionophores incorporated into vesicles (the only study in this area was on valinomycin in perdeuterated DMPC vesicles (57)), it is possible to monitor the effect of ionophores on lipid bilayer phase transition using ^{1}H NMR, differential scanning calorimetry(DSC) and electron spin resonance(ESR) which give information on the lipid bilayer fluidity, orientation of the lipid long chain axes and intervesicular interactions (59). The ^{1}H NMR linewidth measurements of the choline and methylene protons of dipalmitoyl (DPPC) and dimrystiol (DMPC) phosphotidyl choline vesicles impregnated with VM at different lipid: VM ratios indicated that the location of VM, as inferred the above studies, is near the head group of DPPC vesicles and in the hydrophobic core of DMPC vesicles (see Fig.6 in ref.59). Monitoring the ^{31}P NMR chemical shift anisotropy (CSA) parameter ($\Delta\sigma$) and CSA line shapes of phospholipid multibilayers(MLVs) (60,61) on incorporation of ionophores gives information on the orientation and dynamics of head groups and morphology of aggregate structures. The ^{31}P NMR spectra of free DMPC and DPPC vesicles and those with ionophores incorporated studied above the gel-liquid crystalline phase showed a decrease in values upon addition of VM (-45.8 ppm to -35 ppm for DMPC and -50.0 ppm to -40.0 ppm for DPPC). Ionophores, nonactin, lasalocid and A23187

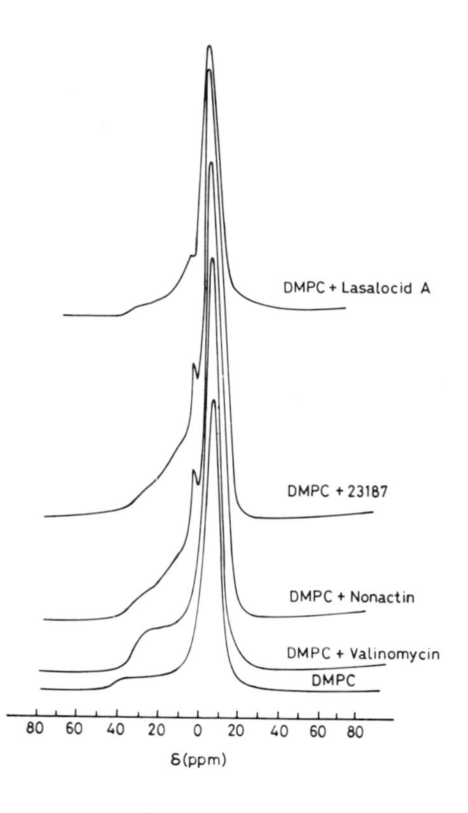

FIGURE 11. 187.5 MHz ^{31}P NMR spectra of free and ionophore incorporated DMPC MLVs at 40°C: Concentration of DMPC was 60mM and lipid: ionophore, 10:1.

broadened the spectral features making it difficult to obtain relative $\Delta\sigma$ values. An interesting observation is the appearence of an additional isotropic signal at about 0 ppm in both lipid systems in the presence of nonactin, lasalocid and A23187 (Fig.11).

Analysis of the data indicated that all the four ionophores studied interact with the polar head groups of lipids or adsorb on the membrane interface (in the case of valinomycin) in MLVs. The sharp signals observed in the ^{31}P NMR spectra of MLVs incorporated with nonactin, lasalocid and A23187 superimposed on the axially symmetric CSA patterns with reduced $\Delta\sigma$ and broadened ends of the $\sigma_{\parallel}$ lines are probably due to species whose motional properties fall into the 'extreme narrowing limit'(62). These species could be small lipid-ionophore aggregates (63,64).

IV. KINETICS OF LANTHANIDE TRANSPORT ACROSS PHOSPHOLIPID VESICLES MEDIATED BY IONOPHORES

The use of paramagnetic lanthanides to distinguish the inner and outer sides of bilayer vesicles (65) has offered a novel method for following the kinetics of lanthanide translocation across vesicles mediated by ionophores. The ^{1}H NMR spectra of phospholipid vesicles contain essentially time averaged signals from the head group cholines, hydrocarbon methylene chains and terminal methyl groups. The paramagnetic lanthanides added to the vesicles bind to the phosphate of the head group jetting out of the side where it is added (Fig.12). The binding of lanthanides gives rise to induced chemical shift in the choline protons in the ^{1}H NMR spectra and phosphate signals in ^{31}P NMR and choline methyl carbon

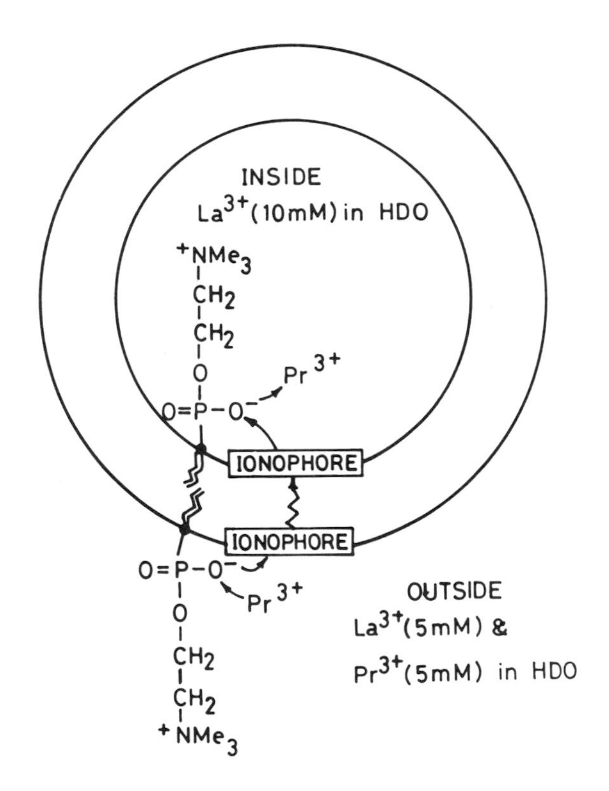

FIGURE 12. Schematic representation of NMR kinetic experiment.

in ^{13}C NMR. The above signals will be either broadened or shifted depending on the nature of the lanthanide ion (66). The strategy of the experiment is to follow the time dependence of the changes in chemical shift, line width and intensity of the signal (choline signals in ^{1}H NMR and phosphate signals in ^{31}P NMR) corresponding to the inner side of the vesicle, which will give information on the actual transport mechanism for the lanthanide transport. A detailed anlysis of such an NMR kinetic experiment has been made by Ting et al. (67).

Several carriers, mainly carboxylic group of ionophores have been studied by ^{1}H and ^{31}P NMR kinetics (68-72). Results on the carrier type ionophores are summarized in Table I. Our detailed anlysis of the transport of lanthanides by A23187 and lasalocid shows that as one goes along the lanthanide series from Pr^{3+} to Lu^{3+} , the rate of transport decreases (73). These results are well correlated with the conformation and cation complexing abilities of ionophores (73). The rates of transport have been found to be faster in the absence of intravesicular La^{3+} as compared with those observed in the presence of intravesicular La^{3+}. Typical time dependence of the choline region of the ^{1}H NMR spectra of A23187 mediated Nd^{3+} transport across DMPC system is shown in Fig.13.

TABLE I. Summary of NMR Kinetics of Carriers

Carrier	cation	lipid	transporting species	Ref.
lasalocid	Pr^{3+}	EYL	2:1	68
	Pr^{3+}	DPPC	-	72
	Pr^{3+}, Eu^{3+} and Nd^{3+}	DPPC, DMPC	2:1	73
	Mn^{2+}	DPPC	1:1,2:1	69
A23187	Pr^{3+}	DPPC	1:1	72
	Pr^{3+}, Eu^{3+} and Nd^{3+}	DPPC, DMPC	2:1	73
Etheromycin	Pr^{3+}	EYL	2:1	70
M139603	Pr^{3+}	EYL	1:1	71
X-14547A	Pr^{3+}	EYL	1:1	71
Tetronomycin	Pr^{3+}	EYL	1:1	71
Narasin	Pr^{3+}	EYL	1:1	71

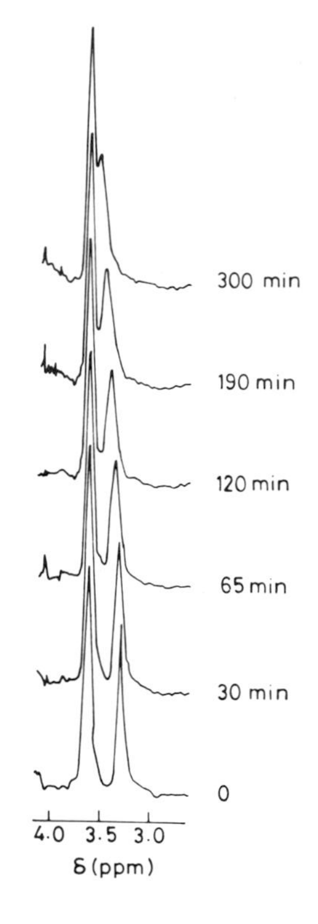

FIGURE 13. Time dependent changes in the choline region for DMPC-A231870-Nd^{3+} system at 40°C (lipid:ionophore, 1000:1).

The initial slopes of the lines obtained by plotting the relative shift ($\Delta\delta$) against time have yielded the apparent rate constants (k) at different concentrations of the ionophore. A double logarithamic plot of the concentration of the ionophore against the apparent rate constants is used to calculate the stoichiometry of the transporting species. The values so obtained (Table I) suggest that the stoichiometry of the transporting species is 2:1 (ionophore:lanthanide ion) for the carboxylic ionophores, lasalocid and A23187. Figure 14 shows the relative shift Vs time plots at various concentrations of the ionophore for Nd^{3+} transport with double logarithamic plot as inset. Our results are in agreement with those reported by earlier workers for Pr^{3+} and Mn^{2+} ions mediated by lasalocid (68,69). However, a 1:1 stiochiometry has been reported for Pr^{3+}-A23187 transporting species (72) as against the 2:1 stoichiometry observed by us.

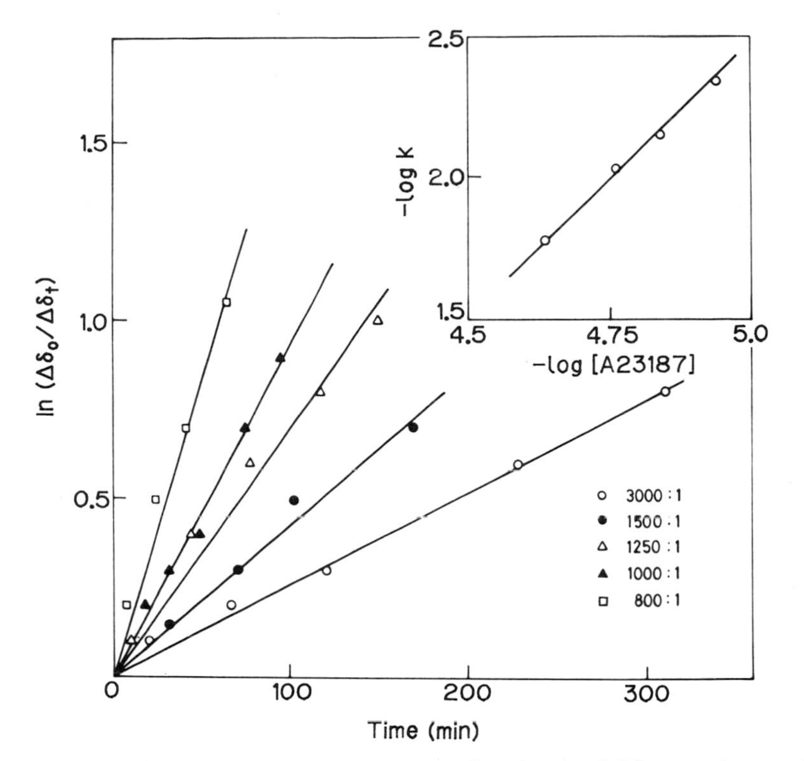

FIGURE 14. Plot of relative chemical shift against time for DPPC-A23187-Nd^{3+} system: Inset , log k vs. log concentration of A23187.

In summary, our studies on the kinetics of lanthanide transport mediated by carboxylic ionophores has shown that (i) A23187 is an efficient carrier as compared to lasalocid, (ii) the rates of transport are dependent on the presence or absence of the intravesicular diamagnetic La^{3+} and the former situation is more realistic in biological membranes, (iii) the stoichiometry of the transporting species is a non-equimolar 2:1 complex for both A23187 and laslocid. Since the paramagnetic lanthanide ions are considered to be the best substituents for the physiologically important divalent cations, the carboxylic ionophores might be bringing about the divalent cation transport via a non-equimolar 2:1 complex in biological membranes also.

Our studies on the kinetics of Pr^{3+} ion transport across vesicles mediated by valinomycin showed that VM does not tarnsport Pr^{3+} across phospholipid vesicles. However, some of the synthetic ion-binding cyclic peptides gave unusual and interesting results. For example, the ^{1}H and ^{13}C NMR experiments on the Pr^{3+} ion transport across DMPC vesicles showed the transport of the lanthanide ion by cyclo $(pro\text{-}gly)_3$. Fig. 15 shows the time dependence of the ^{31}P NMR spectra of DMPC vesicles with cyclo $(pro\text{-}gly)_3$ added. The double logarithamic plot of rate constant of the chemical shift changes as a function of concentration gave a stoichiometry for the transporting species as 5.8. Our results indicate that in the liquid crystalline phase cyclo $(pro\text{-}gly)_3$ probably forms a hexameric stack which acts as a channel for transporting Pr^{3+} across DMPC vesicles, but exhibits carrier like kinetics. The model proposed for the cyclo $(pro\text{-}gly)_3$ channel is shown in Fig.16. It is interesting to note that earlier studies have indicated that divalent cations, Ca^{2+} and Mg^{2+} ions bind strongly to cyclo $(pro\text{-}gly)_3$ forming a 2:1 sandwich type $(pro\text{-}gly)_3 : Ca^{2+}$ complex (74,75). Our studies on the Pr^{3+}

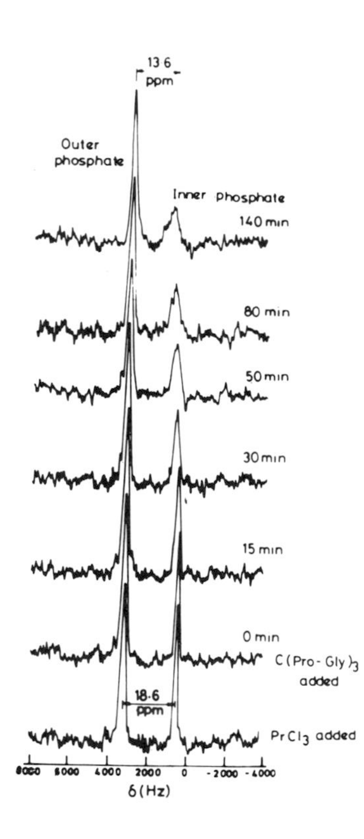

FIGURE 15. ^{31}P NMR spectra of DMPC vesicles at 60°C with cyclo $(pro\text{-}gly)_3$ added. Concentration of DMPC = 0.1 M; of cyclo $(pro\text{-}gly)_3$ = 4.3 mM; of $PrCl_3$ (outside) = 10 mM.

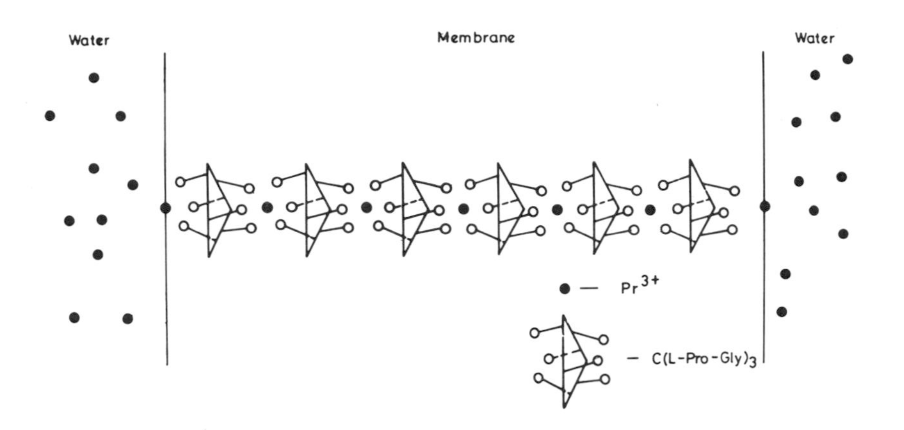

FIGURE 16. Model for the cyclo (pro-gly)$_3$ channel.

transport across vesicles mediated by cyclo (pro-gly)$_3$ are interesting in the sence that this synthetic cyclic peptide can possibly mimic the calcium transporting properties of a calcium ionophore.

The high resolution NMR technique that uses paramagnetic lanthanide cations to follow the kinetics of their transport across vesicles has been quite useful in distinguishing between the two general modes of metal ion transport, i.e whether the ions cross membrane one at a time or in bursts (67). However, it is only limited to studies with small vesicles and hence the method is not suitable for studies with large vesicles or membranes which gives broad resonances. Secondly, the method involves lanthanide ions and hence extrapolating the results to biologically important ions like Ca^{2+} ion need not necessarily be correct. Recent studies (76-78) which involve synthesis of hydrophilic ligands like $Dy(EDTA)^-$, $Dy(DPA)_3^{3-}$, $Dy(PPPi)_2^{3-}$ etc. which complex Na^+ and Li^+ and broaden their signals has permitted use of ^{23}Na, ^{39}K and ^7Li NMR to monitor their transport kinetics. The method has been successfully applied to studies on Na^+ transport across large vesicles (77) and to discriminate intra-and extracellular Na^+, K^+ and Li^+ ions in suspensions of human erythrocytes (78,79), in frog muscle (78) and recently in the studies of perfused rat hearts (80).

V. CONCLUSION

The results discussed in this article clearly demonstrates the importance of the structure and conformation of carrier ionophores and their cation complexes under varied conditions in order to understand the molecular basis for the the transmembrane cation transport. The problem of understanding all possible interactions of the carrier ionophore with model and biological membranes is very complex as can be seen from the reuslts presented. It is hoped that more results will be forthcoming in this direction. The novel NMR method for studying the kinetics of lanthanide transport mediated by carrier ionphores is promising in spite of its limitations.

ACKNOWLEDGMENTS

It is great pleasure to acknowledge the untiring efforts of my associates, Drs Devarajan, Vishwanath, Sankaram and Shastri who have contributed significantly to the studies reported in this article.

REFERENCES

1. Giebisch, G., Tosteson, D. C., and Ussing, H. H., in "Membrane transport in Biology", Vol. III, Springer-Verlag, Berlin, 1980.
2. Scarpa, A., and Azzone, G. F., Eur. J. Biochem. 12:328 (1970).
3. Pressman, B., Ann. Rev. Biochem. 45:501 (1976).
4. Bakker, E. P., in "Antibiotics" (F. E. Hahn, ed.), Vol. V, Springer-Verlag, Berlin, 1979.
5. Ovchinnikov, Yu. A., Ivanov. V. T., and Shkrob, A. M., in "Membrane Active Complexones", Elesevier, Amsterdam, 1974.
6. Harold, F. M., and Baarda, J. R., J. Bacteriol. 94:53 (1967).
7. Lardy, H. A., Graven, S. N., and Estrada, O., Fed. Proc. 26:1355 (1967).
8. Pressman, B. C., Harris, E. J., Jagger, W. S., and Johnson, J. H., Proc. Natl. Acad. Sci. USA 58:1949 (1967).
9. Andreoli, T. E., Tieffenberg, M., and Tosteson, D. C., J. Gen. Physiol. 50:2527 (1967).
10. Harris, E. J., Catlin, G., and Pressman, B. C., Bio-

chemistry 6:1360 (1967).
11. Mueller, P., and Rudin, D. O., Biochem. Biophys. Res. Commun. 26:398 (1967).
12. Szabo, G., Eisenman, G., and Ciani, S., J. Membrane Biol. 1:346 (1969).
13. McLaughlin, S. G. A., Szabo, G., Ciani, S., and Eisenman, G., J. Membrane Biol. 9:3 (1972).
14. Hodgkin, A. L., Huxley, A. F., and Kalz, B., Arch. Sci. Physiol. III:129 (1949).
15. Krough. A., Proc. Roy. Soc. B133:140 (1946).
16. Widdas, W. F., J. Physiol. 118:23 (1952).
17. Ivanov, V. T., Ann, N. Y. Acad. Sci. 264:221 (1975).
18. Ovchinnikov, Yu. A., and Ivanov, V. T., Tetrahedron 31: 2177 (1975).
19. Weiland, T., in "Chemistry and Biology of Peptides" (J. Meinhofer, ed.), p. 377. Ann Arbor Pub., 1973.
20. Madison, V., Atreyi, M., Deber, C. M., and Blout, E. R., J. Amer. Chem. Soc. 96:6725 (1974).
21. Prestegard, J. H., and Chan, S. I., Biochemistry 8:3921 (1969).
22. Grell, E., Funck, T., and Eggers, F., in "Membranes, A Series Advances" (G. Eisenman, ed.), Vol. 111, Chapt. 1, Marcel Dekker, New York, 1974.
23. Grell, E., Funck, Th., and Sauter, H. P., Eur. J. Biochemistry 34:415 (1973).
24. Easwaran, K. R. K., in "Metal Ions in Biological Systems", (H. Sigel, ed.), Vol. 19, Chapt. 5, Marcel Dekker, New York, 1985.
25. Pinkerton, M., Steinrauf, L. K., and Dawkins, P. D., Biochem. Biophys. Res. Commun. 35:512 (1969).
26. Haynes, D. H., Kowalsky., and Pressman, B. C., J. Biol. Chem. 244:502 (1969).
27. Ivanov, V. T., Laine, J. A., Abdulaev, N. D., Senyavina, L. S., Popov, E. M., Ovchinnikov, Yu. A., and Shemayakin, M. M., Biochem. Biophys. Res. Commun. 34:803 (1969).
28. Patel, D. J., and Tonelli, A. E., Biochemistry 12:486 (1973).
29. Glickson, J. D., Gordon, S. L., Pitner, J. P., Agresti, D. G., and Walter, A., Biochemistry 15:5721 (1975).
30. Davis, D. B., and Khalid, Md. A, J. Chem. Soc. Perkin II 1327 (1976)
3]. Grell, E., and Funck, Th., J. Supramol. Struct. 1:307 (1973).
32. Mayers, D. F., and Urry, D. W., J. Amer. Chem. Soc. 94: 77 (1972).
33. Smith, G. D., Duax, W. L., Langs, D. A., DeTitta, G. T., Edmonds, J. W., Rohrer, D. C., and Weeks, C. M, J. Am. Chem. Soc., 97:7242 (1975).
34. Grell, E., Funck, T., and Eggers, F., in "Molecular

and Membranes" (E. Munoz et al., ed), Elsevier, Amsterdam, 1972.
35. Davis, D. G., and Tosteson, D. C., Biochemistry 14:3962 (1975).
36. Sankaram, M. B., and Easwaran, K. R., Bioploymers 21:1557 (1982).
37. Steinrauf, L. K., Hamilton, J. A., Sabesan, M. N., J. Am. Chem. Soc. 104:4085 (1982).
38. Vishwanath, C. K., and Easwaran, K. R., Biochemistry 21: 2612 (1982).
39. Devarajan, S., and Easwaran, K. R., Biopolymers, 20:891 (1981).
40. Devarajan, S., and Easwaran, K. R., Int. J. Pep. Prot. Res. 23:324 (1984).
41. Devarajan, S., and Easwaran, K. R., J. Biosciences 6:1 (1984).
42. Devarajan, S., Nair, C. M., Vijayan, M., and Easwaran, K. R., Nature, 286:640 (1980).
43. Sankaram, M. B., and Easwaran, K. R., Int. J. Pep. Prot. Res. 25:585 (1985).
44. Easwaran, K. R. K., Pease, L. G., and Blout, E. R., Biochemistry 18:61 (1979).
45. Prestegard, J., and Chan, S. I., J. Am. Chem. Soc. 92: 4440 (1970).
46. Viswanath, C. K., and Easwaran, K. R. K., Biochemistry 20:2018 (1981).
47. Vishwanath, C. K., Shamala, N., Easwaran, K. R., Vijayan, M., Acta Crst. C39:1640 (1983).
48. Painter, G. R., and Pressman, B. C., Topics in Curr. Chem. 101:83 (1982).
49. Shen, S., and Patel, D. J., Proc. Natl. Acad. Sci. USA 73:4277 (1976).
50. Patel, D. J., and Shen, S., Proc. Natl. Acad. Sci. USA 73:1796 (1976).
51. Degani, H., and Friedman, H., Biochemistry 13:5022 (1974).
52. Alpha, S. R., and Brady, A. H., J. Am. Chem. Soc. 95:7043 (1973).
53. Vishwanath, C. K., and Easwaran, K. R. K., FEBS Lett. 153:320 (1983).
54. Vishwanath, C. K., and Easwaran, K. R. K., J. Chem. Soc. Perkin Trans II. 65 (1985).
55. Shastri, B. P., and Easwaran, K. R., Int. J. Biol. Macromol. 6:219 (1984).
56. Hsu, M., and Chen S. I., Biochemistry 12:3872 (1972).
57. Feigenson, G. W., and Meers P. R., Nature 283:313 (1980).
58. Walz, D., in "Bioenergetics of Membranes" (L. Packer, ed.), Elsevier, Amsterdam, 1977.
59. Sankaram, M B., and Easwaran, K. R. K., J. Biosciences 6:635 (1984).

60. Jacobs, R. E., and Oldfield, E., Prog. NMR Spect. 14:113 (1981).
61. Seelig, J., Biochim. Biophys. Acta 515:105 (1978).
62. Sankaram, M. B., Shastri, B. P., and Easwaran, K. R. K., Biochemistry (submitted) (1985).
63. Verkleij, A. J., Mombers, C., Leunissen-Bijvelt, J., and Ververgaert, P. H. J. T., Nature 279:162 (1979).
64. Verkleij, A. J., Biochim. Biophys. Acta 779:105 (1978).
65. Bystrov, V. F., Dubronina, N. I., Barsukov, L. I., and Bergelson, L. D., Chem. Phys. Lipids 6:343 (1971).
66. Sankaram, M. B., and Easwaran, K. R. K., in "Magnetic Resonance in Biology and Medicine" (G. Govil et al., ed.), Tata McGraw-Hill Publishing Co., New Delhi 1985.
67. Ting, D. Z., Hagan, P. S., Chan, S. I., Doll, J. P., and Springer, Jr. C. S., Biophys. J. 34:189 (1981).
68. Fernandez, M. S., Celis, H., and Montal, M., Biochim. Biophys. Acta 551:600 (1973).
69. Degani, H., Simon, S., and McLaughlin, A. C., Biochim. Biophys. Acta 646:320 (1981).
70. Donis, J., Grandjean, J., Gresjean, A., and Laszlo, P., Biochem. Biophys. Res. Commun. 102:690 (1981).
71. Grandjean, J., and Laszlo, P., J. Am. Chem. Soc. 106:1472 (1984).
72. Hunt, G. R. A., Tipping, L. R. H., and Belmont, M. R., Biophys. Chem. 8:341 (1978).
73. Shastri, B. P., in "Structural, Kinetic and Mechanistic Aspects of Carboxylic Ionophore Mediated Transmembrane Cation Transport", Ph. D. Thesis, Indian Institute of Science, Bangalore, 1985.
74. Deber, C. M., Torchia, D. A., and Blout, E. R., J. Amer. Chem. Soc. 93:1825 (1971).
75. Kartha, G., Varughese, K. I., and Aimoto, S., Proc. Natl. Acad. Sci. USA 79:4519 (1982).
76. Pike, M., and Springer, Jr. C. S., J. Mag. Res. 46:348 (1982).
77. Pike, M., Simon, S., Balschi, J. A., and Springer, Jr. C. S., Proc. Natl. Acad. Sci. USA 79: 810 (1982).
78. Gupta, R. K., and Gupta, P., J. Mag. Res. 47:344 (1982).
79. Pike, M., Fossel, E. T., Smith, T. W., and Springer, Jr. C. S., Am. J. Physiol. 246:C528 (1984).
80. Pike, M., Frazer, J. C., Dedrick, D. F., Ingwall, J. S., Allen, P. D., Springer, Jr. C. S., and Smith, T. W., Biophys. J. 48:159 (1985).

CARRIERS AND PUMPS: DYNAMIC PROPERTIES AND CATION SPECIFICITY OF NATURAL TRANSPORT MOLECULES.

Ernst Grell
Erwin Lewitzki
Dao Thi Minh Hoa
Achim Gerhard
Horst Ruf
Günther Krause
Gerhard Mager

Max-Planck-Institut für Biophysik
6000 Frankfurt 71
Federal Republic Germany

I. INTRODUCTION

Much effort has been devoted during the last fifteen years to the study of the molecular properties of low-molecular weight carriers for cations such as neutral depsipeptides or peptides and open chain carboxylic acid compounds which can be negatively charged. In addition, related compounds, eg. macrocyclic polyethers (crown ethers) and the bicyclic cryptands have been investigated. Since the selectivity of the facilitated transport of cations through phospholipid bilayer membranes is, among other interactions, related to the stability of the cation complexes of these ligands, it is hoped that their relevant molecular properties are similar to those of transport proteins occurring in natural transport systems. The model compounds mentioned above still represent the only existing class of natural or synthetic compounds that is capable to distinguish between small and large cations, eg., Na^+ and K^+. Most of the ligands that have been investigated until now, select the larger ones out of the alkali or alkaline earth cations, eg., K^+ or Ca^{++}. Very selective membrane carriers for Na^+ or Mg^{++} are not yet existing.

In the following, some common aspects of the low-molecular weight carriers will be derived from the dynamic properties of complex formation between valinomycin and Na^+ and K^+, which have been obtained from detailed kinetic studies in methanol employing ultrasonic absorption and temperature jump relaxation studies (1,2). Here, the fast stepwise substitution of the solvated cation by the coordinating groups of the multidentate ligand, as suggested by Diebler et al. (3), is interrupted due to the formation of a stable precomplex V'-Me^+ (Fig. 1). This intermediate complex is formed by a very fast and nearly diffusion controlled reaction, where only a

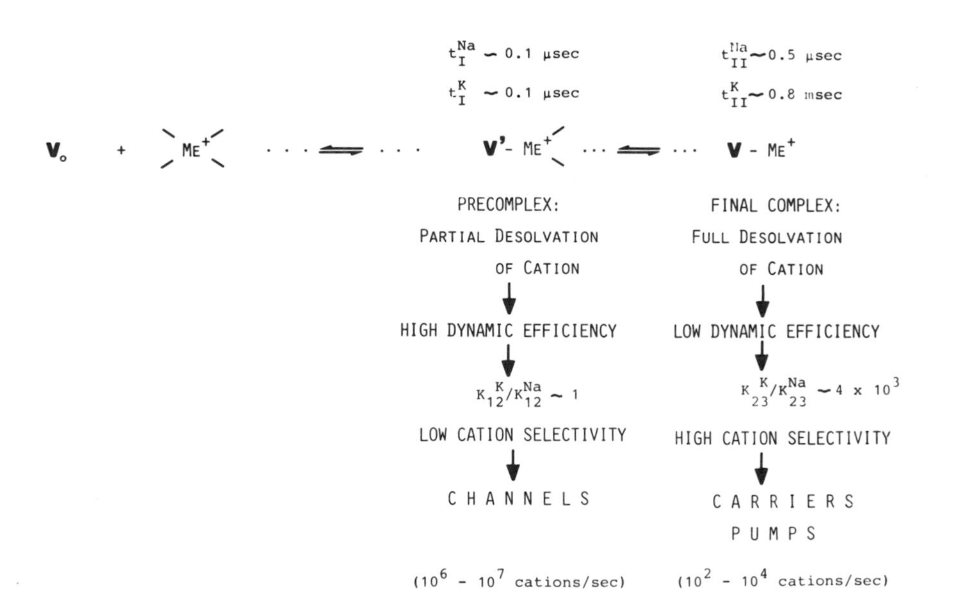

FIGURE 1. Complex formation between valinomycin and monovalent cations: Dynamic aspects and cation selectivity of precomplex and final complex related to natural channels and pumps. K_{12} represents the equilibrium constant between uncomplexed valinomycin and the precomplex, K_{23} that between precomplex and final complex, t represents the life time of a given state.

few solvate molecules are substituted from the cation by the still partially, eg. half-opened conformation of the bound valinomycin molecule. The K^+/Na^+ selectivity at this level of complex formation, expressed by the stability constant K_{12}, is close to 1, and thus unexpectedly low for such a specialized molecule. However, the precomplexes of valinomycin are characterized by very short life times (high dynamic efficiency). The life times of the precomplexes (t_I) with Na^+ and K^+ are both about 0.1 µsec (1,2). If the dissociation of the precomplexes were rate-limiting in cation transport through membranes, turnover numbers around 10^7 cations/sec would be achieved for both alkali ions. Among other reasons, poor cation selectivity results here from partial desolvation; but this disadvantage is abolished due to very high dynamic efficiency. In this respect, the properties of the valinomycin precomplexes closely resemble those of cation channels in membranes which exhibit turnover numbers of 10^6- 10^7 cations/sec.

After the formation of the precomplex, the rate-limiting step of complex formation between valinomycin and the alkali ions occurs, which is attributed to a conformational rearrangement of the precomplex. Subsequent to this rearrangement, the final complex can then be reached after the fast substitution of the still remaining solvate molecules by the last coordinating ester carbonyl groups of the ligand. It is important to note that the coordinated cation is now fully desolvated, and that the unimolecular transition occurring between precomplex and final complex is closely linked to the conformational properties of this flexible ligand. The equilibrium constants for the latter rearrangement process (K_{23}) now reflect a high K^+/Na^+ selectivity as one would expect it for valinomycin (1,2). In contrast to the high cation selectivity of the final complex (cf. Fig. 1), we find a remarkably increased life time of this state (t_{II}) up to 1 msec in case of the K^+ complex (low dynamic efficiency). This reflects the result of full cation desolvation. Now, if the dissociation of K^+ were rate-limiting in a transport process, a turnover number around 10^3 cations/sec could be expected. Typical turnover numbers for carriers in membranes are in the range of 10^2 to 10^5 cations/sec. This analysis of the dynamic aspects of the final complex of valinomycin thus suggests that the relevant properties of low-molecular weight carrier molecules are similar to those of natural cation pump proteins in membranes which are characterized by similar turnover numbers.

Carriers only contribute to the passive transport of cations. Their mechanism of action is related to the diffusion of loaded and unloaded carrier

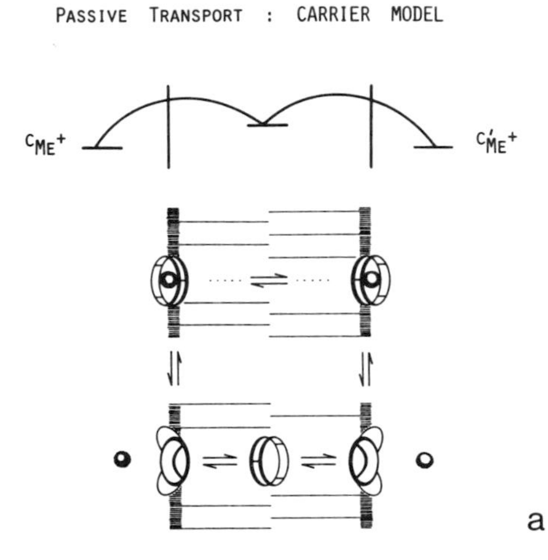

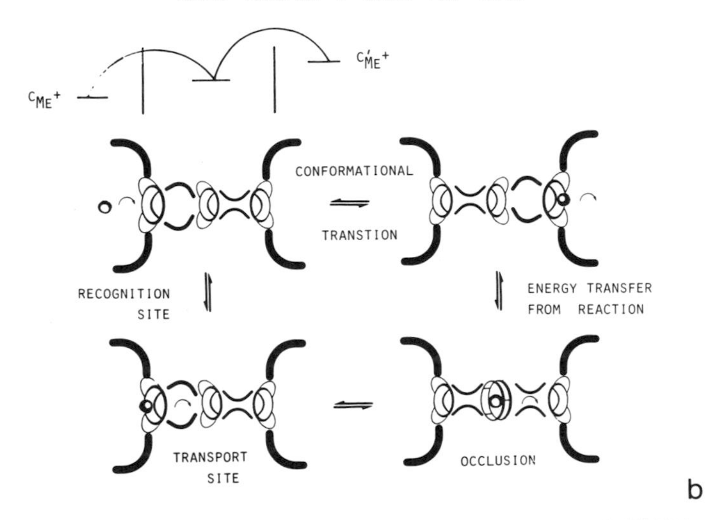

FIGURE 2. Partial models for passive and active transport by a low-molecular weight carrier (a) and a cation pump (b).

molecules across the membranes (Fig. 2a). On the other hand, cation pumps exhibit active transport, where the cation is pumped to the membrane side which is in contact with the higher cation concentration (Fig. 2b). Hydrolysis of ATP generally provides the energy required for such a coupled uphill transport process. In contrast to the carrier molecules, a pump protein must exhibit more than one cation binding site, namely at least one transport and two recognition sites, which are required for the detection of the cation concentrations on both membrane sides. This is due to the fact that the cation pump molecule is not expected to diffuse back and forth between the membrane sides. Since pumps generally do not contribute significantly to the membrane conductivity, there is no evidence supporting the existence of transmembrane-like pores as found in the case of open channels. The individual transport or pump processes must therefore be controlled by a series of successive conformational rearrangements which allow the cations to enter only limited spacial domains of the pump molecule (cf. Fig. 2b). This, however, reveals some striking common properties between low-molecular weight carriers and pump proteins, namely that their functions are controlled by their respective conformational properties.

In the following, certain relevant features such as cation selectivity and conformational properties of some selected carrier molecules and Na^+,K^+-ATPase, which represents a typical example of an E_1/E_2-ATPase, are compared. Some predictions concerning the molecular aspects of pumps will be presented.

II. ALKALI VERSUS ALKALINE EARTH CATION SELECTIVITY

If we neglect, as far as physico-chemical techniques are concerned, the complicated and not yet fully investigated cation binding stoichiometries of Na^+,K^+-ATPase and similar pump molecules, the following principal cation binding reaction scheme (Fig. 3) for a general cation carrier or for a cation pump (AH) is postulated. Since in aqueous solution coordination of alkaline earth cations most likely occurs to negatively charged ligands (A^-), a preceding protolytic reaction step has to be introduced. Monovalent cations can either be coordinated

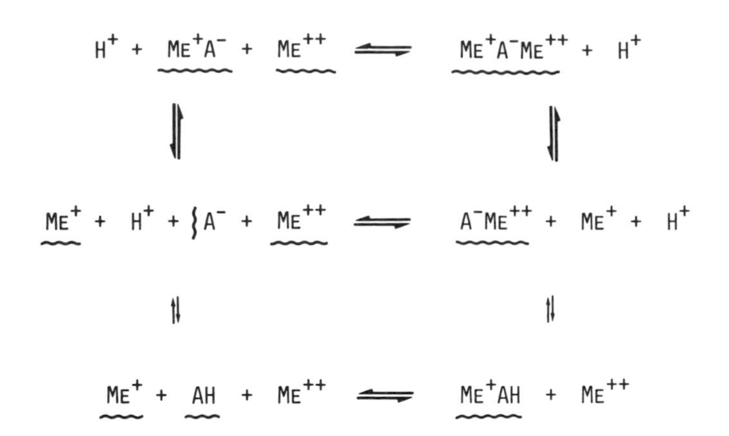

FIGURE 3. Complex formation reaction scheme based on 1:1 stoichiometries for a general cation carrier of pump molecule AH.

to the neutral form AH or also to the monodeprotonated protolytic state of the ligand. In addition, mixed alkali and alkaline earth cation complexes have to be considered too, eg. due to the interaction of a divalent cation with the complex A^-Me^+. Such mixed complexes are relevant as far as cation binding to Na^+,K^+-ATPase is concerned. For the carrier induced transport of divalent cations, the electroneutral complex $(A^-)_2Me^{++}$(4) has also to be considered, but is omitted in Fig. 3.

Since Na^+,K^+-ATPase exhibits enzymatic and transport activity only in the presence of Mg^{++} and Na^+ as well as of K^+, it is of interest to find out how the corresponding high cation selectivities can beachieved and controlled by means of molecular interactions. What is, for example, the reason that the Mg^{++} binding site is not easily occupied by Li^+, which has a similar ionic radius, or by Na^+ and Ca^{++}? The larger Na^+ and Ca^{++} ions, exhibiting similar ionic radii, could be excluded from a Mg^{++} site in a fairly simple way, namely by introducing a small enough cavity.

If we analyse the selectivities between alkaline earth and alkali cations of low-molecular weight carriers, the charge of the ligand seems to play an important role. The neutral, cyclic carrier molecules bind preferentially monovalent cations (1). The valinomycin-Ba^{++} complex in methanol, for example, is about two orders of magnitude lower than that of the K^+ complex (1). Upon transfer of both complexes from methanol to water, the larger free desolvation energy of the divalent cation should lead even to an increase of the K^+/Ba^{++} selectivity. The peptide ligand PV (5) exhibits the same stabilities for monovalent and divalent cations of similar ionic radius (Fig. 4). Here, the reduced conformational mobility of PV, compared with that of valinomycin, may be responsible for this change of cation selectivity. It is important to note that PV forms the most stable cation complexes of all neutral, monocyclic ligands which have been investigated until now. In special cases, as with bicyclic compounds (eg. cryptands), alkaline earth cation complexes with neutral ligands can show even higher stability constants than the corresponding alkali ion complexes (6).

Negatively charged carriers such as virginiamycin S_1 or calcimycin (A23187) are capable of

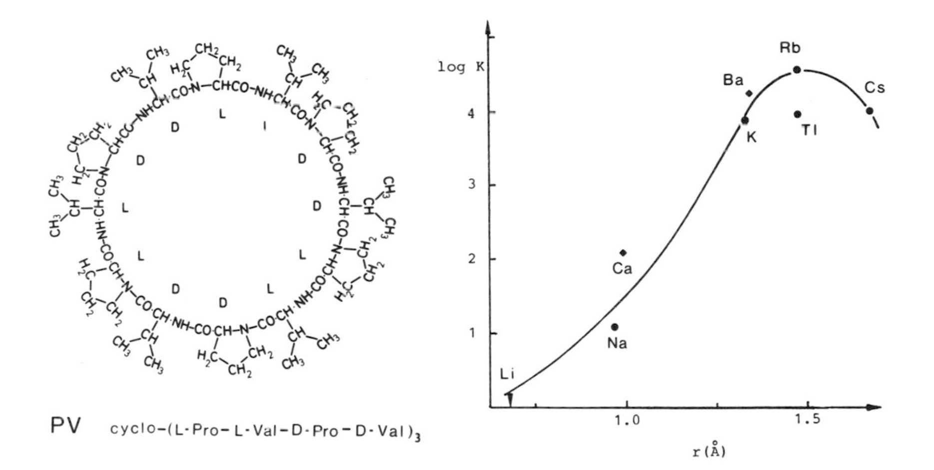

FIGURE 4. Structure and cation selectivity of PV in 55% water in methanol (mole fraction of water ≈ 0.68) at 25°C. Semilogarithmic plot of stability constant K for 1:1 complex formation, as determined by spectrophotometric titration, versus ionic radius r. The PV used here was obtained from B.F. Gisin (New York, USA).

transporting divalent as well as monovalent cations. This is shown by the antibiotic-induced outward flow of Ca^{++} from phospholipid vesicles which is compensated by a corresponding inward transport of Li^{+} (Fig. 5). The structures of virginiamycin S_1, calcimycin and of the similar antibiotic X-14885A together with the cation selectivities of their 1:1 complexes are shown in Fig. 6 . The stability constants have been determined by spectrophotometric titrations at different pH values in 30% water in methanol at 25°C. From the pH dependence of the stability constant K it is concluded that divalent and monovalent cations bind to the negatively charged, deprotonated protolytic states of these ligands.

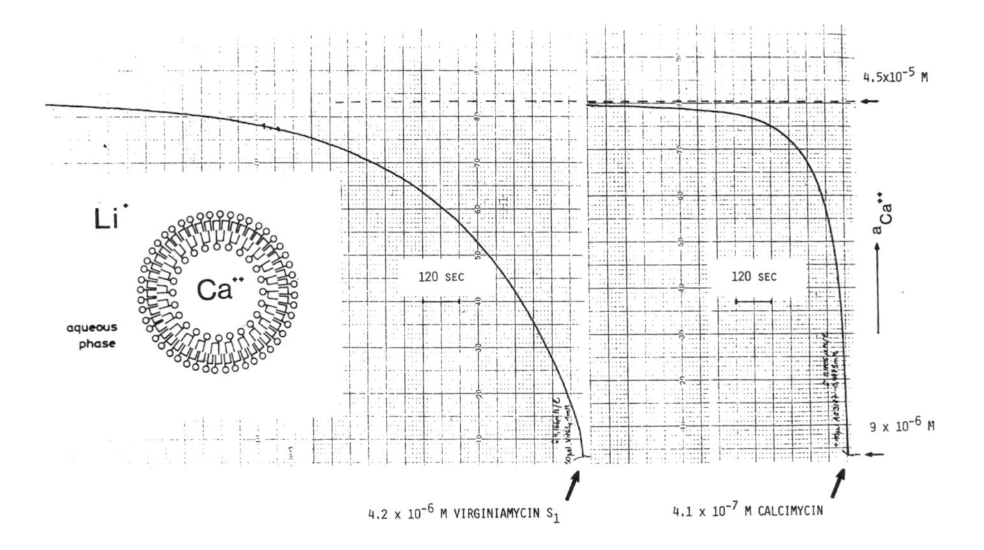

FIGURE 5. Outward transport of Ca^{++} from lecithin vesicles (0.4 mM P_i) containing 20 mM Pipes-LiOH pH 7.2 as buffer and 5 mM $CaCl_2$, 60 mM LiCl. The outer medium contains the same buffer and 102 mM LiCl only upon the addition of calcimycin or virginiamycin S_1 as measured by an external Ca^{++}-selective electrode prepared with ligand ETH 1001 acc. to (7). The calcimycin used here is obtained from Calbiochem (Frankfurt, FRG), and virginiamycin S_1 is isolated acc. to (8), whereas the egg lecithin vesicles are prepared similar to (9).

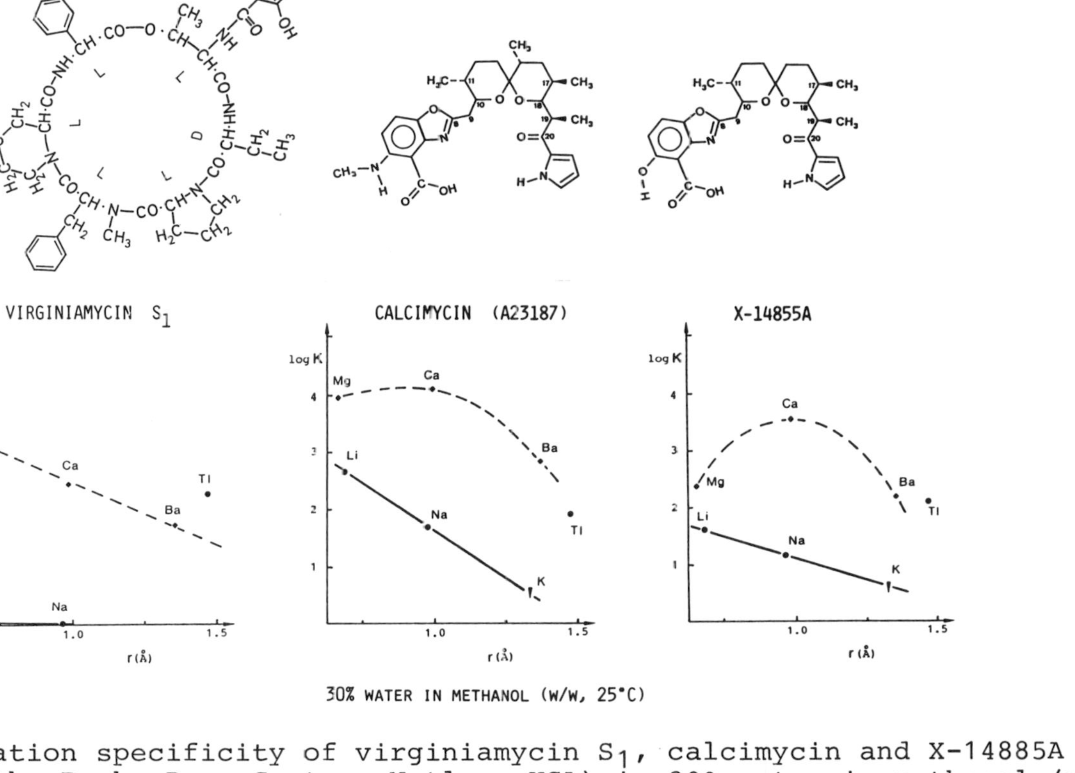

FIGURE 6. Cation specificity of virginiamycin S_1, calcimycin and X-14885A (obtained from the Roche Res. Center, Nutley, USA) in 30% water in methanol (w/w) at 25°C. Semilogarithmic plot of stability constant K versus ionic radius r.

Apparent linear dependences of log K versus the cationic radius are found in case of alkaline earth cation binding to virginiamycin S_1, and of alkali ion binding to both calcimycin and X-14885A. This probably represents a predominant electrostatic contribution to the overall binding interaction. Thus, it is concluded that in case of these complexes no pronounced chelation effect does exist, where the multidentate ligand is wrapped around the cation. This interpretation is also consistent with the observed ionic size dependence of log K. Since the electrostatic interaction of the deprotonated ligand is largest with the smallest cation, the complex stability is expected to decrease with increasing radius, which compares well with the apparent linear dependences shown in Fig. 6 . Virginiamycin S_1 exhibits by far the highest Me^{++}/Me^{+} selectivity of these three ligands, which is due to the surprisingly weak interaction with alkali ions.

With respect to alkaline earth cation binding to calcimycin and X-14885A, the highest stability constants are found for the Ca^{++} complexes. The value of the stability constant decreases with increasing or decreasing ionic radius relative to that of Ca^{++}. This behavior is indicative of a multidentate complex formation, where the open chain ligand is wrapped around the cation and forms a suitable cavity, as it has been suggested for calcimycin (4). X-14885A exhibits the highest Ca^{++}/Mg^{++} selectivity of these three ligands as far as 1:1 complex formation is concerned, although X-14885A has been predicted to be Mg^{++} selective (10). For calcimycin a Ca^{++}/Mg^{++} selectivity close to 1 has also been reported for the higher 2:1 complex (11). On the other hand, X-14885A shows a lower Mg^{++}/Li^{+} selectivity than calcimycin, which may be the result of a larger electrostatic contribution to the binding energy in the case of the twice negatively charged X-14885A molecule in its fully deprotonated state. The structure of the complex formed between virginiamycin S_1 and alkaline earth cations is shown in Fig. 7 . Coordination is assumed to occur to the phenolate oxygen.

In addition to these natural compounds, a synthetic cryptand-like compound has been investigated, which also mimics the specific cation binding site of a transport molecule. It is a derivative of cryptand-222, where an azo-dye fragment has been inserted (Fig. 8). Similar to virginiamycin S_1, this dye-

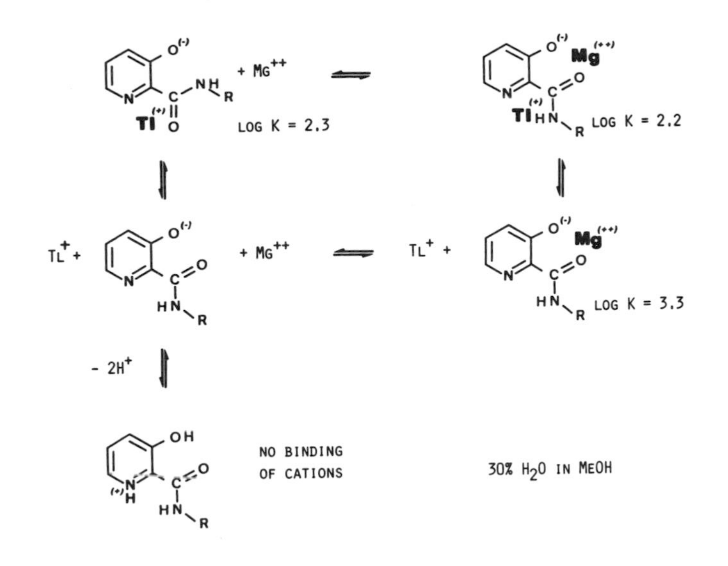

FIGURE 7. Structures of cation complexes of virginiamycin S_1 (R: peptide lactone residue).

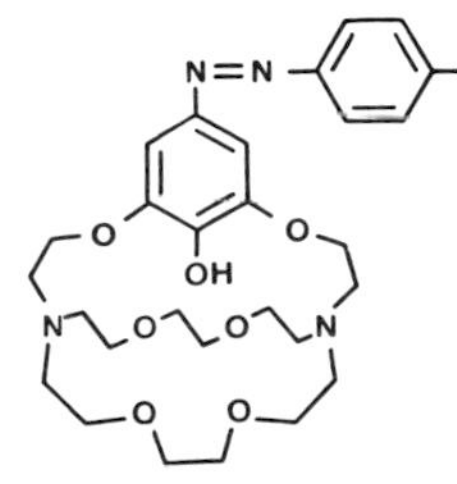

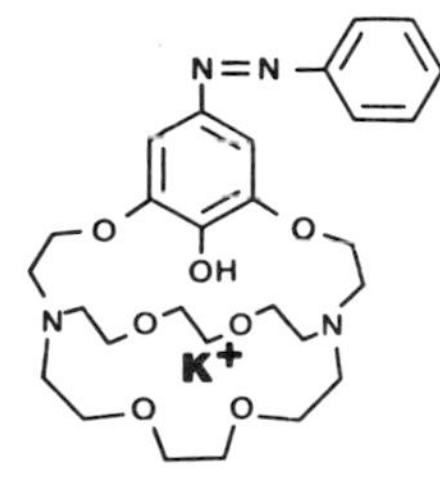

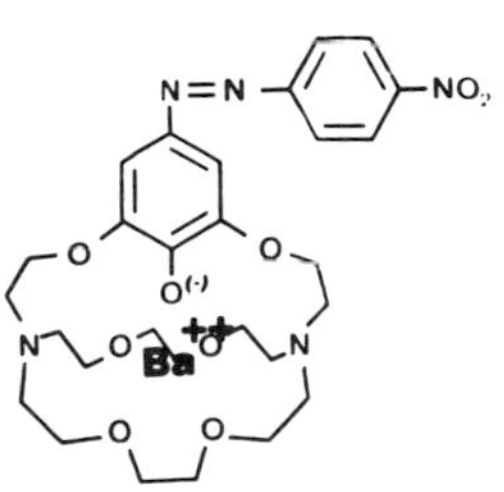

log K_{OH} = 8.9 log K_{K^+} = 3.5 log $K'_{Ba^{++}}$ = 1.2

(pH = 7.4)

FIGURE 8. Structure of dye-cryptand and of its K^+ and Ba^{++} complex. In aqueous solution at 25°C, the log K_{OH} value of protolysis is 8.9; the log K value of the K^+ complex is 3.5, and that of the Ba^{++} complex is 1.2 in 30 mM Tes-tetramethylammonium hydroxide pH 7.4 . The dye-cryptand was obtained from Merck (Darmstadt, FRG).

cryptand has a phenolic OH group which can be deprotonated so that a negatively charged ligand is formed. Complex formation studies are carried out in aqueous solution at pH 7.4 with Ba^{++} and K^+. Both cations have a similar ionic radius. A surprising result is observed, namely that K^+ forms a considerably stronger complex with this ligand than Ba^{++}. From the spectral changes observed during the spectrophotometric titration experiments, it is concluded that K^+ is coordinated almost entirely to the neutral state whereas Ba^{++} is coordinated preferentially to the negatively charged state of the ligand (Fig. 8). Although the phenolic OH group may somewhat disturb the cavity of the ligand, the result is consistent with the general rule expressed before, namely that neutral cyclic molecules can select alkali ions whereas negatively charged ones often interact stronger with alkaline earth cations. This type of coordination observed here is considered to represent an example for the binding of a protonated ligand AH to a monovalent cation, as predicted in the general binding scheme of Fig. 3 .

III. Tl^+/K^+ SELECTIVITY OF CARRIERS AND Na^+,K^+-ATPase

According to the previous arguments, it is understandable how a Mg^{++} selective binding site can be created in a pump molecule by allowing the cation to interact with one or two deprotonated carboxylic acid groups, originating from amino acid residues,in a cavity of suitable size. On the other hand, for principal reason it seems to be more difficult to build up an alkali ion, eg. K^+ selective binding site which excludes the coordination of divalent cations such as Mg^{++} and Ca^{++}. From the results obtained with low-molecular weight carriers one would expect that an uncharged site with amide carbonyl groups as coordinating ligands could fulfill this aim. However, in case of a pump molecule like Na^+,K^+-ATPase, the nature of the coordinating groups that form the K^+ binding site is not yet known. In order to characterize further the alkali ion binding sites of low-molecular weight carriers and Na^+,K^+-ATPase, their Tl^+/K^+ selectivities are investigated. It has been suggested earlier that alkali ion selectivities can be considered to act as valuable criteria for

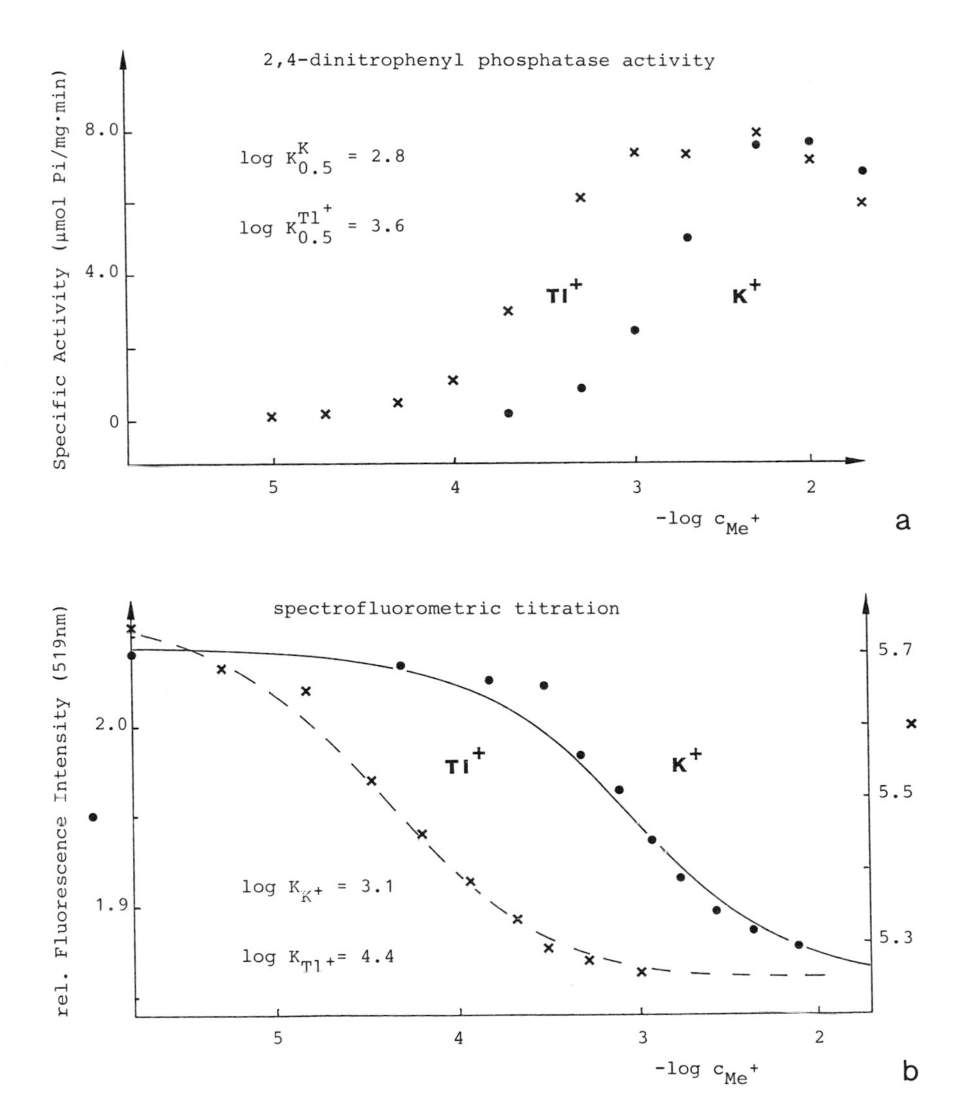

FIGURE 9. (a) Dinitrophenyl phosphatase activity of FITC-Na^+,K^+-ATPase in 25 mM triethanolamine-$HClO_4$ pH 7.5 containing 5 mM $Mg(ClO_4)_2$ at various $KClO_4$ and $TlClO_4$ concentrations at 37°C. (b) Spectrofluorometric titration (λ_{exc} = 489 nm) of 0.015 µM FITC-Na^+,K^+-ATPase in 25 mM triethanolamine-$HClO_4$ pH 7.5 with $KClO_4$ and of 0.045 µM FITC-enzyme with $TlClO_4$ at 37°C.

the characterization of the corresponding binding sites (12).

Since large Tl^+/K^+ selectivities are found for several carriers (Fig. 6), it is of interest to find out whether Na^+,K^+-ATPase also exhibits similar binding properties. To carry out such binding studies with Na^+,K^+-ATPase, a fluorescent derivative of the protein seems to be more suitable for this purpose, as its fluorescence properties have been reported to be sensitive to the presence of alkali ions (13). The modification is achieved by labelling with fluorescein isothiocyanate according to (13).

Prior to a titration experiment, the activity of the FITC-derivative of Na^+,K^+-ATPase (isolated from pig kidney according to (14)) with K^+ and Tl^+, employing 2,4-dinitrophenylphosphate (15) as a K^+-sensitive substrate, is investigated. Tl^+ induces the same degree of activation as K^+ (Fig. 9a), however in a concentration range which is about ten times lower than that of K^+. This result suggests that Tl^+ is bound much more strongly than K^+ which can be confirmed by the results of direct spectrofluorometric titrations (Fig. 9). The titrations are analyzed on the basis of a 1:1 complex formation stoichiometry. A log K value of 4.4 is found for the Tl^+, and 3.1 for the K^+ complex. Thus, the Tl^+/K^+ selectivity of FITC-Na^+,K^+-ATPase is about 10.

Rather unexpectedly high Tl^+/K^+ selectivities have been reported above for some of the low-molecular weight carriers. Whereas the neutral carriers valinomycin and enniatin B or dibenzo-30-crown-10 exhibit Tl^+/K^+ selectivities between 1 and 0.1 (cf. data in methanol acc. to (1,16)), the Ca^{++} carriers shown in Fig. 6, on the other hand, reach values even higher than 20. For virginiamycin S_1, where K^+ complex formation is nearly no more observable in the medium 30% water in methanol, the complex between the deprotonated form of this ligand and Tl^+ reaches a stability as high as that of the Ca^{++} complex. It is thus likely that Tl^+ doesn't bind to the alkaline earth cation site. Therefore, it is suggested that Tl^+ binds essentially to the pyridine nitrogen of the chromophoric group of this antibiotic (Fig. 7). This takes into account that Tl^+ has higher affinities to nitrogen or sulfur atoms than to oxygen atoms, which are much less polarizable. This interpretation is further supported by the observation that Mg^{++} can even bind to the 1:1 Tl^+ complex of virginiamycin S_1 by interacting with the phenolate oxygen (Fig. 7). However, the stability constant is about ten times lower than that observed in the absence of Tl^+. This mixed complex formally can be considered to represent a molecular model related to the mixed complexes formed between Na^+,K^+-ATPase and alkali and alkaline earth cations. In the case of this enzyme, evidence has been presented that the distance between Tl^+ and the divalent cation is in the range of 4 Å (17), which is surprisingly short. The high affinity observed also for Tl^+ binding to calcimycin and X-14885A cannot only be attributed to the interaction of Tl^+ with the negative charge of the ligand, but must to a substantial degree be due to the interaction with one or both nitrogen atoms of the ligand. The similar Tl^+/K^+ selectivities observed for the titratable binding site of Na^+,K^+-ATPase as well as for some of the low-molecular weight carriers may imply the existence of similarities between the nature of the coordinating groups in both types of molecules.

In order to account for the experimentally observed Tl^+/K^+ selectivity of Na^+,K^+-ATPase of about 10, it seems likely that a nitrogen atom (eg. from

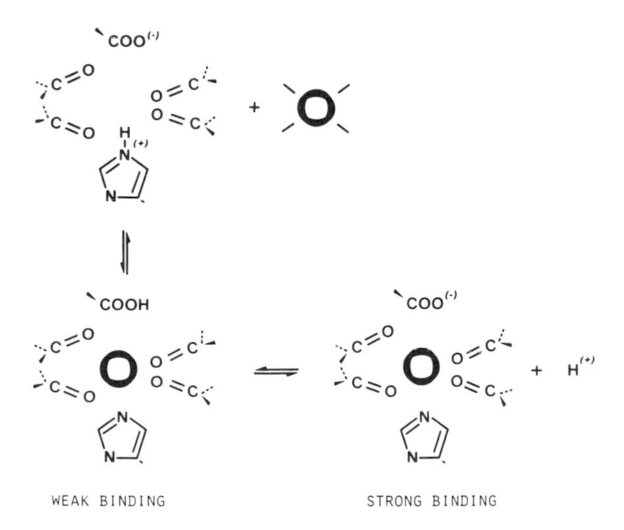

FIGURE 10. Hypothetical structure of the binding site for K^+ in Na^+,K^+-ATPase.

a histidine residue) is involved in the titratable binding site for K^+ or Tl^+. Similar selectivities are reported above for some of the deprotonated Ca^{++} carrier molecules. For general reasons it is probable that also one carboxylic acid group from a glutamic or an aspartic acid residue belongs to this alkali ion binding site. A hypothetical binding site is indicated in Fig. 10. If two or more deprotonable carboxylic acid groups were involved, the alkali ion binding site would be expected to exhibit very high affinities for alkaline earth cations, which is concluded from results obtained with model compounds. However, such high affinities for these divalent cations have not been observed until now.

The remaining coordinating groups of the site for K^+ or Tl^+ are attributed to amide carbonyl groups of the peptide back bone arranged in a regular way due to suitable backfolding of the chain (cf. Fig. 10). Since the coordination number for K^+ would have to be at least six (eg. octahedral arrangement), a minimum number of four such amide groups have to be postulated. The molecular model shown in Fig. 10 is also capable to explain the protection of an alkali ion binding site within Na^+,K^+-ATPase, either by protonation of the nitrogen atom, which could lead to the formation of a salt bridge, or by a conformational rearrangement, leading to the displacement of one or several of the coordinating groups.

Furthermore, this hypothetical model allows to explain the pH dependence of alkali ion binding such as the transition from low affinity to high affinity binding. This transition could, for instance, be due to a deprotonation of the coordinated carboxylic acid group (cf. Fig. 10), which would enable a stronger electrostatic binding of the monovalent cation to its site. These predictions clearly indicate that binding studies on low-molecular weight carriers can provide interesting suggestions concerning the nature of the alkali and alkaline earth cation binding sites in transport proteins.

IV. OCCLUSION OF CATIONS AND CONFORMATIONAL PROPERTIES

In this chapter, occlusion of alkali ions by Na^+,K^+-ATPase is compared with the coordination of alkali ions to valinomycin, where a molecular reac-

MODELS FOR OCCLUSION PROCESSES

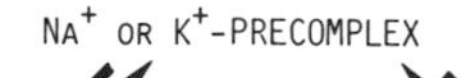

FIGURE 11. Precomplex and final complex between valinomycin and Na^+ or K^+ as a molecular model for occlusion of cations in Na^+,K^+-ATPase.

tion mechanism has been derived from the results of kinetic studies (2,18). In polar solvent systems, valinomycin predominantly exists in opened conformational states (1,2,18,19) which are significantly changed upon binding of eg. K^+ due to the formation of a bracelet-type, closed structure which is stabilized by the formation of six intramolecular hydrogen bonds (19,20). This conformation of the cation complex represents the final state of complex formation (cf. Fig. 1). The transition between the precomplex of valinomycin, where the coordination of an only partially desolvated cation to an half opened conformational state occurs (weak binding, short life time), and its final complex (strong binding, long life time) may be regarded as a molecular model for alkali ion occlusion in Na^+,K^+-ATPase (cf. Fig. 11). This transition between precomplex and final complex is essentially attributed to a conformational transition of the depsipeptide already bound to the cation.

Under suitable conditions (E_2 state for K^+, E_1P state for Na^+), Na^+ or K^+ is bound very tightly to Na^+,K^+-ATPase, implying that the bound alkali ion is not capable to exchange quickly with the unbound cations on either side of the membrane (cf. Fig. 2b). For this situation, the term occluded binding has been introduced (21). Whereas the closed or occluded state of the final valinomycin K^+ complex exists in solvents of different polarities, the corresponding Na^+ complex (Fig. 11) is only detectable in solvents that are less polar than methanol (2,18).

For the alkali ion transport action of Na^+,K^+-ATPase it has been suggested that cation deocclusion may be rate-limiting (22). Also in this respect, a striking analogy is observed between the properties of the depsipeptide carrier and the transport protein. Both processes, the deocclusion of alkali ions from Na^+,K^+-ATPase as well as the dissociation of these cations from valinomycin are related to conformational properties of the corresponding ligands. In the case of valinomycin, the transition from the final cation complex to the precomplex represents the rate-limiting step of cation dissociation (1,2). This transition is again attributed to a conformational rearrangement of the bound ligand to the half-opened precomplex, from where cation dissociation can occur. In conclusion we can say that many properties of valinomycin alkali ion complexes compare well with cation occlusion by Na^+,K^+-ATPase.

ACKNOWLEDGMENTS

The antibiotic X-14885A was made available by Dr. J.W. Westley, Hoffmann La Roche (Nutley, USA). The dye-cryptand was obtained from Dr. R. Klenk and Dr. E.G. Ross, E. Merck (Darmstadt, FRG). The peptide PV was obtained from Prof. B.F. Gisin, The Rockefeller University (New York, USA). The help of Dr. D. Ammann and Prof. W. Simon concerning the preparation of ion selective electrodes is acknowledged. The authors wish to thank Dr. A.-M. Albrecht, Dr. E. Bamberg, Prof. F.J. Kayne, Prof. H. Kessler and Prof. P.L. Jørgensen for many helpful suggestions and discussions during this work. The excellent technical assistance of Mrs. A. Ifftner and Mr. G. Schimmack is appreciated. This work was partially supported by the Deutsche Forschungsgemeinschaft (SFB 169).

REFERENCES

1. Grell, E., Funck, Th., and Eggers, F., in "Molecular Mechanism of Antibiotic Action on Protein Biosynthesis and Membranes" (E. Munoz, F. Garcia-Ferrandiz, and D. Vasquez, eds.), p. 645. Elsevier, Amsterdam, 1972.
2. Grell, E., Funck, Th., and Eggers, F., in "Membranes Vol. III" (G. Eisenman, ed.), p.1. Dekker, New York, 1975.
3. Diebler, H., Eigen, M., Ilgenfritz, G., Maaß, G., and Winkler, R., Pure Appl. Chem. 20:93 (1969).
4. Pfeiffer, D.R., Taylor, R.W., and Lardy, H.A., Annals New York Acad. Sci. 307:402 (1978).
5. Gisin, B.F., Ting-Beall, H.P., Davis, D.G., Grell, E., and Tosteson, D.C., Biochim. Biophys. Acta 509:201 (1978).
6. Dietrich, B., Lehn, J.M., and Sauvage, J.P., Chemie in unserer Zeit 7:120 (1973).
7. Simon, W., Ammann, D., Oehme, M., and Morf, W.E., Annals New York Acad. Sci. 307:52 (1978).
8. Oberbäumer, I., Grell, E., Raschdorf, F., and Richter, W.J., Helv. Chim. Acta 65:2280 (1982).
9. Brunner, J., Skrabal, P., and Hauser, H., Biochim. Biophys. Acta 453:322 (1976).
10. Liu, C.-M., Chin, M., Prosser, La T., Palleroni, N.J., Westley, J.W., and Miller, P.A., J. Antibiotics 36:1118 (1983).
11. Krause, G., Grell, E., Albrecht-Gary, A.M., Boyd, D.W., and Schwing, J.P., in "Physical Chemistry of Transmembrane Ion Motions" (G. Spach, ed.), p. 255. Elsevier, Amsterdam, 1983.
12. Eisenman, G., and Krasne, S., in "MTP International Review of Science, Biochemistry Series Vol. 2" (C.F. Fox, ed.), p. 27 . Butterworths, London, 1973.
13. Hegyvary, C., and Jørgensen, P.L., J. biol. Chem. 256:6296 (1981).
14. Jørgensen, P.L., Biochim. Biophys. Acta 356:36 (1974).
15. Gache, C., Rossi, B., Leone, F.A., and Lazdunski, M., in "Na^+,K^+-ATPase, Structure and Kinetics" (J.C. Skou, and J.G. Nørby, eds.), p. 301. Academic Press, London, 1979.
16. Chock, P.B., Proc. Nat. Acad. Sci. USA 69:1939 (1972).

17. Grisham, Ch.M., and Mildvan, A.S., J. Supramol. Struct. 3:304 (1975).
18. Grell, E., and Oberbäumer, I., in "Molecular Biology, Biochemistry and Biophysics, Vol. 24" (I. Pecht and R. Rigler, eds.), p. 371 . Springer, Heidelberg, 1977.
19. Ivanov, V.T., Laine, I.A., Abdulaev, N.D., Senyavina, L.B., Popov, E.M., Ovchinnikov, Yu.A., and Shemyakin, M.M., Biochem. Biophys. Res. Comm. 34:803 (1969).
20. Pinkerton, M., Steinrauf, L.K., and Dawkins, P., Biochem. Biophys. Res. Comm. 35:512 (1969).
21. Beaugé, L.A., and Glynn, I.M., Nature 280:510 (1979).
22. Forbush, B., Anal. Biochem. 140:495 (1984).

STRUCTURAL BASES OF MEMBRANE PROTEIN FUNCTIONING

Yuri A. Ovchinnikov

Shemyakin Institute of
Bioorganic Chemistry
USSR Academy of Sciences
Moscow, USSR

During the last two decades much attention has been paid not only to the study of the composition and structure of membranes as such but also to the understanding of the structure and function of membrane proteins. So far, the light-sensitive membrane proteins of animal and bacterial origins, namely, bacteriorhodopsin and rhodopsin, are best characterized.

Light absorption by rhodopsin leads to its activation that subsequently triggers a cascade of enzymatic reactions resulting in a fast decrease of the cGMP level in rod outer segments. The activated molecule of rhodopsin is then inactivated by phosphorylation (1,2). Until recently rhodopsin has been the only known light-transducing protein with retinal as a chromophore.

Halophilic microorganisms of the *Halobacterium* family utilize solar radiation energy due to the presence of bacteriorhodopsin, a light-driven primary proton translocase. Bacteriorhodopsin is a relatively small protein (about 250 amino acid residues) containing protonated aldimine of the retinal as a prosthetic group. Each working cycle of bacteriorhodopsin induced by the light quantum absorption is accompanied with transfer of at least one proton across the membrane, the retinal aldimine being reversibly deprotonated in the course of this cycle as judged from spectral data (3). Henderson and Unwin determined the three-dimensional structure of bacteriorhodopsin to a resolution of 7 Å within the membrane plane and about 14 Å perpendicular to the plane. According to these data the bacteriorhodopsin molecule consists of seven roughly parallel segments each spanning the membrane (4).

Our interest is focused on chemical and biochemical aspects of this unique membrane protein functioning. No doubt, such studies are necessary for elucidation of both the mechanism of proton translocation by bacteriorhodopsin and the dynamics of functioning of even more complex membrane proteins. A recent review from this laboratory (5) presents data concerning the structural basis for bacteriorhodopsin and rhodopsin function. A model which correlates the bacteriorhodopsin amino acid sequence and three-dimensional structure was elaborated, but it requires further improvement.

We started a detailed immunological investigation of bacteriorhodopsin which proved to be useful in studying the orientation and surface topology of the protein. Khorana et al. used this approach to identify distinct antibody binding sites of the cytoplasmic surface of bacteriorhodopsin (6) that allowed identification of a peptide loop between α-helical segments 3 and 4 and confirmation of the cytoplasmic location of the exposed C-terminal tail. In the course of immunological studies we obtained five hybridomas producing monoclonal antibodies to different membrane-exposed parts of the polypeptide chain. Specificity of these antibodies was established using modified derivatives of bacteriorhodopsin and a number of overlapping peptides yielded from enzymatic or chemical cleavages of the protein.

The antigenic determinants are situated on the following exposed parts of bacteriorhodopsin (7): Glu^{1}-Glu^{9} with three N-terminal amino acids; Gly^{35}-Met^{56} including Asp^{36} and/or Asp^{38} and Phe^{42}; Phe^{156}-Met^{163} with Phe^{156}; Glu^{194}-Leu^{207} including residue Glu^{194}; Pro^{200}-Leu^{207}. Thus bacteriorhodopsin fragments 4-65 and 156-231 have membrane-exposed peptide regions. All the data, experimentally obtained and earlier available, concerning the membrane location of fragments 66-72 and 231-248 evidence that each of the sequences 4-65 and 156-231 traverses the membrane at least twice (Fig. 1).

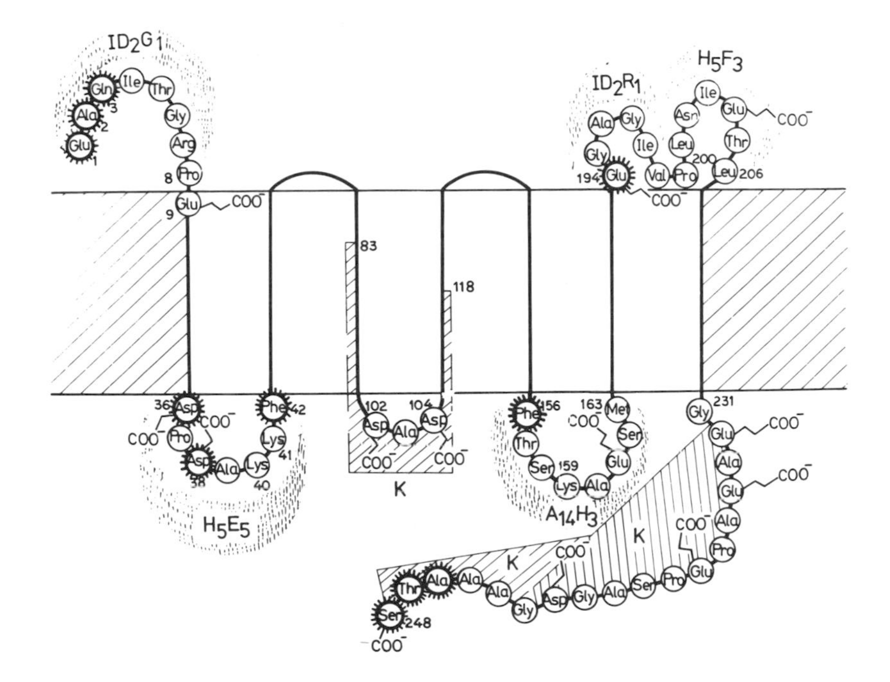

FIGURE 1. Antigenic determinants of bacteriorhodopsin.

Of even more importance is accessibility of Glu[194] to a monoclonal antibody. Upon the study of the chromophore orientation in bacteriorhodopsin by cross-linking using the photosensitive p-diaziridinophenyl analog of retinal, Ser[193] and Glu[194] were shown to be the sites of cross-linking with the diaziridine group located at the phenyl ring (8). Based on this finding a structural model with Glu[194] well in the membrane was put forward. Our result, on the contrary, proved that the residue, as a part of an antigenic determinant, should be located on the membrane surface. Obviously, only further studies will establish the real topography of bacteriorhodopsin and its chromophore in the membrane.

The structure-functional studies of bacteriorhodopsin pose a number of questions of fundamental significance. Here we would like to know: a) what is the nature of the marked spectral shift of retinal in bacteriorhodopsin (380-570 nm), b) is it possible to prove or disprove the models underlying the active sites of light-sensitive proteins, i.e. how valid is the point-charge model, c) are there any connections between the spectral intermediates of pigment bleaching and different stages of proton translocation in bacteriorhodopsin, d) what is the role of membrane-bound and exposed parts of the protein molecule in the pigment function, e) how valid is the conception of proton translocation through the chain of hydrogen bonds and, most important, could the proton pumping activity of bacteriorhodopsin be inhibited by replacing the potential "participants" of this pathway by means of site-directed mutagenesis - the technique with many advantages compared with direct chemical modification.

Naturally, all the questions can be solved only by gene manipulation utilizing the recombinant DNA technique; of vital importance here is finding of conditions for high level expression of the bacteriorhodopsin gene. Several recombinant plasmids were constructed to achieve the expression of bacterioopsin in *E. coli* (Fig. 2). To detect the level of expression we decided to exploit the well known fact that β-galactosidase of *E. coli* forms enzymatically functional hybrids if its several N-terminal amino acids are exchanged for some other polypeptide. It is easy to test expression of any foreign protein in *E. coli* if its C-terminal part is fused to the gene coding for β-galactosidase with several N-terminal codons cut off. Such a hybrid gene placed under promoter control programs synthesis of the hybrid protein possessing galactosidase activity and containing the polypeptide chain of the foreign protein of interest.

The bacterioopsin gene was fused to the lac Z gene (gene of β-galactosidase) and this hybrid gene was placed downstream the *E. coli* tryptophan promoter containing also the ribosome-binding site. This plasmid POG contains the full bacterioopsin gene. It means that this gene encodes an opsin precursor which is 13 amino acids longer than the mature opsin. Judging by β-galactosidase activity the level of expression is rather low.

We removed the signal peptide and changed the system of expression regulation. The POG 1 plasmid was reconstructed as follows. The precursor region of opsin gene was removed and a strong promoter of

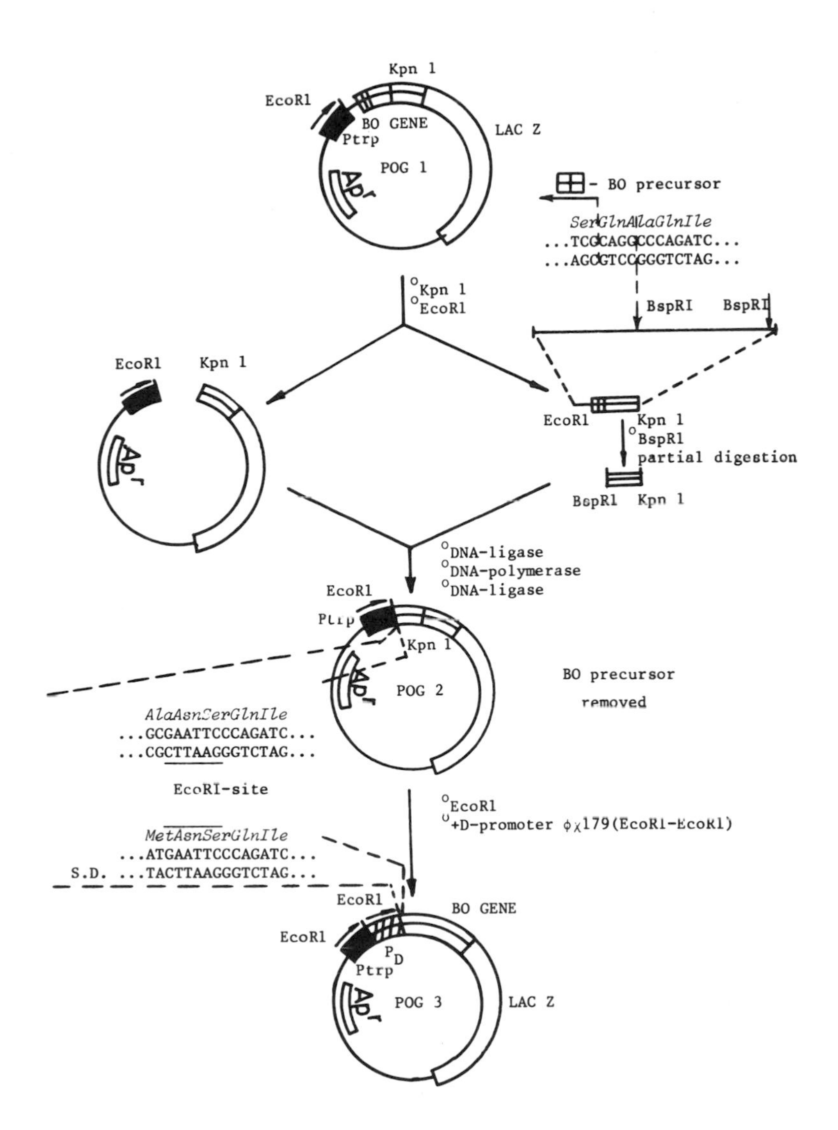

FIGURE 2. Construction of the POG 3 recombinant plasmid for expression of bacterioopsin-β-galactosidase fusion protein.

phage φχ 179 containing its own ribosome-binding site and initiating ATG-codon was placed upstream the gene. High galactosidase activity implies a high level of hybrid protein expression (up to 1% of total cellular protein). The same level of expression was achieved using plasmid pCP3 under the control of the λp_L promoter. This opens wide perspectives for investigation into site-directed mutagenesis of bacterioopsin gene to produce mutated bacterioopsins and to study their functioning.

Experimental methods developed for the bacteriorhodopsin research were applied to its analog - visual pigment rhodopsin. We determined the complete amino acid sequence of rhodopsin and showed that the polypeptide chain of the protein consisted of 348 amino acid residues (5). Related results were obtained in the USA (9) by sequencing a structural gene of bovine rhodopsin. Nucleotide sequence analysis of the cloned DNA provided an intron-exon map of the gene.

The characteristic feature of the amino acid sequence of rhodopsin is the presence of extended regions of the polypeptide chain made up of nonpolar amino acid residues interrupted by comparatively small polar sites. These hydrophobic extended regions compose the membrane part of the protein molecule.

The primary structure of rhodopsin underlies elucidation of the polypeptide chain arrangement in the membrane. The model building demands the combination of two approaches: a) analysis of the regions of the protein polypeptide chain located in the aqueous phase and accessible to the action of proteolytic enzymes; b) localization of the protein regions containing the least number of polar amino acid residues and capable of spanning the lipid bilayer. Besides, two considerations are taken into account. First, membrane regions of the molecule have the α-helical conformation and are situated perpendicularly to the membrane plane; secondly, the N- and C-terminal regions of rhodopsin are located on opposite sides of the membrane and, consequently, the polypeptide chain of the protein molecule should traverse the membrane uneven number of times. Now let us follow the path of the polypeptide chain in the membrane. Thirty amino acid residues in the N-terminal region of the protein molecule are accessible to the action of various proteolytic enzymes upon the treatment of inside-out photoreceptor disks. The N-terminal region of 30 amino acid residues is localized in the intradisk space. The region accessible to the chymotrypsin action is identified on the outer surface of photoreceptor disks (residues Phe^{146}-Arg^{147}). The region of 27 amino acid residues of the polypeptide chain in the α-helical conformation can traverse the lipid bilayer of the membrane (to span the entire membrane width 26-30 residues are necessary). Taking this finding into account in the analysis of distribution of polar and nonpolar residues of the polypeptide chain between two regions of proteinolysis (Tyr^{30}-Phe^{146}), we identified three membrane segments separated by two small clusters of hydrophilic residues (regions 62-73 and 92-101). Thus region Tyr^{30}-Phe^{146} of the polypeptide chain traverses the photoreceptor membrane three times. The region accessible to the papain action (Ser^{186}-Cys^{187}) is located on the intradisk membrane surface. Apparently, the region of 40 amino acid residues can span the membrane only once; it contains membrane segment Ile^{154}-Ser^{176}.

On the outer surface of photoreceptor disks there is rather extended region Gln^{236}-Lys^{245} accessible to the action of different enzymes. Since this region and the papain-accessible one are on different membrane surfaces, the polypeptide chain site between them (55 amino acid residues) spans the membrane uneven number of times. This region appeared to contain the only membrane segment (Phe^{203}-Phe^{228}).

The C-terminal region of rhodopsin is situated on the outer disk surface as region Gln^{236}-Lys^{245}. The polypeptide chain part connecting these two regions traverses the membrane even number of times. From the study of distribution of polar and nonpolar amino acid residues of this part of the protein molecule (75 residues) two membrane segments (Met^{235}-Phe^{276} and Phe^{283}-Met^{309}) were identified.

So seven segments of the polypeptide chain in the α-helical conformation spanning the photoreceptor membrane width compose the membrane moiety of the rhodopsin molecule (Fig. 3). The N- and C-terminal regions are the largest sites exposed to the water phase.

There are three functionally important domains in the rhodopsin molecule: retinal-binding site, C-terminal region and polypeptide region connecting the V and VI membrane segments.

Retinal and rhodopsin have absorption maxima of ~380 nm and 498 nm, respectively; for such a considerable bathochromic shift (120 nm) the charged groups should be incorporated into the retinal binding site.

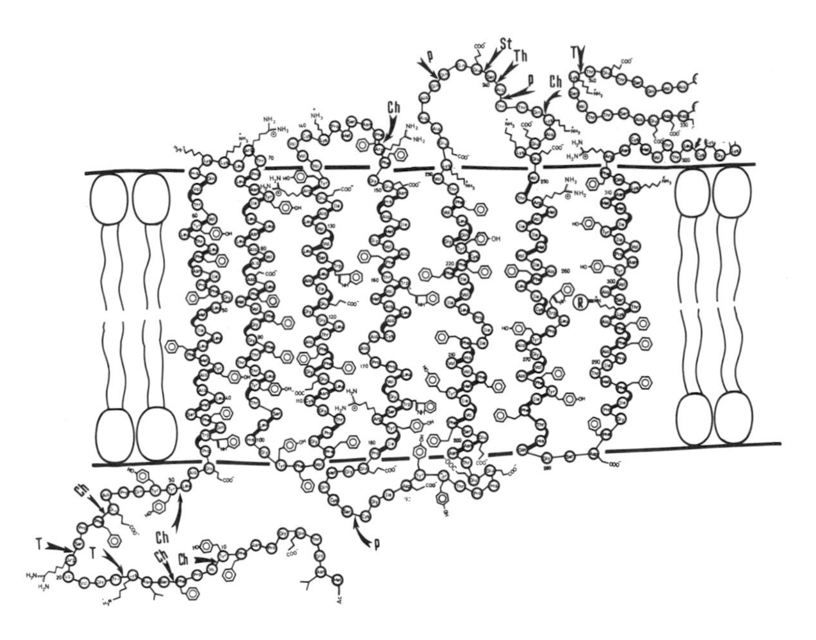

FIGURE 3. Topography of the visual rhodopsin molecule in the photoreceptor membrane.

Besides, spectral investigations show that the Trp residue should be located in the active site of retinal. Lys^{296}, responsible for the retinal binding, is located in the VII segment in the middle of the lipid bilayer in accord with data obtained from the study of the energy transfer between the chromophore and terbium ions. It is noteworthy that Ala precedes the retinal binding Lys residue in all the studied retinal binding proteins. The absence of the three-dimensional model of the rhodopsin arrangement precludes localization of other amino acid residues in the active site of the protein. As known retinal is located at the 16-23° angle to the membrane plane, i.e. almost parallel to its surface. Upon studying various chromophore analogs the length of the retinal binding region was established to be 10.1-10.9 A. The β-ionone ring of retinal, or at least its two methyl groups, are extremely important for the prosthetic group binding.

The study of the model compounds shows that the bathochromic shift (120 nm) is possible if a carboxyl group of aspartic or glutamic acid is at the distance of 3 Å from the protonated Schiff base. The second negative charge should be localized within the region of double bond C(11)-C(12) in the chromophore polyene chain. According to the proposed model three negatively charged amino acid residues Asp^{83}, Gln^{113} are in the lipid matrix of the photoreceptor membrane, one or two of them can regulate the bathochromic shift of the visual pigment at the interaction with its chromophore. In addition, negatively charged amino acid residues, such as Glu^{134} or Glu^{150}, located in the vicinity of the membrane surface, can be included in the retinal binding site.

According to the model of the rhodopsin polypeptide chain arrangement, four of five Trp residues (positions 126, 161, 175 and 265) are in the lipid matrix of the photoreceptor membrane. Each of them can be involved in the active site of the protein, however, Trp^{265} located in the VI segment near the aldimine bond seems to be the most probable.

The active site of the protein includes the His residue and its imidazole ring can be involved in the proton transfer upon the rhodopsin functioning. In the lipid matrix of the membrane the only residue, His^{211}, is situated. The data on the tertiary rhodopsin structure, chemical midification of the protein and application of the photoaffinity retinal analog provide detailed information on the retinal binding region of the visual pigment. A second functionally important rhodopsin domain is the C-terminal protein region exposed to the water phase. This region of the polypeptide chain is phosphorylated that results in inhibition of the action of the photolyzed molecule of the visual pigment. Determination of the primary structure of the C-terminal region reveals seven possible sites of kinase action.

There exist data on participation of the C-terminal domain in formation of its complex with GTPase; except for the finding that cleavage of the 12-membered peptide at the short time thermolysine action (region Val^{337}-Ala^{348}) does not change the pigment-enzyme binding.

The C-terminal region of rhodopsin contains three of ten Cys residues (positions 316, 322, and 323). Cys^{316} is accessible to the action of SH-blocking reagents upon the treatment of the native protein in the photoreceptor membrane, whereas residues Cys^{322} and Cys^{323} are not modified even after the pigment bleaching. The study of the chemistry of rhodopsin sulfhydryl groups by covalent chromatography on the basis of thiol-disulfide exchange shows that after immobilization of the protein with reduced disulfide bonds and cyanogen bromide cleavage the C-terminal fragment (containing residues Cys^{322} and Cys^{323}) is covalently bound to the carrier. If disulfide bonds are not reduced, cyanogen bromide fragment forms no covalent bond with the carrier. Being isolated and incubated with dithiotreitol this peptide is capable of immobilization. The data obtained indicate the presence of the disulfide bond between residues Cys^{322} and Cys^{323} in the C-terminal region of the rhodopsin molecule. The study of the fragment by the resonance Raman spectroscopy in the region of the S-S valence vibrations supports this. That is the first case of discovering a disulfide bond between the adjacent Cys residues in the naturally occurring sample, up to now such a bond has been found only in synthetic peptides.

Though the disulfide bond does not change upon the rhodopsin bleaching, its participation in the thiol-disulfide exchange when forming the pigment-transducin complex is possible, as in case of the receptor-insulin complex. Modification of transducin with SH-blocking reagents precludes formation of the photolyzed pigment-enzyme complex (the α-subunit of transducin is modified).

The third functionally important domain is the polypeptide chain region connecting the V and VI membrane segments. Apparently, it participates in complex formation with transducin, since at the thermolysin cleavage of this region into two membrane-bound fragments the photoreceptor disks considerably decrease its ability to the enzyme binding. If so, substantial homology in this region among rhodopsins from different sources should be observed. However, careful analysis of sequence data on bovine, human, fruit fly rhodopsins shows no homology. Thus this region seems to be less, if not at all essential for binding amplifier proteins. At the same time in the sequence of octopus rhodopsin, we are now studying, there is certain homology in the second cytoplasmic peptide loop.

O	Met	Ile	Ser	Ile	Asp	Arg	Tyr	Asn	Val	Ile	Gly	Arg	Pro	Met
D	Met	Ile	Ser	Leu	Asp	Arg	Tyr	Gln	Val	Ile	Val	Lys	Gly	Met
B	Val	Leu	Ala	Ile	Glu	Arg	Tyr	Val	Val	Val	Cys	Lys	Pro	Met

O - octopus, D - *Drosophila*, B - bovine (130-143)

Close similarity in this region may indicate its importance for binding the enzymes involved in signal transduction.

As mentioned above the C-terminal region of rhodopsin contains seven potential sites for the kinase action. However, 9 moles of Pi can be included into 1 mole of the protein upon its bleaching. One more site of the kinase action is probably located on the polypeptide chain between the V and VI membrane segments containing three possible sites of the enzyme action, Ser^{240}, Thr^{242}, and Thr^{243}. So incorporation of two phosphate groups in this region after the pigment illumination might regulate the complex formation of photolyzed rhodopsin with GTPase. Site-directed mutagenesis should hopefully enable us to find out functional roles for these and other parts of rhodopsins.

Another trend, our laboratory intensely develops nowadays, is research into structural bases of functioning of the systems of the active ion transport across biological membranes or ion pumps. Na^+,K^+-activated adenosine triphosphatase of plasma membranes from animal cells translocates sodium and potassium ions against electrochemical potential gradients (10); this process coupled with the ATP hydrolysis provides nonequilibrium distribution of the ions between the cell and medium. The Na^+,K^+-ATPase molecule consists of equimolar amounts of two subunits, α and β (for the studied enzyme from the outer medulla of pig kidney their Mr ~110 and ~47 kDa, respectively). Its carbohydrate moiety (Mr ~7 kDa) consists of two or three chains linked to the β-subunit by N-glycosidic bonds (11). The α-subunit is a functionally basic component, the ATP-hydrolyzing catalytic site being located in the cytoplasmic region of its polypeptide chain. Regions of protein molecule exposed on the membrane outer side make up the binding site for cardiac glycosides - enzyme specific inhibitors (12,13).

By immunochemical methods the exposed domains of the α-subunit were found at both membrane sides, whereas the hydrophilic part of the β-subunit involving the N-terminal glycosylated region was only on the outer membrane surface (11,14). The transmembrane organization of the α-subunit testifies to participation of its polypeptide chain in formation of a cation-conducting pathway in the Na^+,K^+-ATPase molecule. At present we know next to nothing about a functional role of β-glycoprotein. To elucidate the function mechanism of the active transport of monovalent cations the structural organization of the enzyme molecule should be analyzed from every point of view.

The minimal structural unit - protomer $\alpha\beta$ in a solubilized form completely preserves its ability for the ATP hydrolysis (15). On the other hand, there are contradictory data on the number of protomers in a functionally active complex of the native membrane. Our results on affinity modification of the enzyme active site with the alkylating ATP analog (16) and on freeze-fracture electron microscopy of the membrane-bound enzyme (17) evidence in favor of the tetrameric structure ($\alpha_4\beta_4$) of the Na^+-pump in the membrane.

Electron microscopy of two-dimensional crystals gave a general outlook on the three-dimensional enzyme structure. Three forms of these crystals were obtained by equilibrium cooling of the membrane in the presence of specific ligands (Mg^{2+}, VO_3^-, PO_4^{3-}) (18).

One of these forms was used to establish the three-dimensional structure of Na^+,K^+-ATPase (19,20). The cell unit of these crystals has the following parameters: a=72 Å; b=123 Å and γ=77° and the thickness is ~100 Å. The resolution is ~20 Å.

The cell unit is formed by two identical protomers $\alpha\beta$. Each protomer protrudes from the membrane and the major part of the protein is exposed on its surface. The protomers are somewhat narrowed in the central part and they contact on the one side of the membrane. The height of the contact region is ~20 Å.

These general ideas on the organization of the Na^+,K^+-ATPase molecule can be defined at the molecular level only by the analysis of complete chemical structures of the subunits. At present we bring this work to an end. The subunit primary structure is studied according to the conventional scheme involving the parallel application of genetic engineering and protein chemistry techniques (21-24). Since the α-subunit was known to have large hydrophilic regions exposed to both sides of the membrane and to be sensitive to the protease action in comparison with the glycosylated β-subunit, we decided to analyze first the peptides from the α-subunit hydrophilic regions. The selective tryptic digestion of the native membrane-bound Na^+,K^+-ATPase under the conditions for extensive hydrolysis of exposed regions of the α-subunit with complete retention of glycoprotein

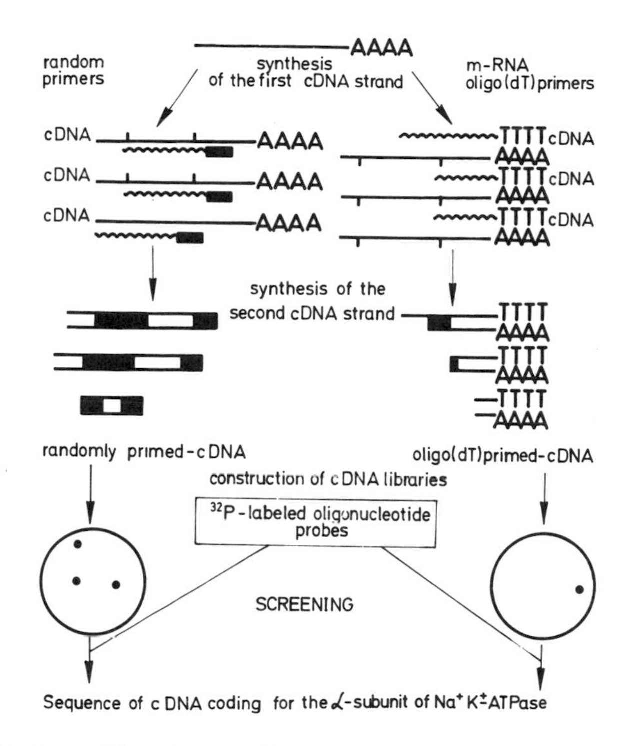

FIGURE 4. Cloning of cDNA coding for the α-subunit.

intactness was performed. The approach revealed amino acid sequences covering 33% of the α-subunit polypeptide chain (21). The data gave us information necessary for the synthesis of several specific 17-membered oligonucleotide probes.

To isolate RNA from the outer medulla of pig kidney use was made of selective precipitation with lithium chloride in the presence of urea and the vanadyl ribonucleoside complex. The fraction of mRNA resulted from two cycles of chromatography on oligo-(dT)-cellulose. Then it was ultracentrifuged in the sucrose concentration gradient. Fractions with mRNA of the α-subunit (25,26) were identified by hybridization with specific oligonucleotide probes (22,23).

To obtain cDNA corresponding *a fortiori* to all sites of this large mRNA not only oligo$(dT)_{12-18}$ but also "random" primers (mean-statistic hydrolyzate of calf thymus DNA) were used to initiate the synthesis of the first cDNA chain (Fig. 4). A second chain was synthesized by DNA polymerase I and RNAase H. Double-

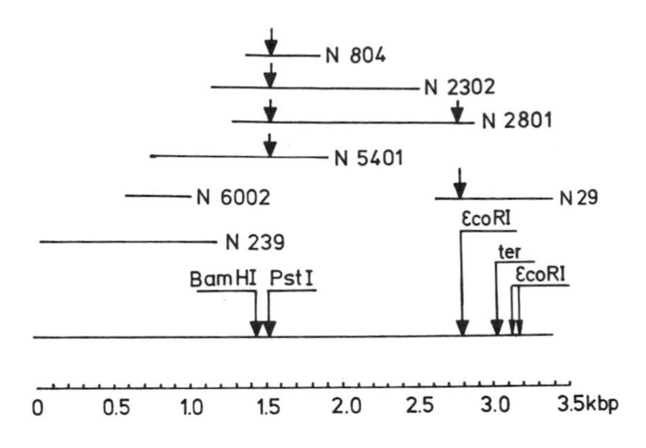

FIGURE 5. Map of cDNA coding for the α-subunit from pig kidney Na^{+},K^{+}-ATPase. Overlapping independent clones used to derive the complete cDNA structure.

stranded DNA was cloned in plasmid pBR 322 cleaved with Pst I by the method of dG:dC cohesive ends. To screen the clone libraries created by means of "random" ($2 \cdot 10^5$ colonies) and oligo(dT) ($5 \cdot 10^4$ colonies) primers, use was made of nucleotide probe 5'-TT(T,C) TC(A,G)AAGGC(A,G)TC(T,C)TT-3', corresponding to peptide Lys-Asp-Ala-Phe-Gln-Asn. As a result ~50 positive clones were found. Some of them were hybridized also with probes 5'-CC(T,C)TCNGG(A,G)AA(T,C)TG(T,C) TC-3' and 5'-TT(T,C)TC(T,C)TG(A,G)TTNGC(T,C)GT-3' corresponding to peptides Glu-Gln-Phe-Pro-Glu-Gly and Gln-Ala-Asn-Gln-Glu-Asn, respectively (Fig. 5) (22,23). In such a way clones with insertions, covering all regions of the α-subunit mRNA, were identified from the nucleotide sequence (25) determined by the Maxam - Gilbert and Sanger techniques. As a result the primary structure of the Na^{+},K^{+}-ATPase α-subunit was deduced (Fig. 6) (some of these data were published in (26)). All peptide structures established by the protein chemistry methods are completely in accord with that derived from the nucleotide sequence.

Sequence analysis of the mature protein N-terminal part (11) revealed the presence of five addition-

Position	Sequence
1-15	AGGACCCGGCGCGGGACACAGACCACCGCCACT ATG GGG AAG GGG GTT
(-5)-(-1)	Met-Gly-Lys-Gly-Val-
16-90	GGA CGC GAT AAA TAT GAG CCC GCA GCC GTG TCA GAG CAT GGC GAC AAA AAG AAG GCC AAG AAG GAG AGG GAT ATG
1-25	Gly-Arg-Asp-Lys-Tyr-Glu-Pro-Ala-Ala-Val-Ser-Glu-His-Gly-Asp-Lys-Lys-Lys-Ala-Lys-Lys-Glu-Arg-Asp-Met-
91-165	GAT GAG CTG AAG AAG GAA GTT TCT ATG GAT GAC CAT AAA CTT AGC CTT GAT GAG CTT CAT CGC AAA TAC GGA ACG
26-50	Asp-Glu-Leu-Lys-Lys-Glu-Val-Ser-Met-Asp-Asp-His-Lys-Leu-Ser-Leu-Asp-Glu-Leu-His-Arg-Lys-Tyr-Gly-Thr-
166-240	GAC TTG AGC CGA GGC TTA ACA CCT GCT CGA GCT GCT GAG ATC CTA GCC CGA GAC GGT CCC AAT GCC CTG ACA CCC
51-75	Asp-Leu-Ser-Arg-Gly-Leu-Thr-Pro-Ala-Arg-Ala-Ala-Glu-Ile-Leu-Ala-Arg-Asp-Gly-Pro-Asn-Ala-Leu-Thr-Pro-
241-315	CCA CCC ACA ACC CCT GAA TGG GTC AAG TTC TGT CGG CAG CTC TTC GGA GGC TTC TCC ATG TTA CTG TGG ATC GGA
76-100	Pro-Pro-Thr-Thr-Pro-Glu-Trp-Val-Lys-Phe-Cys-Arg-Gln-Leu-Phe-Gly-Gly-Phe-Ser-Met-Leu-Leu-Trp-Ile-Gly-
316-390	GCG ATT CTT TGT TTC TTG GCC TAT GGC ATT CAA GCT GCT ACA GAA GAG GAA CCT CAA AAT GAT AAT CTG TAC CTT
101-125	Ala-Ile-Leu-Cys-Phe-Leu-Ala-Tyr-Gly-Ile-Gln-Ala-Ala-Thr-Glu-Glu-Glu-Pro-Gln-Asn-Asp-Asn-Leu-Tyr-Leu-
391-465	GGT GTG GTG CTC TCC GCC GTC GTC ATC ATA ACT GGC TGT TTC TCC TAC TAT CAA GAA GCG AAA AGC TCA AAG ATC
126-150	Gly-Val-Val-Leu-Ser-Ala-Val-Val-Ile-Ile-Thr-Gly-Cys-Phe-Ser-Tyr-Tyr-Gln-Glu-Ala-Lys-Ser-Ser-Lys-Ile-
466-540	ATG GAA TCC TTC AAA AAC ATG GTT CCT CAG CAA GCC CTC GTG ATT CGA AAT GGT GAA AAG ATG AGC ATA AAT GCA
151-175	Met-Glu-Ser-Phe-Lys-Asn-Met-Val-Pro-Gln-Gln-Ala-Leu-Val-Ile-Arg-Asn-Gly-Glu-Lys-Met-Ser-Ile-Asn-Ala-
541-615	GAG GAA GTC GTC GTC GGG GAT CTG GTG GAG GTG AAG GGA GGG GAT CGA ATC CCT GCT GAC CTC AGG ATC ATA TCT
176-200	Glu-Glu-Val-Val-Val-Gly-Asp-Leu-Val-Glu-Val-Lys-Gly-Gly-Asp-Arg-Ile-Pro-Ala-Asp-Leu-Arg-Ile-Ile-Ser-
616-690	GCG AAC GGC TGC AAG GTG GAC AAC TCC TCC CTC ACT GGT GAA TCA GAA CCG CAG ACC AGG TCT CCA GAT TTC ACC
201-225	Ala-Asn-Gly-Cys-Lys-Val-Asp-Asn-Ser-Ser-Leu-Thr-Gly-Glu-Ser-Glu-Pro-Gln-Thr-Arg-Ser-Pro-Asp-Phe-Thr-
691-765	AAT GAG AAC CCC CTG GAG ACT AGG AAC ATC GCC TTT TTT TCA ACC AAC TGC GTT GAA GGC ACT GCA CGT GGT ATT
226-250	Asn-Glu-Asn-Pro-Leu-Glu-Thr-Arg-Asn-Ile-Ala-Phe-Phe-Ser-Thr-Asn-Cys-Val-Glu-Gly-Thr-Ala-Arg-Gly-Ile-
766-840	GTG GTG TAC ACT GGC GAT CGC ACC GTG ATG GGC AGA ATC GCT ACC CTT GCT TCC GGG CTG GAA GGG GGC CAG ACT
251-275	Val-Val-Tyr-Thr-Gly-Asp-Arg-Thr-Val-Met-Gly-Arg-Ile-Ala-Thr-Leu-Ala-Ser-Gly-Leu-Glu-Gly-Gly-Gln-Thr-
841-915	CCC ATC GCT GCG GAG ATT GAA CAT TTT ATC CAC ATC ATC ACG GGC GTG GCC GTG TTC CTG GGC GTG TCC TTC TTC
276-300	Pro-Ile-Ala-Ala-Glu-Ile-Glu-His-Phe-Ile-His-Ile-Ile-Thr-Gly-Val-Ala-Val-Phe-Leu-Gly-Val-Ser-Phe-Phe-
916-990	ATC CTT TCT CTG ATC CTC GAG TAC ACC TGG CTC GAG GCC GTC ATC TTC CTC ATC GGG ATC ATT GTA GCC AAC GTG
301-325	Ile-Leu-Ser-Leu-Ile-Leu-Glu-Tyr-Thr-Trp-Leu-Glu-Ala-Val-Ile-Phe-Leu-Ile-Gly-Ile-Ile-Val-Ala-Asn-Val-
991-1065	CCT GAA GGT TTG CTG GCC ACC GTC ACG GTG TGC TTG ACC CTG ACT GCC AAG CGC ATG GCC AGG AAG AAC TGC CTT
326-350	Pro-Glu-Gly-Leu-Leu-Ala-Thr-Val-Thr-Val-Cys-Leu-Thr-Leu-Thr-Ala-Lys-Arg-Met-Ala-Arg-Lys-Asn-Cys-Leu-
1066-1140	GTG AAG AAC TTG GAG GCT GTG GAG ACC CTG GGG TCC ACA TCC ACC ATC TGC TCA GAC AAA ACC GGA ACC CTC ACC
351-375	Val-Lys-Asn-Leu-Glu-Ala-Val-Glu-Thr-Leu-Gly-Ser-Thr-Ser-Thr-Ile-Cys-Ser-Asp-Lys-Thr-Gly-Thr-Leu-Thr-
1141-1215	CAG AAC CGA ATG ACA GTG GCC CAC ATG TGG TTC GAC AAT CAA ATC CAC GAG GCT GAC ACG ACG GAA AAT CAG AGC
376-400	Gln-Asn-Arg-Met-Thr-Val-Ala-His-Met-Trp-Phe-Asp-Asn-Gln-Ile-His-Glu-Ala-Asp-Thr-Thr-Glu-Asn-Gln-Ser-
1216-1290	GGT GTC TCA TTC GAC AAG ACT TCG GCC ACC TGG CTT GCT CTG TCC AGA ATT GCA GGT CTT TGT AAC AGG GCA GTG
401-425	Gly-Val-Ser-Phe-Asp-Lys-Thr-Ser-Ala-Thr-Trp-Leu-Ala-Leu-Ser-Arg-Ile-Ala-Gly-Leu-Cys-Asn-Arg-Ala-Val-
1291-1365	TTC CAG GCC AAC CAG GAA AAC CTA CCT ATC CTG AAG CGG GCA GTG GCG GGC GAC GCC TCC GAG TCC GCG CTC TTA
426-450	Phe-Gln-Ala-Asn-Gln-Glu-Asn-Leu-Pro-Ile-Leu-Lys-Arg-Ala-Val-Ala-Gly-Asp-Ala-Ser-Glu-Ser-Ala-Leu-Leu-
1366-1440	AAG TGC ATC GAG CTG TGC TGT GGG TCC GTG AAG GAG ATG AGG GAG CGA TAC ACC AAG ATC GTC GAG ATT CCC TTC
451-475	Lys-Cys-Ile-Glu-Leu-Cys-Cys-Gly-Ser-Val-Lys-Glu-Met-Arg-Glu-Arg-Tyr-Thr-Lys-Ile-Val-Glu-Ile-Pro-Phe-
1441-1515	AAC TCC ACC AAC AAG TAC CAG CTG TCC ATC CAC AAG AAC CCC AAC ACG GCT GAG CCC CGG CAC CTG CTG GTG ATG
476-500	Asn-Ser-Thr-Asn-Lys-Tyr-Gln-Leu-Ser-Ile-His-Lys-Asn-Pro-Asn-Thr-Ala-Glu-Pro-Arg-His-Leu-Leu-Val-Met-
1516-1590	AAA GGT GCT CCA GAA AGG ATC CTG GAC CGC TGC AGC TCC ATC CTC ATC CAC GGC AAG GAG CAG CCC CTA GAC GAG
501-525	Lys-Gly-Ala-Pro-Glu-Arg-Ile-Leu-Asp-Arg-Cys-Ser-Ser-Ile-Leu-Ile-His-Gly-Lys-Glu-Gln-Pro-Leu-Asp-Glu-
1591-1665	GAG CTG AAG GAC GCC TTT CAG AAC GCC TAC CTG GAG CTG GGT GGC CTC GGG GAA CGC GTG CTG GGT TTC TGC CAC
526-550	Glu-Leu-Lys-Asp-Ala-Phe-Gln-Asn-Ala-Tyr-Leu-Glu-Leu-Gly-Gly-Leu-Gly-Glu-Arg-Val-Leu-Gly-Phe-Cys-His-
1666-1740	CTT TTC CTG CCG GAC GAG CAG TTC CCC GAA GGC TTC CAG TTT GAC ACC GAC GAT GTG AAT TTC CCT CTC GAT AAT
551-575	Leu-Phe-Leu-Pro-Asp-Glu-Gln-Phe-Pro-Glu-Gly-Phe-Gln-Phe-Asp-Thr-Asp-Asp-Val-Asn-Phe-Pro-Leu-Asp-Asn-
1741-1815	CTC TGC TTC GTT GGG CTC ATC TCC ATG ATT GAC CCA CCG CGA GCG GCC GTC CCG GAT GCC GTG GGC AAA TGT CGA
576-600	Leu-Cys-Phe-Val-Gly-Leu-Ile-Ser-Met-Ile-Asp-Pro-Pro-Arg-Ala-Ala-Val-Pro-Asp-Ala-Val-Gly-Lys-Cys-Arg-
1816-1890	AGC GCT GGC ATT AAG GTC ATC ATG GTC ACC GGC GAT CAT CCC ATC ACA GCC AAA GCC ATT GCC AAA GGT GTG GGC
601-625	Ser-Ala-Gly-Ile-Lys-Val-Ile-Met-Val-Thr-Gly-Asp-His-Pro-Ile-Thr-Ala-Lys-Ala-Ile-Ala-Lys-Gly-Val-Gly-
1891-1965	ATC ATC TCG GAA GGC AAT GAA ACG GTC GAA GAC ATC GCT GCC CGC CTC AAC ATC CCA GTG AGC CAG GTG AAC CCC
626-650	Ile-Ile-Ser-Glu-Gly-Asn-Glu-Thr-Val-Glu-Asp-Ile-Ala-Ala-Arg-Leu-Asn-Ile-Pro-Val-Ser-Gln-Val-Asn-Pro-
1966-2040	AGG GAT GCC AAG GCC TGC GTG GTC CAT GGA AGC GAT CTG AAA GAC ATG ACC TCG GAG CAG CTG GAT GAC ATC TTG
651-675	Arg-Asp-Ala-Lys-Ala-Cys-Val-Val-His-Gly-Ser-Asp-Leu-Lys-Asp-Met-Thr-Ser-Glu-Gln-Leu-Asp-Asp-Ile-Leu-
2041-2115	AAG TAC CAC ACG GAG ATC GTG TTT GCC CGG ACG TCT CCT CAG CAG AAG CTC ATC ATT GTG GAA GGC TGC CAG AGA
676-700	Lys-Tyr-His-Thr-Glu-Ile-Val-Phe-Ala-Arg-Thr-Ser-Pro-Gln-Gln-Lys-Leu-Ile-Ile-Val-Glu-Gly-Cys-Gln-Arg-
2116-2190	CAG GGC GCC ATC GTG GCC GTG ACT GGC GAC GGT GTC AAT GAC TCT CCC GCT CTG AAG AAG GCA GAC ATC GGG GTT
701-725	Gln-Gly-Ala-Ile-Val-Ala-Val-Thr-Gly-Asp-Gly-Val-Asn-Asp-Ser-Pro-Ala-Leu-Lys-Lys-Ala-Asp-Ile-Gly-Val-

```
2191-2265  GCC ATG GGG ATT GCT GGC TCG GAC GTG TCA AAG CAA GCT GCT GAC ATG ATC CTC CTG GAT GAC AAC TTC GCC TCC
 726- 750  Ala-Met-Gly-Ile-Ala-Gly-Ser-Asp-Val-Ser-Lys-Gln-Ala-Ala-Asp-Met-Ile-Leu-Leu-Asp-Asp-Asn-Phe-Ala-Ser-

2266-2340  ATT GTG ACG GGA GTA GAG GAA GGT CGT CTG ATC TTT GAT AAC TTG AAG AAA TCC ATT GCC TAC ACC CTC ACC AGT
 751- 775  Ile-Val-Thr-Gly-Val-Glu-Glu-Gly-Arg-Leu-Ile-Phe-Asp-Asn-Leu-Lys-Lys-Ser-Ile-Ala-Tyr-Thr-Leu-Thr-Ser-

2341-2415  AAC ATT CCA GAG ATC ACC CCC TTC CTG ATA TTT ATT ATT GCG AAC ATT CCA CTG CCC CTG GGC ACC GTC ACC ATC
 776- 800  Asn-Ile-Pro-Glu-Ile-Thr-Pro-Phe-Leu-Ile-Phe-Ile-Ile-Ala-Asn-Ile-Pro-Leu-Pro-Leu-Gly-Thr-Val-Thr-Ile-

2416-2490  CTC TGC ATC GAC TTG GGC ACA GAC ATG GTT CCT GCC ATC TCC CTG GCG TAT GAG CAG GCG GAG AGC GAC ATC ATG
 801- 825  Leu-Cys-Ile-Asp-Leu-Gly-Thr-Asp-Met-Val-Pro-Ala-Ile-Ser-Leu-Ala-Tyr-Glu-Gln-Ala-Glu-Ser-Asp-Ile-Met-

2491-2565  AAG AGG CAG CCC CGA AAC CCC AAG ACA GAC AAA CTC GTC AAT GAG CAG CTC ATC AGC ATG GCC TAC GGA CAG ATA
 826- 850  Lys-Arg-Gln-Pro-Arg-Asn-Pro-Lys-Thr-Asp-Lys-Leu-Val-Asn-Glu-Gln-Leu-Ile-Ser-Met-Ala-Tyr-Gly-Gln-Ile-

2566-2640  GGT ATG ATC CAG GCC CTG GGC GGC TTC TTC ACT TAC TTT GTG ATC CTG GCT GAG AAC GGC TTC CTC CCG ATT CAC
 851- 875  Gly-Met-Ile-Gln-Ala-Leu-Gly-Gly-Phe-Phe-Thr-Tyr-Phe-Val-Ile-Leu-Ala-Glu-Asn-Gly-Phe-Leu-Pro-Ile-His-

2641-2715  CTG CTG GGC CTC CGG GTG AAC TGG GAT GAC CGC TGG ATC AAC GAC GTG GAG GAC AGC TAC GGG CAG CAG TGG ACC
 876- 900  Leu-Leu-Gly-Leu-Arg-Val-Asn-Trp-Asp-Asp-Arg-Trp-Ile-Asn-Asp-Val-Glu-Asp-Ser-Tyr-Gly-Gln-Gln-Trp-Thr-

2716-2790  TAC GAA CAG AGG AAG ATC GTG GAG TTC ACC TGC CAC ACG GCC TTC TTT GTC AGC ATC GTG GTG GTG CAG TGG GCC
 901- 925  Tyr-Glu-Gln-Arg-Lys-Ile-Val-Glu-Phe-Thr-Cys-His-Thr-Ala-Phe-Phe-Val-Ser-Ile-Val-Val-Val-Gln-Trp-Ala-

2791-2865  GAC TTG GTC ATC TGC AAG ACC CGG AGG AAT TCC GTC TTC CAG CAG GGG ATG AAG AAC AAA ATC TTG ATC TTT GGC
 926- 950  Asp-Leu-Val-Ile-Cys-Lys-Thr-Arg-Arg-Asn-Ser-Val-Phe-Gln-Gln-Gly-Met-Lys-Asn-Lys-Ile-Leu-Ile-Phe-Gly-

2866-2940  CTC TTC GAA GAG ACG GCC CTG GCT GCT TTC CTC TCC TAC TGC CCC GGA ATG GGC GTG GCC CTG AGG ATG TAC CCC
 951- 975  Leu-Phe-Glu-Glu-Thr-Ala-Leu-Ala-Ala-Phe-Leu-Ser-Tyr-Cys-Pro-Gly-Met-Gly-Val-Ala-Leu-Arg-Met-Tyr-Pro-

2941-3015  CTC AAA CCT ACC TGG TGG TTC TGT GCC TTC CCC TAC TCG CTC CTC ATC TTC GTC TAT GAC GAA GTC AGG AAG CTC
 976-1000  Leu-Lys-Pro-Thr-Trp-Trp-Phe-Cys-Ala-Phe-Pro-Tyr-Ser-Leu-Leu-Ile-Phe-Val-Tyr-Asp-Glu-Val-Arg-Lys-Leu-

3016-3066  ATC ATC AGG CGA CGC CCT GGC GGC TGG GTG GAG AAG GAA ACC TAC TAC TAG ACCCCCTCCTGCACGCCG
1001-1016  Ile-Ile-Arg-Arg-Arg-Pro-Gly-Gly-Trp-Val-Glu-Lys-Glu-Thr-Tyr-Tyr
```

FIGURE 6. Nucleotide sequence of cDNA and deduced amino acid sequence of Na^+,K^+-ATPase α-subunit. a) Structures established by peptide sequencing (solid lines). b) Sequence which demans verification (dotted line).

al residues in the primary translation product. The C-terminal sequence found by carboxypeptidases coincides with that foregoing the termination codon (TAG). So the α-subunit polypeptide chain consists of 1016 amino acid residues.

Two known sites of the α-subunit involved in formation of the ATP-hydrolyzing catalytic site are located in the polypeptide chain as follows. A functionally important aspartic acid residue phosphorylated upon the acylphosphate intermediate formation is localized in position 369. The Cys-Ser-Asp-Lys sequence completely coincides with that reported in (27). It is noteworthy that the 368-375 region is a common element for catalytic site structures of all known transport ATPases (including F_0F_1-ATPases) (24).

Another region (496-506) of the α-subunit is, apparently, a part of the enzyme catalytic site containing the lysine residue, which can be modified with an irreversible inhibitor of the enzyme - fluorescein isothiocyanate (28,29).

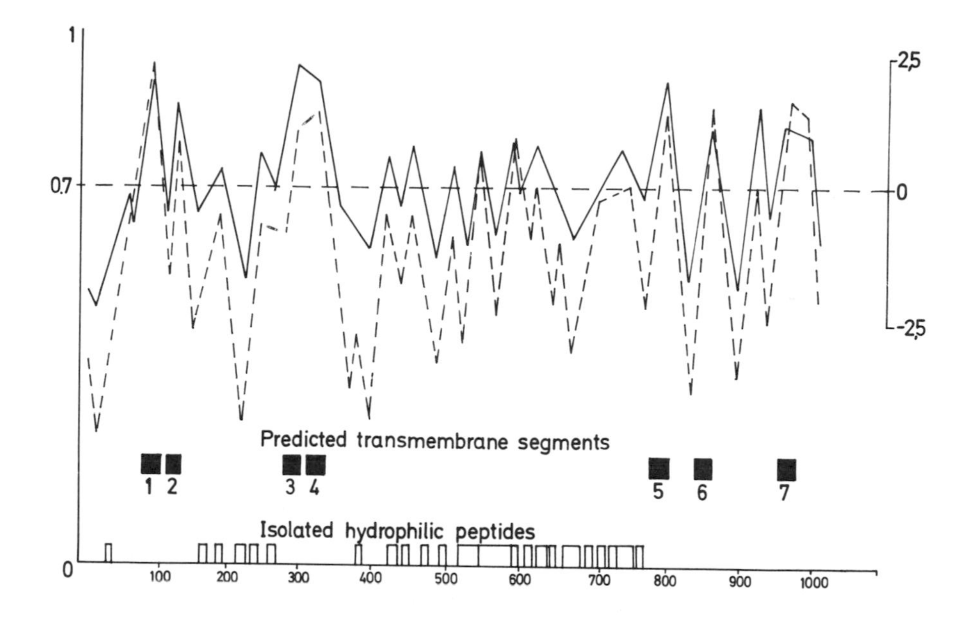

FIGURE 7. Hydrophobicity profile of Na^+,K^+-ATPase α-subunit. Dotted line - (5,32). Solid line - (33).

The first tentative model of the α-subunit polypeptide chain arrangement originated from the following data. The cytoplasmic region comprising the major part of the protein molecule includes the catalytic site (12,30); here the N-terminus is localized (30,31). To elucidate potential transmembrane segments of the α-subunit, local hydrophobicity of its polypeptide chain (Fig. 7) was calculated by two methods (5,32,33). Eleven hydrophobic regions, which could serve, in principle, as intramembrane rods, were found. However comparing the above data with those on amino acid sequences of the peptides resulted from tryptic hydrolysis of the α-subunit within the membrane-bound enzyme (21), we positioned regions 89-114, 123-142, 284-306 and 313-341, but not 530-553 and 569-597, inside the membrane. Sequences of isolated extramembrane fragments cover mainly the 143-283 and 342-778 regions, one of the peptides being localized in the N-terminal part (1-88). All the mentioned hydrophilic regions are really exposed to the cytoplasm.

It is hard to predict the arrangement of the remaining portion of the polypeptide since hydrophilic peptides are not isolated from this region. Positioning of only three of five segments (780-804, 842-867 and 946-972) in the membrane in order to expand an extracellular portion of the protein seems reasonable. If so, the polypeptide chain of the α-subunit transverses the lipid bilayer seven times and its C-terminus is exposed on the outer membrane site as shown in Fig. 8. Numbered arrows indicate the most probable localization of the points of primary proteolytic splits (12,30).

Limited tryptic hydrolysis of the membrane-bound Na^+,K^+-ATPase revealed the trypsin-unspecific splits of some bonds (Ile-Val, Ala-Ala, Ile-Met-Val, etc.) (21,25,26). These sequences are, apparently, situated on the surface of the protein globula in bends of the polypeptide chain. Positions of these bonds are marked by arrows in Fig. 8. Within the membrane-bound enzyme the C-terminus of the α-subunit is inaccessible to carboxypeptidases. Evidently it is protected, to a large extent, by the glycosylated β-subunit like all other outer regions. The proposed organization of the α-subunit is rather tentative, for instance, the ratio of inner to outer hydrophilic regions is about 6:1. In this case the outer portion of the protein is smaller than it has been predicted from chemical modification with impermeable reagents (20,30,34,35). The extracellular domain of the α-subunit is to be large enough since it forms the binding site for almost all parts of the cardiac glycoside and the acceptor site for K^+-ions (12).

So the α-subunit may be of more complex organization than it is shown in Fig. 8. One cannot exclude, for example, the presence of unidentified hydrophilic stretches necessary to form cation-conducted pathway(s) inside the α-helical bundle. According to the model only three residues of glutamic acid (327, 953 and 954) are situated in the membrane that seems to be insufficient for ion transport. Another possibility is the cation being transported through the contact regions of αβ protomers within the functionally active tetramer (16).

As to the β-subunit we have already shown that its major portion is on the outer surface of the membrane (14). Here the N-terminus with the glycosylated residue Asn^2 is localized (11). The β-subunit regions exposed to cytoplasm were not revealed (14,

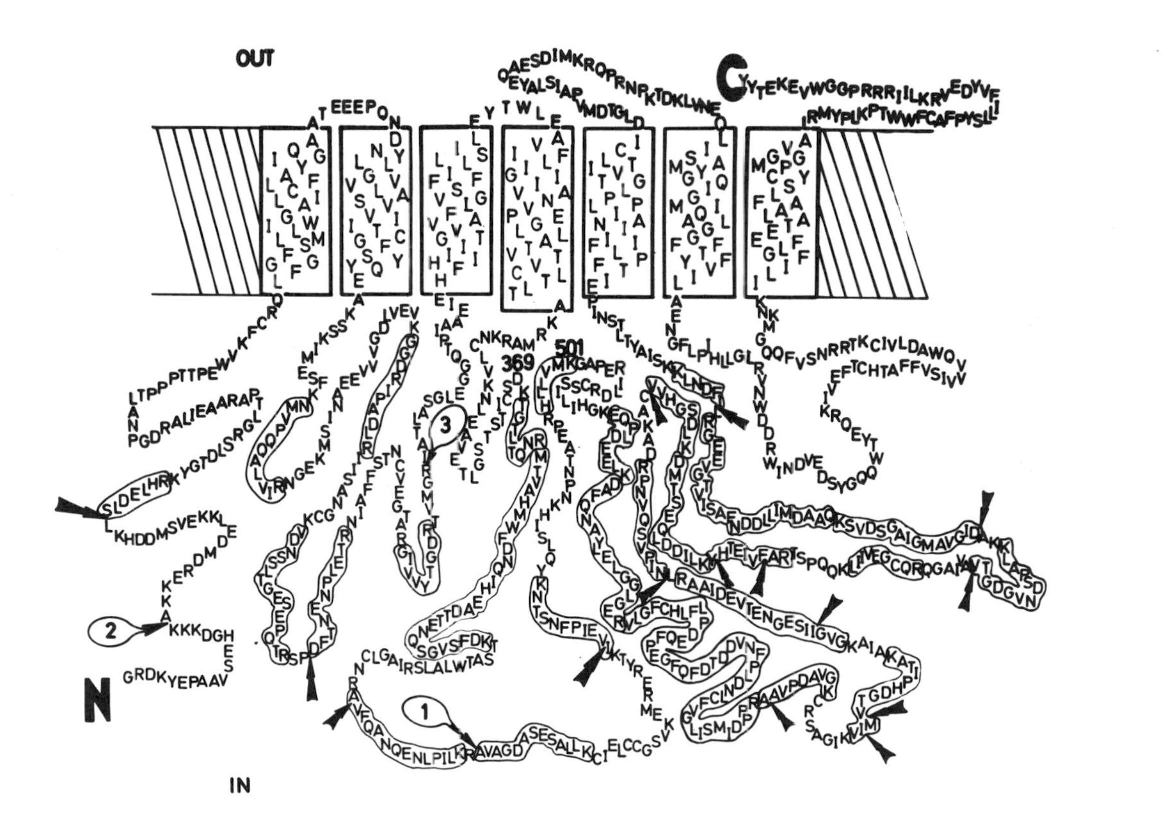

FIGURE 8. Model of Na^+,K^+-ATPase α-subunit arrangement in the lipid bilayer (details in the text).

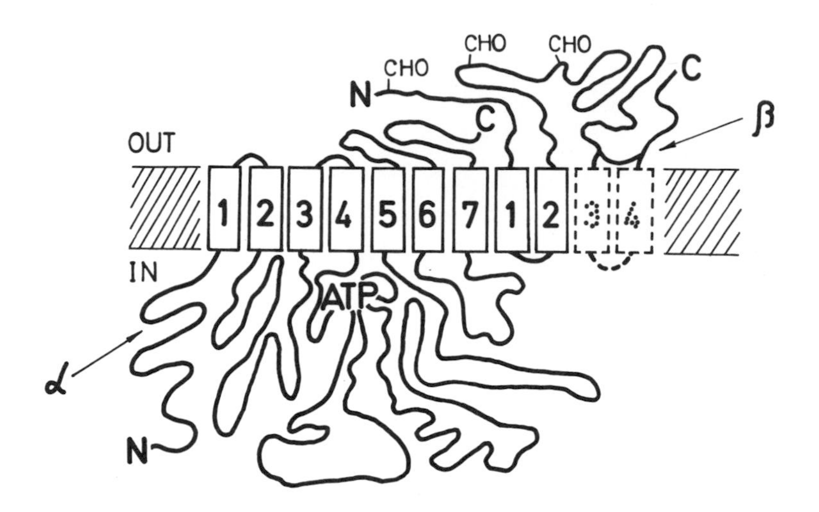

FIGURE 9. Proposed arrangement of Na^+,K^+-ATPase $\alpha\beta$ protomer in the membrane.

20,34). The C-terminus of the β-subunit accessible to the carboxypeptidase action in the intact enzyme is, obviously, situated on the outer membrane surface. So the β-subunit must traverse the membrane an even number of times. Maximum four stretches of the β-subunit together with those of the α-subunit fit the cross-section of the hydrophobic area of the three-dimensional model (19,20).

At present we sequenced over a half of the β-subunit polypeptide chain and determined partial structure of the carbohydrate moiety: SA_{2-3}(Galβ1→4GlcNAc)$_{2-3}$ $Man_3GlcNAc_2$-Asn (bi-and three-antennary chains) and SA_x (Galβ1→4GlcNAc)$_5$$Man_3$GlcNAc β1→4(Fucα1→6)GlucNAc-Asn. We almost completed the analysis of the cDNA nucleotide sequence. Figure 9 outlines the Na^+, K^+-ATPase organization in the membrane.

Deciphering of the primary structure of Na^+,K^+-ATPase and other ion-transporting ATPases will reveal homologous elements of the structure showing common function principles of ion pumps and their structural peculiarities predetermining unique ion transport selectivity.

We have discussed results on the primary structure of bacteriorhodopsin, rhodopsin and Na^+,K^+-ATPase derived from the analysis of both proteins and corresponding genes. Membrane arrangement of these proteins was established by various approaches: limited proteolysis, immunochemical technique and determination of the distribution of polar and non-polar amino acids along the respective polypeptide chains. The clear clustering of hydrophobic and hydrophilic residues made it possible to propose a model of the most probable spatial organization of Na^+,K^+-ATPase. This work has advanced our understanding of biochemical mechanisms which govern the functioning of membrane proteins.

REFERENCES

1. Stryer, L., Hurley, J.B., and Fung, K.-K., Trends in Biochem. Sci. 6:245 (1981).
2. Chabre, M., Annual Rev. Biophys. and Biophys. Chem. 14:331 (1985).
3. Oesterhelt, D., and Stoeckenius, W., Nature New Biol. 233:149 (1947).
4. Henderson, R., and Unwin, P.N.T., Nature 257:28 (1975).
5. Ovchinnikov, Yu.A., FEBS Letters 148:179 (1982).
6. Kimura, R., Nason, T.L., and Khorana, H.G., J. Biol. Chem. 257:2859 (1982).
7. Ovchinnikov, Yu.A., Abdulaev, N.G., Vasilov, R. G., Vturina, I.Yu., Kuryatov, A.B., and Kiselev, A.V., FEBS Letters 179:343 (1985).
8. Huang, K.S., Radhakrishnan, R., and Khorana, H. G., J. Biol. Chem. 257:13616 (1982).
9. Nathans, J., and Hoggnes, D., Cell 35:807 (1982).
10. Skou, J.C., Biochim. Biophys. Acta 23:394 (1957).
11. Dzhandzhugazyan, K.N., Modyanov, N.N., and Ovchinnikov, Yu.A., Bioorgan. Khim. 7:847 (1981).
12. Jørgensen, P.L., Biochim. Biophys. Acta 694:27 (1982).
13. Cantley, L.C., Curr. Top. Bioenerg. 11:201 (1981).
14. Dzhandzhugazyan, K.N., Modyanov, N.N., and Ovchinnikov, Yu.A., Bioorgan. Khim. 7:1790 (1981).
15. Brotherus, J.R., Jacobsen, L., and Jørgensen, P. L., Biochim. Biophys. Acta 731:290 (1983).
16. Dzhandzhugazyan, K.N., Modyanov, N.N., and Mustaev, A.A., Biol. Membranes 1:823 (1984).

17. Demin, V.V., Barnakov, A.N., Dzhandzhugazyan, K.N., and Vasilova, L.A., Bioorgan. Khim. 7:1783 (1981).
18. Demin, V.V., Barnakov, A.N., Lunev, A.V., Dzhandzhugazyan, K.N., Kuzin, A.P., Modyanov, N. N., Hovmoller, S., and Farrants, G., Biol. Membranes 1:831 (1984).
19. Demin, V.V., Kuzin, A.P., Barnakov, A.N., Lunev, A.V., Dzhandzhugazyan, K.N., Modyanov, N.N., and Ovchinnikov, Yu.A., Special FEBS Meeting, p. 74 Abstracts, Algarve, Portugal, 1985.
20. Ovchinnikov, Yu.A., Demin, V.V., Barnakov, A.N., Kuzin, A.P., Lunev, A.V., Modyanov, N.N., and Dzhandzhugazyan, K.N., FEBS Letters 190:73 (1985).
21. Arzamazova, N.M., Arystarkhova, E.A., Shafieva, G.I., Nazimov, I.V., Aldanova, N.A., and Modyanov, N.N., Bioorgan. Khim. 11:1598 (1985).
22. Petrukhin, K.E., Broude, N.E., Arsenyan, S.G., Grishin, A.V., Dzhandzhugazyan, K.N., and Modyanov, N.N., Bioorgan. Khim. 11:1607 (1985).
23. Petrukhin, K.E., Grishin, A.V., Arsenyan, S.G., Broude, N.E., Grinkevich, V.A., Filippova, L.Yu., Severtsova, I.V., and Modyanov, N.N., Bioorgan. Khim. 11:1636 (1985).
24. Modyanov, N.N., Arzamazova, N.M., Arystarkhova, E.A., Gevondyan, N.M., and Ovchinnikov, Yu.A., Biol. Membranes 8:844 (1985).
25. Ovchinnikov, Yu.A., Monastyrskaya, G.S., Arsenyan, S.G., Broude, N.E., Petrukhin, K.E., Grishin, A.V., and Modyanov, N.N., Dokl, Acad. Nauk SSSR, in press (1985).
26. Ovchinnikov, Yu.A., Monastyrskaya, G.S., Arsenyan, S.G., Broude, N.E., Petrukhin, K.E., Grishin, A.V., Arzamazova, N.M., Severtsova, I. V., and Modyanov, N.N., Dokl. Acad. Nauk SSSR 283:1278 (1985).
27. Walderhaug, M.O., Post, R.L., Saccomani, G., Leonard, R.T., and Briskin, D.P., J. Biol. Chem. 260:3852 (1985).
28. Farley, R.A., Tran, C.M., Carilli, C.T., Hawke, D., and Shively, J.E., J. Biol. Chem. 259:9532 (1984).
29. Kirley, T.L., Wallick, E.T., and Lane, L.K., Biochem. Biophys. Res. Commun. 125:767 (1984).
30. Jørgensen, P.L., Skriver, E., Hebert, H., and Maunsbauch, A.B., Ann. N.Y., Acad. Sci. 402:207 (1982).
31. Chin, G., and Forgac, M., Biochemistry 22:3405

(1983).
32. Capaldi, R.A., and Vanderkooi, G., Proc. Natl. Acad. Sci. USA 69:930 (1972).
33. Kyte, J., and Doolittle, R.F., J. Mol. Biol. 157:105 (1982).
34. Dzhandzhugazyan, K.N., and Jørgensen, P.L., Biochim. Biophys. Acta 817:165 (1985).
35. Sharkey, R., Biochim. Biophys. Acta 730:327 (1983).

ION MOVEMENT THROUGH CHANNELS WITH CONFORMATIONAL SUBSTATES

P. Läuger

Department of Biology
University of Konstanz
Konstanz, F.R.G.

I. INTRODUCTION

Ions permeate through cellular membranes by special mechanisms different from simple diffusion through the lipid bilayer. In the discussion of possible passive transport pathways, two alternatives are usually considered: carrier and channel mechanism. A carrier may be defined as a transport system with a binding site that is exposed alternately to the left and to the right side(but not to both sides simultaneously). A channel, on the other hand, consists of one or several binding sites arranged in a transmembrane sequence and is accessible from both sides at the same time.

Clear-cut examples of carrier and channel mechanisms in ion transport have been obtained from the study of certain small or medium-sized peptides and depsipeptides. Cyclodepsipeptides, such as valinomycin, have been shown to act by a translatory carrier mechanism which involves a movement of the whole carrier molecule with respect to the lipid matrix of the membrane. A well-characterized ion channel is the channel formed by the linear pentadecapeptide gramicidin A. In these cases the distinction between a channel which is more or less fixed within the membrane and a carrier moving within the lipid matrix is unambiguous.

The discrimination between carrier and channel mechanisms becomes less obvious in the case of large transport proteins spanning the cell membrane. Such a protein is unlikely to move as a whole within the membrane. It still can act as carrier (according to the definition given above), however, if a conformational change within the protein switches the binding site from a left-exposed to a right-exposed state. A channel, on the other hand, does not necessarily have a fixed, time-independent structure. Proteins may assume many conformational substates and move from one state to the other. Accordingly, a channel may carry out conformational transitions between states differing in the height of the energy barriers that restrict the movement of the ion. It can be shown that such a channel with multiple conformational states may approach the kinetic behaviour of a carrier. Channel and carrier models should therefore not be regarded as mutually exclusive possibilities, but rather as limiting cases of a more general mechanism.

II. THE GRAMICIDIN CHANNEL AS A MODEL CHANNEL

The finding that gramicidin A, a hydrophobic peptide with known primary structure, forms alkali-ion permeable channels in lipid bilayer membranes (1) opened up the possibility of studying ion permeation through channels in a simple model system. Gramicidin A is a linear pentadecapeptide with the sequence HCO-L-Val-Gly-L-Ala-D-Leu-L-Ala-D-Val-L-Val-D-Val-L-Trp-D-Leu-L-Trp-D-Leu-L-Trp-D-Leu-L-Trp-$NHCH_2CH_2OH$. Evidence that gramicidin A forms channels (and does not act as a mobile carrier) has been obtained in experiments in which very small amounts of the peptide were added to a planar bilayer membrane (1). Under this condition the membrane current under a constant applied voltage fluctuates in a step-like manner. The size of the single conductance step is about 90 pS in 1 M Cs^+, corresponding to a transfer of 6×10^7 Cs^+ ions per sec. This transport rate is larger by a factor of about one thousand than the turnover number of a mobile carrier of the valinomycin type, making a translatory carrier mechanism highly unlikely.

A structural model of the gramicidin channel has been proposed by Urry (2). According to this model which is now supported by many experimental findings (3), the channel consists of a helical dimer that is formed by head-to-head (formyl end to formyl end) association of two gramicidin monomers and is stabilized by intra- and inter-molecular hydrogen bonds. The central hole along the axis of the helix has a diameter of about 0.4 nm and is lined with oxygen atoms of the peptide carbonyls, whereas the hydrophobic amino-acid residues lie on the exterior surface of the helix. The total length of the dimer is about 2.5-3.0 nm, the lower limit of the hydrophobic thickness of the lipid bilayer.

The entry of the ion into the channel is made energetically favorable by interaction with the peptide carbonyls. In the

0.4 nm wide channel, water molecules can precede and follow the ion through the channel so that probably only part of the primary hydration shell is striped off. The interaction with the ligand groups creates a series of potential energy minima along the pathway of the ion. Superimposed onto this potential is the dielectric interaction of the ion with the water phase and the membrane lipid which gives rise to a broad energy barrier with the peak in the middle of the membrane. This picture predicts the existence of an energetically favorable "binding" site for the ion at either end of the channel.

III. CHANNELS WITH MULTIPLE CONFORMATIONAL STATES

Ion movement through a transmembrane protein channel may be described as a series of thermally activated jumps over energy barriers. The potential wells and barriers along the transport pathway are determined by the structure of the protein, i.e., by the spatial distribution of ligands such as oxygen atoms of carbonyl groups. In the traditional treatment of ion transport in channels the energy profile is considered to be fixed, i.e., independent of time and independent of the movement of the ion. Such a description corresponds to an essentially static picture of protein structure. During the last years evidence has been accumulated, however, that proteins can assume many conformational substates and at physiological temperatures rapidly move from one substate to the other (4,5). Support for the dynamic nature of protein structure comes from x-ray diffraction and Mössbauer studies, fluorescence depolarization experiments and nuclear magnetic resonance measurements. These and other studies have shown that internal motions in proteins occur in a wide time range, from picoseconds to seconds.

Direct evidence that ionic channels may assume different conformational states comes from single-channel records obtained by the patch-clamp technique (6). Intermediate conductance levels between the fully open and the fully closed state have been observed, for instance, with acetylcholine-activated endplate channels (7). In these systems the lifetimes of the substates were sufficiently long so that transitions could be observed directly in the current records. The detection of fast transitions between substates is limited, however, by the finite bandwidth of the measurement. This means that in many cases the observed single-channel current represents merely an average over unresolved conductance states. As will be discussed below, the existence of such hidden substates may strongly influence the observable properties of the channel, such as the current-voltage characteristic or the concentration dependence of conductance.

Of particular interest is the possibility that ion translocation in the channel becomes coupled to conformational transitions (8). In this case the conductance of the channel explicitly depends on the rate constants of conformational transitions. Such coupling may occur when the average lifetimes of conformational states are of the same order or lower than the dwelling times of ions in the binding sites. The channel may then exhibit unusual flux-coupling behavior in experiments with more than one permeable ion species.

A. Two-State Channel with a Single Binding Site

We consider a channel that (in the conducting state) fluctuates between two conformations A and B. We assume that the rate of ion flow through the channel is limited by two (main) barriers on either side of a single (main) binding site (Fig. 1). In series with the rate-limiting barriers, smaller barriers may be present along the pathway of the ion. This model corresponds to a channel consisting of a wide, water-filled pore and a narrow part, acting as a selectivity filter, in which the ion interacts with ligand groups. Since the binding site may be empty or occupied, the channel may exist in four substates (Fig. 2): A^o, conformation A, empty; A^*, conformation A, occupied; B^o, conformation B, empty; B^*, conformation B, occupied. The rate constants for transitions between A and B depend, in general, on whether the binding site is empty or occupied (i.e., $k^o_{AB} \neq k^*_{AB}$ and $k^o_{BA} \neq k^*_{BA}$). The coulombic field around the ion tends to polarize the neighbourhood by reorienting dipolar groups of the protein. In this way, the probability of a given transition may be strongly affected by the presence of the ion in the binding site.

The ohmic conductance Λ of the channel under the condition $c' = c'' = c$ has the form (8):

$$\Lambda(c) = \frac{z^2 e_o^2}{kT} \cdot \frac{c(\alpha + \beta c)}{\gamma + \delta c + \varepsilon c^2} \qquad [1]$$

z: valency of the permeable ion; e_o: elementary charge; k: Boltzmann's constant; T: absolute temperature.
The parameters α, β, δ, and ε are concentration-independent

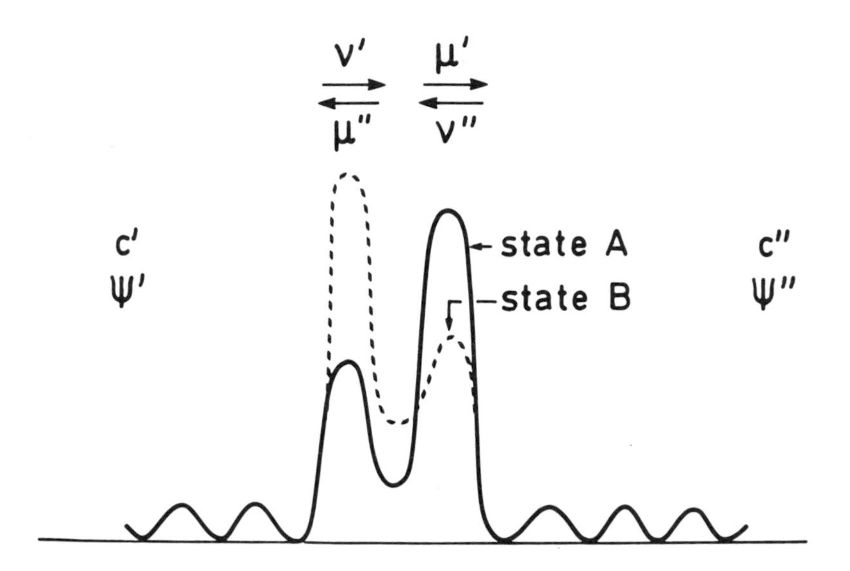

FIGURE 1. Energy profile of a channel with two conformational states: ν' and ν'' are the frequencies of jumps from the soultions into the empty site; μ' and μ'' are the jumping frequencies from the occupied site into the solutions; c', c'' and ψ', ψ'' are the ion concentrations and the electrical potentials in the left and right aqueous solutions.

combinations of all rate constants.

Thus, $\Lambda(c)$ is a nonlinear function of ion concentration containing terms that are quadratic in c. This behavior may be compared with the properties of a one-site channel with fixed barrier structure, which always exhibits a simple saturation characteristic of the form

$$\Lambda(c) = \frac{z^2 e_o^2}{kT} \cdot \frac{\rho c}{\mu + \rho c} \cdot \frac{\mu' \mu''}{\mu} \qquad [2]$$

$(\mu \equiv \mu' + \mu''; \ \nu \equiv \nu' + \nu'' = c\rho)$

It can easily be shown that for certain combinations of rate constants $\Lambda(c)$ goes through a maximum with increasing ion concentration. Such a nonlinear concentration dependence

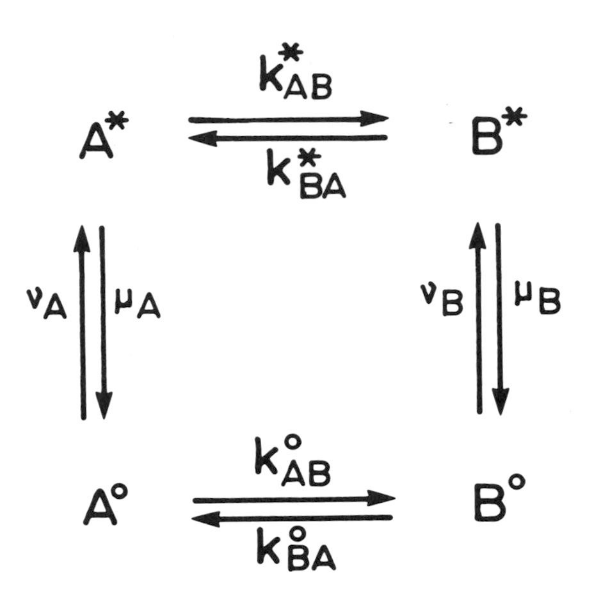

FIGURE 2. Transitions among four substates of a channel with one binding site. A°, conformation A, empty; A*, conformation A, occupied; B°, conformation B, empty; B*, conformation B, occupied.

of conductance is usually taken as evidence for ion-ion interaction in the channel or for the existence of regulatory binding sites. In the channel model discussed here, the nonlinearity of $\Lambda(c)$ is a direct consequence of the coupling between ion flow and conformational transitions.

For further discussion of equation [1], it is useful to consider two limiting cases in which conformational transitions are either much slower or much faster than ion translocation.

$$\text{Case 1: } k^o_{AB},\ k^*_{AB},\ k^o_{BA},\ k^*_{BA} \ll \nu_A,\ \nu_B,\ \mu_A,\ \mu_B$$

Under these conditions, the mean lifetime of a given state is much longer than the average time an ion spends in the energy well, which may be as short as 10^{-11} sec. Since many ions pass through the channel during the lifetime of the individual state, a well-defined conductance can be assigned to each state. On the other hand, the frequency of transitions between state A and B may still be much too high to be resolved in a single-channel record. The observed current is then averaged over the two rapidly interconverting conductance states. Under the conditions given above, equation [1] reduces to

$$\Lambda(c) = p_A \Lambda_A + (1 - p_A)\Lambda_B \quad [3]$$

where Λ_A and Λ_B are the conductances of the channel in states A and B, respectively, which have the form of equation [2]; p_A is the probability of finding the channel in state A (A^o or A^*) which, in the vicinity of equilibrium, is given by (with $K_A \equiv \rho_A/\mu_A$; $K_B \equiv \rho_B/\mu_B$; $\kappa^* K_A = \kappa^o K_B$; $\kappa^* = k^*_{AB}/k^*_{BA}$; $\kappa^o = \kappa^o = k^o_{AB}/k^o_{BA}$):

$$p_A = \frac{1 + cK_A}{1 + \kappa^o + (1 + \kappa^*)cK_A} \quad [4]$$

According to equation [3], Λ is equal to the weighted average of the conductances in states A and B. Since not only Λ_A and Λ_B but also p_A contains the ion concentration c, the concentration dependence of Λ is different from the simple saturation characteristic given by equation [2].

If the frequency of conformational transitions is so low that discrete conductance states can be observed in a single-channel current record, the mean lifetime τ_A and τ_B of the two conductance states can be determined. If $\tilde{p}^o_A = p^o_A/p_A$ is the conditional probability that the channel is in state A^o (given that it is in state A^o or A^*), the transitions frequency $1/\tau_A$ is equal to $\tilde{p}^o_A k^o_{AB} + (1 - \tilde{p}^o_A)k^*_{AB}$. This yields, in the limit of slow conformational transitions,

$$\tau_A = \frac{\mu_A + \nu_A}{\mu_A k^o_{AB} + \nu_A k^*_{AB}} \qquad \tau_B = \frac{\mu_B + \nu_B}{\mu_B k^o_{BA} + \nu_B k^*_{BA}} \quad [5]$$

Thus, the mean lifetimes depend on ion concentration (through ν_A and ν_B). Only when the transition frequencies are unaffected by the presence of the ion in the binding site ($k^o_{AB} = k^*_{AB}$, $k^o_{BA} = k^*_{BA}$) are the lifetimes given by the usual concentration indpendent relationships $\tau_A = 1/k^o_{AB}$ and $\tau_B = 1/k^o_{BA}$.

$$\text{Case 2: } k^{o}_{AB},\ k^{*}_{AB},\ k^{o}_{BA},\ k^{*}_{BA} >> \nu_A,\ \nu_B,\ \mu_A,\ \mu_B$$

When interconversion of states A and B is much faster than ion transfer between binding site and water, coupling between ion translocation and conformational transitions is lost, since states A and B are always in equilibrium with each other, even for nonzero ion flow through the channel. This equilibrium may be described by introducing the probability $\bar{p}^{o}_{A}$ that an empty channel is in state A and the probability $\bar{p}^{*}_{A}$ that an occupied channel is in state A:

$$\bar{p}^{o}_{A} = \frac{1}{1+\kappa^{o}} \qquad \bar{p}^{*}_{A} = \frac{1}{1+\kappa^{*}} \qquad [6]$$

Under the condition of fast conformational transitions, equation [1] reduces to the simple form of equation [2] when the following substitutions are introduced:

$$\rho = \bar{p}^{o}_{A}\rho_A + (1-\bar{p}^{o}_{A})\rho_B \qquad [7]$$

$$\mu' = \bar{p}^{*}_{A}\mu'_A + (1-\bar{p}^{*}_{A})\mu'_B \qquad [8]$$

$$\mu'' = \bar{p}^{*}_{A}\mu''_A + (1-\bar{p}^{*}_{A})\mu''_B \qquad [9]$$

This means that in the limit of fast interconversion of states the equation for Λ becomes formally identical to the corresponding equation derived for a channel with fixed barrier structure, provided that the rate constants are replaced by weighted averages of the rate constants in the two states. This result [which can be generalized to multisite channels with more than two conformational states (Läuger et al. (8))] has to be expected since under the above conditions the lifetime of a given conformation is much shorter than the time an ion spends in the binding site, and therefore, the ion "sees" an average barrier structure. Despite the formal identity of the conductance equation with equation [2], the interpretation of the transport process in a channel with variable barrier structure is different, since an ion will preferably jump over the barrier when the barrier is low; this means that the jump rate largely depends on the frequency with which conformational states with low barrier heights are assumed.

B. Carrier-like Behavior of Channels

A special situation (with strong coupling) occurs when in state A the barrier to the right is very high (binding site mainly accessible from the left) and in state B the barrier to the left is very high (binding site mainly accessible from the right). In this case, neither state is ion conducting, but ions may pass through the channel by a cyclic process in which binding of an ion in state A from the left is followed by a transition from A to B and release of the ion to the right. In other words, the channel approaches the kinetic behavior of a carrier. (A carrier is defined as a transport system with a binding site that is exposed alternately to the left and to the right external phase.) Indeed, in the limit $\mu_A' \approx 0$, $\rho_A'' \approx 0$, $\mu_B'' \approx 0$, $\rho_B' \approx 0$, equation [1] reduces to the expression for the conductance of a carrier with a single binding site (Läuger, 1980). This means that channel and carrier mechanisms are not mutually exclusive possibilities; rather, a carrier may be considered as a limiting case of a channel with multiple conformational states.

C. Single-Channel Currents with Rectifying Behavior

An example of an ionic transport system that exhibits a strongly asymmetric current-voltage characteristic in symmetrical electrolyte solutions is the so-called inwardly rectifying potassium channel, which has been found in a number of biological membranes (9). It has recently been demonstrated that the open-channel current itself shows a rectifying characteristic (9). A possible explanation of this finding is the assumption that the channel fluctuates between different conductance states with voltage-dependent transition frequencies that are too high to be resolved in a single-channel record.

In order to illustrate the possibility that hidden conformational states give rise to rectifying single-channel currents, we assume that in the channel model described above the ion translocation rates in conformation B vanish ($\nu_B = \mu_B = 0$), so that this state becomes nonconducting. Furthermore, the mean lifetime of state A is assumed to be much longer than the mean dwell time of an ion in the binding site. The average single-channel current I is then given by

$$I = p_A I_A \qquad [10]$$

where p_A is the probability of finding the channel in conformation A,

$$p_A = \frac{\mu_A + \nu_A}{\mu_A(1+\kappa^o) + \nu_A(1+\kappa^*)} \qquad [11]$$

and I_A is the current through the channel in state A,

$$I_A = ze_o[1-\exp(zu_o - zu)]\,\frac{\nu'_A \mu'_A}{\nu_A + \mu_A} \qquad [12]$$

$$u \equiv \frac{\psi' - \psi''}{kT/e_o} \qquad [13]$$

$$zu_o = \ln(c''/c') \qquad [14]$$

In order to simplify the analysis further we assume that the equilibrium constants κ^o and κ^* for the conformational states are identical ($\kappa^o = \kappa^* \equiv \kappa$) so that $p_A = 1/(1+\kappa)$. The voltage dependence of the transitions between conformations A and B may be formally described by introducing a gating charge of magnitude $a \cdot e_o$, so that

$$p_A = \frac{1}{1+\bar{\kappa}\cdot\exp(au)} \qquad [15]$$

where $\bar{\kappa}$ is the value of κ for $u = 0$. Thus, for negative voltages u, the probability p_A approaches unity but vanishes for increasing positive values of u. This means that the single-channel current I (equation [10]) exhibits a strongly rectifying behavior for sufficiently large values of a.

IV. CONFORMATIONAL TRANSITIONS DRIVEN BY AN EXTERNAL ENERGY SOURCE: ION PUMPS

The concept of channels with multiple conformational states may be used as a basis for the description of active ion transport. An ion channel functions as a pump when the energy profile of the channel is transiently modified in an appropriate way by an energy-supplying reaction (10-12). Absorption of a light-quantum, transitions to another redox state, or phosphorylation of the channel protein may alter the binding constant of an ion-binding site in the channel and, at the same time, change the height of adjacent barriers. In this way an ion may be preferentially released to one side of the membrane, while during the transition back to the original state of the channel another ion is taken up from the opposite side. As a specific example we consider a proton pump driven by the hydrolysis of ATP, such as the proton-translocating ATPases in fungi and higher plants (13,14). A minimum model of the pumping cycle is depicted in Fig. 3. It is assumed that in the dephosphorylated state (HA) of the pump a proton is located in a binding site which is accessible from the left-hand (cytoplasmic) medium but separated from the right-hand (extracellular) medium by a high barrier. Phosphorylation creates a state HB in which the barrier heights are changed in such a

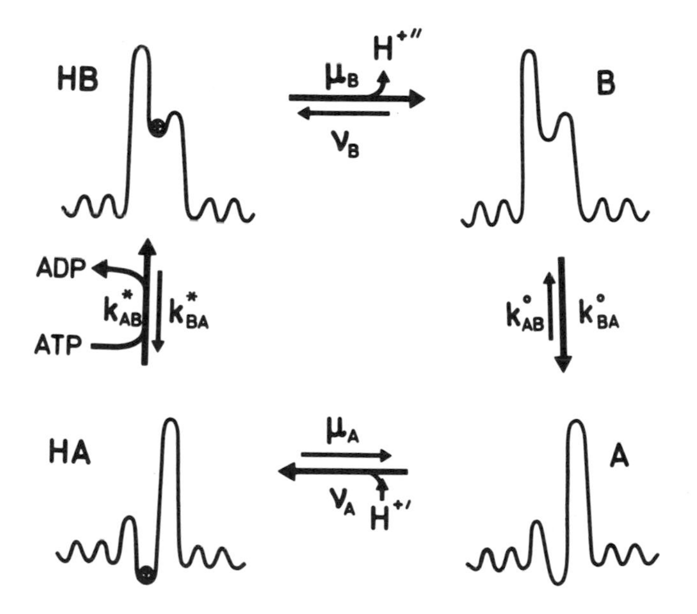

FIGURE 3. Channel mechanism for an ATP-driven proton pump. The energy profile of the channel is transiently modified by phosphorylation of the channel protein. In the dephosphorylated state HA/A the proton binding site is exposed to the left-hand (cytoplasmic) medium (') and in the phosphorylated state HB/B to the right-hand (extracellular) medium ("). During the cycle HA→HB→B→A→HA a proton is translocated from the cytoplasmic to the extracellular medium.

way that the proton is released preferentially to the external medium. After dissociation of H^+ the protein is dephosphorylated and relaxes back to a conformation with a low barrier on the left (cytoplasmic) side ($B \rightarrow A$). The original state is restored by uptake of H^+ from the cytoplasmic side ($A \rightarrow HA$). During the cycle a proton is translocated from the cytoplasm (phase ') to the extracellular medium (phase ").

The rate constants for the transition between HA and HB (Fig. 3) may be written as

$$k^*_{AB} = pc_T + qc_P \qquad [16]$$

$$k^*_{BA} = rc_D + s \qquad [17]$$

where c_T, c_D, and c_P are the concentrations of ATP, ADP, and inorganic phosphate (P_i), respectively. In Eqs. [16] and [17] it is assumed, for generality, that HA may be phosphorylated not only by ATP but also by direct reaction with P_i and that dephosphorylation may take place either by transfer of P_i from the protein to ADP or by release of P_i to the aqueous medium. It may easily be shown that the rate constants p, q, r, and s are connected by

$$qr/ps = \bar{c}_T/\bar{c}_D\bar{c}_P = 1/K \qquad [18]$$

where $\bar{c}_T$, $\bar{c}_D$, and $\bar{c}_P$ are equilibrium concentrations and K is the equilibrium constant of ATP hydrolysis. For the rate constants of uptake and release of the proton the same notation is used as in Fig. 1.

If the pump starts to work at zero initial voltage ($V \equiv \psi' - \psi'' = 0$) and equal proton concentrations in both aqueous phases ($c' = c''$), a difference in the electrochemical potential μ_H of H^+ builds up, which consists partly in a voltage V and partly in a pH difference. With increasing $\Delta\tilde{\mu}_H$ the rates of the reverse processes ($B + H^{+''} \rightarrow HB$, $HA \rightarrow A + H^{+'}$) are enhanced so that eventually a limiting value of $\Delta\tilde{\mu}_H$ is reached at which the net rate of proton translocation vanishes. The limiting value of $\Delta\tilde{\mu}_H/F$ (F is the Faraday constant) is the so-called protomotive force (pmf) of the pump which is obtained as (15)

$$\mathrm{pmf} = \left(\frac{\Delta\tilde{\mu}_H}{F}\right)_{max} = -\left[\frac{RT}{F}\ln\frac{c'}{c''} + (\psi' - \psi'')\right]_{max}$$

$$= \frac{RT}{F}\ln\frac{X + \theta + \mu'_A\mu''_B/\mu''_A\mu'_B}{1 + \theta + \mu'_A\mu''_B/\mu''_A\mu'_B} \qquad [19]$$

$$\theta \equiv \frac{s}{rc_D} \left(1 + \frac{\mu'_A \mu''_B}{\mu''_A \mu'_B}\right) + \frac{\nu''_A}{\nu''_B rc_D k^o_{AB}} \left(\mu_B k^o_{BA} + k^o_{BA} k^*_{BA} + \nu_B k^*_{BA}\right)$$

$$+ \frac{\mu''_B}{\mu''_A} \left[\frac{qc_P}{s} \left(1 + \frac{\nu_A}{k^o_{AB}}\right) \left(X + \frac{s}{rc_D}\right) + \frac{\mu_A}{rc_D}\right] \qquad [20]$$

The quantity X is related to the free energy ΔG of ATP hydrolysis:

$$X \equiv \frac{c_T/c_D c_P}{\bar{c}_T/\bar{c}_D \bar{c}_P} = K \frac{c_T}{c_D c_P} = \exp(-\Delta G/RT) \qquad [21]$$

For a pump with ideally asymmetric barriers which has the binding site in state A accessible only from the left and in state B only from the right ($\mu'_A = \mu''_B = \nu''_A = \nu'_B = 0$), Eq. [19] reduces to

$$\text{pmf} = \frac{RT}{F} \ln \frac{X + s/rc_D}{1 + s/rc_D} \qquad [22]$$

If, in addition, phosphorylation - dephosphorylation and proton translocation are perfectly coupled, i.e., if the rate of spontaneous dephosphorylation is small ($s \ll rc_D$), the protomotive force becomes

$$\text{pmf} = (RT/F) \ln X = - \Delta G/F \qquad [23]$$

Under these conditions the pump works with maximum thermodynamic efficiency, converting chemical free energy completely into (electro)osmotic energy.

V. CONCLUSION

The notion that the conformation of ionic channels fluctuates in time is suggested by recent studies on the dynamics of proteins. Whereas long-lived substates may be directly observed in records of single-channel currents, fast transitions may merely contribute to open-channel noise or may escape detection altogether. Such "hidden" substates nevertheless influence the measurable properties of channels such as the dependence of conductance on ion concentration.

Since ions interact with the ligand system of the channel, transitions between different conformational substates will depend in general on whether a binding site is empty or occupied. If the rate of conformational transitions is comparable to the jumping rates of ions in the channel, coupling between conformational transitions and ion flow occurs. In this case the permeability of the channel depends explicitly on the rate constants of conformational transitions. A two-state channel that in one state has the binding site accessible only from the left and in the other state only from the right exhibits carrier-like behavior. Many biological transport systems are likely to function by mechanisms intermediate between a "pure" carrier and a "pure" channel mechanism.

The concept of a channel with multiple conformational substates may also be applied to ion pumps. A channel acts as a pump when the barrier structure of the channel is transiently modified by an external energy source in such a way that an ion binding site is switched from a left-exposed to a right-exposed state. In this way cyclic conformational changes driven by an external energy source are coupled to vectorial ion translocation.

REFERENCES

1. Hladky, S. B., and Haydon, D. A., Biochim. Biophys. Acta 274:294 (1977).
2. Urry, D. W., Proc. Natl. Acad. Sci. USA 68:672 (1971).
3. Läuger, P., Angew, Chemie, Intern. Ed. 1985 (in press).
4. Frauenfelder, H., Petsko, G. A., and Tsernoglu, D., Nature (Lond.) 280:558 (1979).
5. Karplus, M., and McCammon, J. A., Ann. Rev. Biochem. 52:263 (1983).
6. Sakmann, B., and Neher, E. (eds), in Single Channel Recording, Plenum Publishing Comp., New York, 1983.
7. Hamill, O.P., and Sakmann, B., Nature (Lond.) 294:462 (1981).
8. Läuger, P., Stephan, W., and Frehland, E., Biochim. Biophys. Acta 602:167 (1980).
9. Sakmann, B., and Trube, G., J. Physiol. (Lond.) 347:641 (1984).
10. Patlak, C. S., Bull. Math. Biophys. 19:209 (1957).
11. Jardetzky, O., Nature (Lond.) 211:969 (1966).
12. Läuger, P., Biochim. Biophys. Acta 553:143 (1979).
13. Goffeau, A., and Slayman, C. W., Biochim. Biophys. Acta 639:197 (1981).

14. Poole, R. J., Ann. Rev. Plant Physiol. 29:437 (1978).
15. Läuger, P., Biochim. Biophys. Acta 779:301 (1984).

ELECTRICAL SIGNS OF RAPID FLUCTUATIONS IN THE ENERGY PROFILE OF AN OPEN CHANNEL

George Eisenman[1]

Department of Physiology
UCLA Medical School
Los Angeles, california, USA

I. INTRODUCTION

Remarkable progress has been made in elucidating the static structures of membrane proteins which provided the energetically favorable paths, called channels, by which ions cross the apolar interior of the cell membrane (1-3). An important interface between the structure of a channel and its function exists at the level of the energy profile generated by the structure and "felt" by the ion; for this energy profile determines the functional electrical behavior (e.g. selectivity, conductance, and binding) of the channel (4,5). The energy profile is a direct consequence of the structure (6-8) and is, in principle, calculable from it; although such calculations are presently restricted almost entirely to static structures (but see Ref. 9). Because of this, virtually all interpretations of ion permeation phenomena in channels are dominated by concepts based on a static picture of the channel's structure (and energy profile) during the permeation process (cf. 4,5, 10-16). The situation is hardly surprising in view of the astronomical number of parameters involved when trying to predict fluctuating protein structures to say nothing of the proliferation implied by the consequences of such a structure for the already numerous parameters of even the simplest barrier representation of it (17,18).

---

[1]Supported by NSF (BNS 84-11033) and USPHS (GM 24749).

Static pictures are useful and will be with us for some time to come; but we must be prepared to be flexible in the interpretations we draw from them; for they dominate our thinking in ways of which we are not always aware. Consider the following two examples. We presently interpret deviations from simple Michaelis-Menten dependence of conductance upon concentration as a sign of multiple binding sites (5,13, 15,19) or surface charge effects (20-22); but such behavior will be shown here to be a direct consequence in a 1-site, 1-ion channel, of structural fluctuations at an appropriate frequency, as first pointed out by Lauger et al. (23). It has also become common to accept "anomalous mole fraction effects" (5,24), such as a minimum in conductance in mixtures of two permeant ions at constant total concentration, as a sign of multiple occupancy (25-27). This is certainly correct from our present static perspective, for it can be deduced from Lauger's treatment (28) that a static 1-ion channel cannot show such an effect. However, this conclusion need not hold for a fluctuating 1-ion channel whose energy profile fluctuates on the time scale of ion hopping; for the form of Ciani's (29) expression for the conductance of a such a channel shows the distinct possibility for a minimum in conductance.

The present restriction to static structures is clearly an approximation since regions of proteins involved in ligand binding and catalysis characteristically exhibit particularly large thermal displacements (30), with smaller scale fluctuations acting as the "lubricant" for larger scale displacements (31). It seems likely that regions of a channel protein do likewise. In particular, the frequency of structural fluctuations is diagnostic of different kinds of motion. For example, individual ligands librate on a time scale of picoseconds. More concerted motions of groups of ligands, coupled via the H-bonding of water molecules in the channel or of the channel forming peptides, occur in the nanosecond range. Cooperative motions of even larger assemblies such as globular subunits or regions of subunits linked by flexible segments can occur in the 10-100 nanosecond range; and electrical signs of fluctuations in the submillisecond range have been identified (32). The most concerted motion of a channel, the gating process by which it is opened or closed, has been proposed to occur by tilting of whole subunits (3,33) on a time scale of milliseconds to seconds, depending on the channel.

Smaller motions of the subunits, occurring on a faster time scale, could account for the general properties of the fluctuations inferred here because such fluctuations will couple to ion hopping and will strongly influence the observable (time-averaged) electrical behavior (17,18,23,29,34,35) even though the fluctuations themselves are too fast for the present (microsecond) level of time resolution for electrical measurements. Indeed, it is not difficult to conceive of substates of the full gating process taking place in the 10-100 nanosecond range; and a teleological argument can be made for the likelihood that channel structures will fluctuate on the time scale of ion hopping because such fluctuations enable a higher flux to be maintained for a given selectivity than can occur in a system that is static or fluctuates at a frequency which does not couple to ion hopping. Since a higher flux for a given selectivity could confer an evolutionary advantage, natural selection should favor its occurrence.

It is, therefore, becoming increasingly apparent that the present static picture of a channel will need to be replaced by a more dynamic one; and it is interesting that a functional importance of both dynamic structural transformations and diffusional processes was suggested as early as 1943 by Schrodinger (36).

The present paper examines some consequence for the current-voltage and conductance-concentration behaviors of a channel when significant fluctuations in the energy profile occur on the same time scale as ion hopping. The basic strategy is the following. The acetylcholine receptor (AChR) channel is chosen as a prototype because it is the functionally simplest biological channel of known structure for which sufficiently detailed electrical data exist to confront with a fluctuating energy profile model. The simplest possible fluctuating profile model is developed for this channel, a fluctuating version of the conventional, static 2-barrier, 1-site model (10-12,20,37,38). In this "fluctuating 2B1S model" the energy profile for ion permeation is assumed to fluctuate between two different conformational states, each having a static 2-barrier, 1-site profile.

A preliminary exploration with static models is carried out first, from which the decision is made to reduce the parameters in the fluctuating model by 6 by fixing the locations of the barriers and well at physically reasonable locations which are found not to be critical for the fixed barrier model. Experimental current-voltage (I-V) relations (20, 22,38,39) and chord conductances (G) calculated from these are then used, together with a variable metric error minimization algorithm (40) on a VAX 11/730 computer to obtain a set of permeation rate constants and fluctuation frequencies that adequately describe the data without imposing any external constraints on the fluctuation frequency or on the rate constants for ion hopping. Best fits are obtained for various locations of site and barriers, the rate constants for which are expressed in the convenient form of energy profiles for permeation and fluctuation. The shared features of the fits are noted and a particular case for a centrally located site is examined further. Not surprisingly, in view of the previous general theory (23,29,34,35), the permeation profiles turn out to be highly asymmetrical with high and low barriers alternatively facing one side of the channel and then the other and with the fluctuations occurring on the same general time scale as ion hopping.

Next, the fluctuation frequency is then deliberately varied away from its optimal value, again under the constraints imposed by the experimental data, to assess how the frequency influences the theoretically predicted behavior. A surprisingly complex dependence of conductance on concentration (and voltage) is found, which appears to have no precedent in previous studies on static models, even those with multiple occupancy. The most unusual aspect of this behavior is its prediction of two different kinds of "2-site" behavior. The first kind of behavior is a deviation from linear Eadie-Hofstee (or Scatchard) plots at low voltages in the direction of a suppression of conductance at high concentration (which appears as a flattening or even a maximum of conductance). The second is an opposite deviation which shows up at low concentrations for high voltages of either sign as an enhancement of conductance, or "foot" on such plots. Although critical experiments have yet to be done to test these new expectations, particularly at low voltages and high concentrations, the theory can be confronted in a limited way by using a polynomial fit to the existing I-V data to generate the crucial chord conductance data at low voltages. This is done and the tentative conclusion reached that the best representation of the data occurs for a frequency of fluctuation between 16-30 megahertz. Some speculations about the implications of this frequency are then presented.

II. THEORY

A. The Energy Profiles

Figure 1 (top) defines the energy profile for ion translocation for a channel with 2 conformations o and *. States 0 and X represent the empty and occupied channel in conformation o. States 0* and X* represent the empty and occupied channel in conformation *. The transitions between these two conformational states are represented at the bottom; and the state diagram for the transitions among the 4 substates of the channel is given in Fig. 2. Rate constants from outer to inner solution are denoted by k's; those from inner to outer by b's. Subscripts 1, 2, 3 refer to the successive energy maxima or minima encountered from out to in.

The rate for an ion to jump into the site is proportional to the aqueous ionic activity (a) times the entry rate constants k_1, b_3, k^*_1, b^*_3, and the rate for an ion to leave the site is proportional to the exit rate constants b_1, k_3, b^*_1, k^*_3, (For simplicity, the solutions are assumed to be symmetrical in the present treatment). These 8 rate constants are

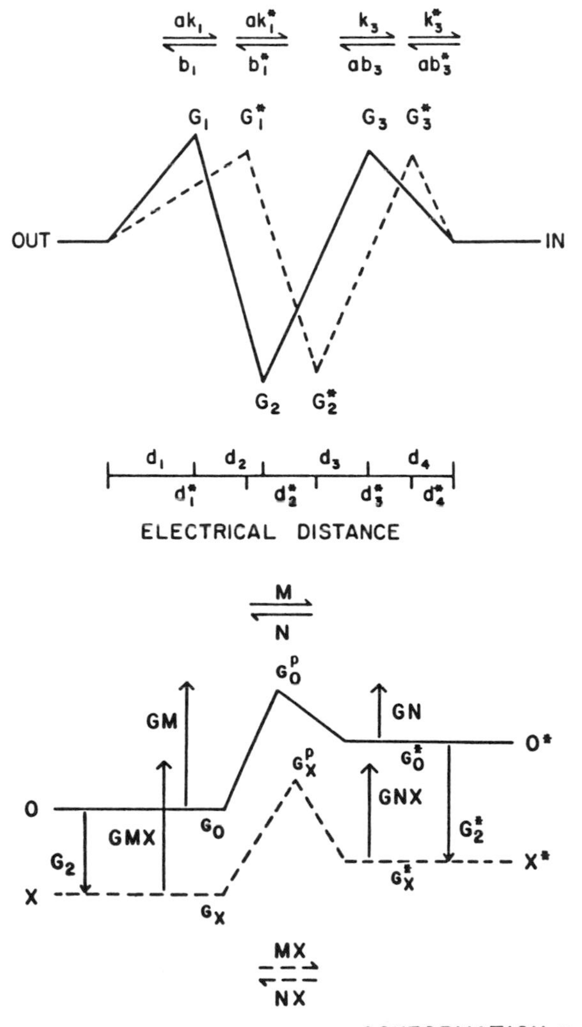

FIGURE 1. Energy profiles for translocation (top) and fluctuation (bottom).

defined in terms of the 6 Gibbs free energies G_1, G_2, G_3, G^*_1, G^*_2, G^*_3, for the peaks and wells through Eqs. 12-19. The voltage dependences of the rate constants are defined in the figure through the electrical distances which represent the fraction of the potential traversed during a jump under the constraint that the total potential drop occurs across the length of the channel (Eqs. 21-22).

The rates of fluctuation between the conformational states are defined through the activation energies GM and GN governing the forward and backward rates of transition between conformations o and * for the empty channel and GMX and GNX governing the forward and backward rates of transition between conformations o and * for the occupied channel (Eqs. 23-29).

The energy profiles of the top and bottom parts of Fig. 1 are interrelated. The energy levels for the empty and occupied o conformations (states 0 and X), labelled G_0 and G_X, are separated by the energy G_2; and the energy levels for the empty and occupied * conformations (states 0* and X*), labelled G^*_0 and G^*_X, are separated by the energy G^*_2

$$G_X - G_0 = G_2 \qquad [1]$$

$$G^*_X - G^*_0 = G^*_2 \qquad [2]$$

G_2 fixes the relative levels of the empty and occupied channels in conformation o through Eq. 1 and enables all the parameters of the model to be inferred from the energy dia-

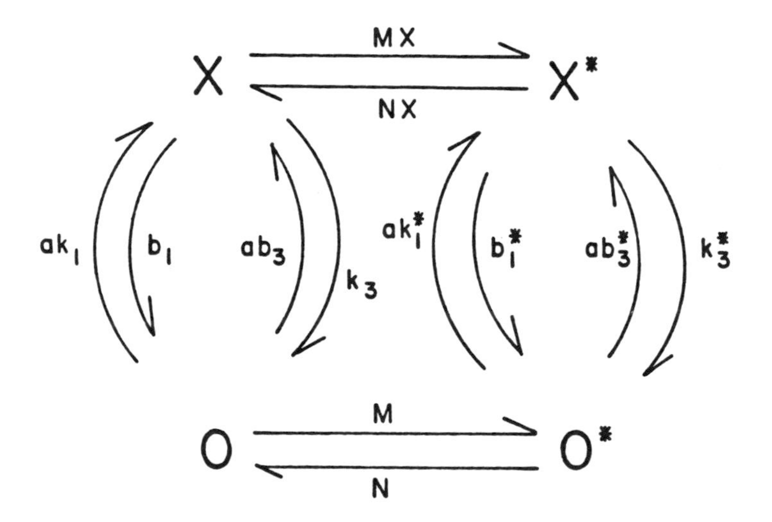

FIGURE 2. State diagram for transitions among the 4 possible channel states.

grams presented as inserts to later figures. G^*_2 imposes a similar constraint on the levels in conformation *. These two constraints link the fluctuation rate constants (cf. Eqs. 23-26) with those for translocation (cf. Eqs. 12-19) and thereby make one of these a dependent variable through the requirement that

$$-G_2 + GM - GN + G^*_2 + GNX - GNX = 0 \qquad [3]$$

obtained by summing the energies around a cycle in Fig. 1 (bottom), thus imposing microscopic reversibility.

The fluctuation energies in Fig. 1 are defined by

$$GM = G^P_0 - G_0 \qquad [4]$$

$$GN = G^P_0 - G^*_0 \qquad [5]$$

$$GMX = G^P_X - G_X \qquad [6]$$

$$GNX = G^P_X - G^*_X \qquad [7]$$

The free energy of the system is defined as zero when the channel is unoccupied and in state 0. The rate constants of translocations (Eqs. 12-19) or fluctuations (Eqs. 23-26) are inversely related to the exponential of the barrier heights by the Eyring (41) frequency factor, A,

$$\text{rate constant} = A/\exp\ (\text{barrier height}) \qquad [8]$$

where $kT/h = 6.21 \times 10^{12}\ \text{sec}^{-1}$, the rate constant is in reciprocal seconds, and the barrier height is in kT units.

B. Steady-State Current

The net steady-state current is the same across any barriers in series. The outward ionic flux across the left hand barrier at the top of Fig. 1 is given by the sum of outward fluxes for both the * and o conformations

$$\text{outward flux} = (b_1 P(X) + b^*_1 P(X^*))$$

where $(P(X), P(X^*))$ are the probabilities that the channel is occupied in conformation o or conformation * and b_1 and b^*_1 are the corresponding exit rate constants. The inward flux across this barrier is given by

$$\text{inward flux} = a.(k_1 P(0) + k^*_1 P(0^*))$$

which is the sum for both conformations of the probability that the channel is empty multiplied by the entrance rate.

The net current (I) is the difference of these contributions:

$$I/Q = (b_1 P(X) + b^*_1 P(X^*)) - a.(k_1 P(0) + k^*_1 P(0^*)) \qquad [9]$$

where Q is the value of the elementary charge (1.602×10^{-19} coulombs).

C. Matrix Method for Calculating Occupancy

The probability of occupancy in each state is defined by the vector |P,

$$|P = P(0), |P(X), P(0^*), P(X^*)| \qquad [10]$$

which can be evaluated numerically using the matrix methods (cf. 16,18) as

$$|P = |M^{-1} \cdot |B \qquad [11]$$

where |B is a 4-element vector (see (18) for details),

$$|B = |0, 0, 0, 1|$$

and |M is a 4 by 4 matrix obtained, as described below (also see Ref. 18), by eliminating the third row of the 5 × 4 matrix constructed from the state diagram by the following procedure.

The matrix |M is written with the help of the state diagram (Fig. 2) by expressing the rate constants for leaving the states of the header of the columns as the diagonal elements (these have negative signs). This header corresponds to the |P vector. The other elements represent the fluxes to the states labelled at the left from the states in the header. |M is:

STATES	P(0)	P(X)	P(0*)	P(X*)
0	$-a(k_1+b_3)$ $-M$	b_1+k_3	N	0
X	$a(k_1+b_3)$	$-b_1-k_3$ $-MX$	0	NX

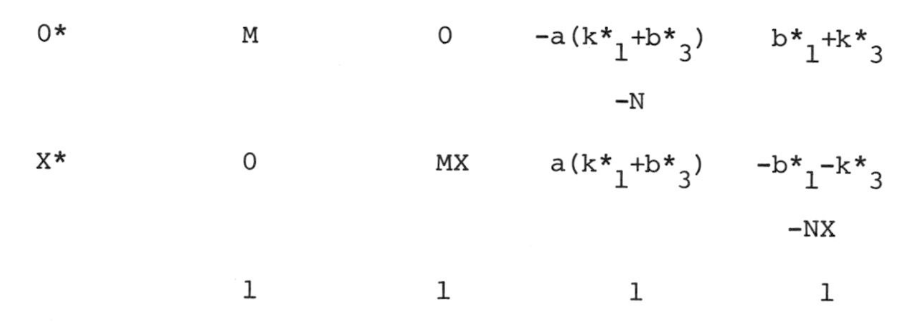

0*	M	0	$-a(k^*_1+b^*_3)$ $-N$	$b^*_1+k^*_3$
X*	0	MX	$a(k^*_1+b^*_3)$	$-b^*_1-k^*_3$ $-NX$
	1	1	1	1

Where, for example, the first row of $|M$ stands for the transition fluxes from and toward state 0, the total sum being zero (the corresponding element in the B vector):

$$-(a\ (k_1 + b_3) + M)\ P(0) + (b_1 + k_3)\ P(X) + NP(0^*) = 0$$

Here, the first term represents all transitions leaving state 0 in Fig. 2 (two ionic fluxes going to state X and one fluctuation going to state 0*). The second term represents the transition to state 0 from state X, which occurs by way of 2 ionic fluxes leaving state X toward state 0. The third term represents the transition to state 0 from state 0*, which is proportional to the backward rate constant N.

The first 4 rows of this 5 × 4 matrix are linearly related to each other, as seen by adding all these rows columnwise, thus producing a zero-vector. The fifth row is the normalization equation which expresses the requirement that the sum of all probabilities of possible states be equal to 1. This row is used, and any of the first 4 rows is eliminated to produce a 4 × 4 matrix.

Eliminating the 3rd row gives for $|M$:

STATES	P(0)	P(X)	P(0*)	P(X*)
0	$-a(k_1+b_3)$ $-M$	b_1+k_3	N	0
X	$a(k_1+b_3)$	$-b_1-k_3$ $-MX$	0	NX
X*	0	MX	$a(k^*_1+b^*_3)$	$-b^*_1-k^*_3$ $-NX$
	1	1	1	1

Using $|M$ to compute the probabilities through Eq. 11, the I-V behavior can be calculated from Eq. 9, using the following definitions of the rate constants.

D. Rate Constants for Ionic Translocation

The rate constants for ionic translocations and their voltage dependences are given in terms of the energy levels and distances of Fig. 1 by Eqs. 12-19:

$$k_1 = A \exp(-G - d_1 U) \quad [12]$$

$$b_1 = A \exp(G_2 - G_1 + d_2 U) \quad [13]$$

$$k_3 = A \exp(G_2 - G_3 - d_3 U) \quad [14]$$

$$b_3 = A \exp(-G_3 + d_4 U) \quad [15]$$

$$k^*_1 = A \exp(-G^*_1 - d^*_1 U) \quad [16]$$

$$b^*_1 = A \exp(G^*_2 - G^*_1 + d^*_2 U) \quad [17]$$

$$k^*_3 = A \exp(G^*_2 - G^*_3 - d^*_3 U) \quad [18]$$

$$b^*_3 = A \exp(-G^*_3 + d^*_4 U) \quad [19]$$

where A is the Eyring frequency factor and the voltage U is expressed in reduced (kT/Q = RT/F) units, defined in terms of the potential difference (in volts) between the inner and outer solutions as:

$$U = (V_{in} - V_{out})/(kT/Q) \quad [20]$$

The voltage dependences are expressed in terms of the fractional distances in the potential field

$$d_1 + d_2 + d_3 + d_4 = 1 \quad [21]$$

$$d^*_1 + d^*_2 + d^*_3 + d^*_4 = 1 \quad [22]$$

The rate constants for the fluctuations (M, N, MX, NX) are assumed, for simplicity, to have no intrinsic voltage dependence and are defined as:

$$M = A \exp(-GM) \quad [23]$$

$$N = A \exp(-GN) \quad [24]$$

$$OMX = A \exp(-GMX) \quad [25]$$

$$ONX = A \exp(- GNX) \qquad [26]$$

where OMX and ONX are the voltage independent portions of the fluctuation rate constants MX and NX for the occupied channel. Note that MX and NX can depend on voltage through Eqs. 27 and 28 because a voltage dependence due to the fluctuations occurs if a site occupied by an ion changes its location between the o and * conformations, according to:

$$MX = OMX \exp(-AA\ U/2) \qquad [27]$$

$$NX = ONX \exp(+AA\ U/2) \qquad [28]$$

due to the displacement of a site carrying an ion through the distance AA in the potential field.

$$AA = (d^*_1 + d^*_2) - (d_1 + d_2) \qquad [29]$$

Although there appear to be 21 parameters defining the behavior of this system (12 rate constants and 9 distances including AA); the Peak-Well formalism (14) shows that only 15 of the parameters of the model are actually independent (6 distances and 9 energies). This can be seen by counting the energies in Fig. 1.

III. FITTING THE MODEL TO EXPERIMENTAL I-V DATA FOR THE AChR CHANNEL

The theory can now be confronted with experimental data; and the results can be expressed in a compact form as energy profiles, assuming that A = kT/h and that the partial molar volume of water is the same in the channel as in bulk solution. Neither of these assumptions affects the comparison of data with theory nor the values of rate constants; but for different assumptions the energy levels in the figures would be different.

A. The Static 2B1S Model

Figure 3A characterized the I-V behavior of the AChR channel using the experimental data of Dani and Eisenman (20, 22,38,39) for Cs, which are plotted as the jagged curves at the 5 indicated symmetrical concentrations. Voltage is defined with reference to the inner solution, and outward current is defined as positive. In this, and in all subsequent figures, the horizontal scale represents 100 mV and the vertical scale is 10 pA for the current and 10 kT for the energy profiles. The smooth theoretical curves are produced by the static 2B1S model (10,12,20,38) with the indicated energy profile. The simple model fits the data surprisingly well, though not perfectly (Hamilton R-factor (42) = 0.08, corresponding to a cumulative relative misfit of 8% (see Appendix of Ref. 18).

The imperfections are not due to an inherent inability of the static 2B1S model with its 6 parameters to produce the appropriate shape at a given concentration. For example, Fig. 3B demonstrates that the model can provide an excellent fit at any single concentration, even with the locations of site and barriers fixed at the positions of Fig. 3A, provided that the steady-state energy levels are allowed to vary with concentrations as indicated for the 5 concentration-specific energy profiles indicated above. (Each energy profile cor-

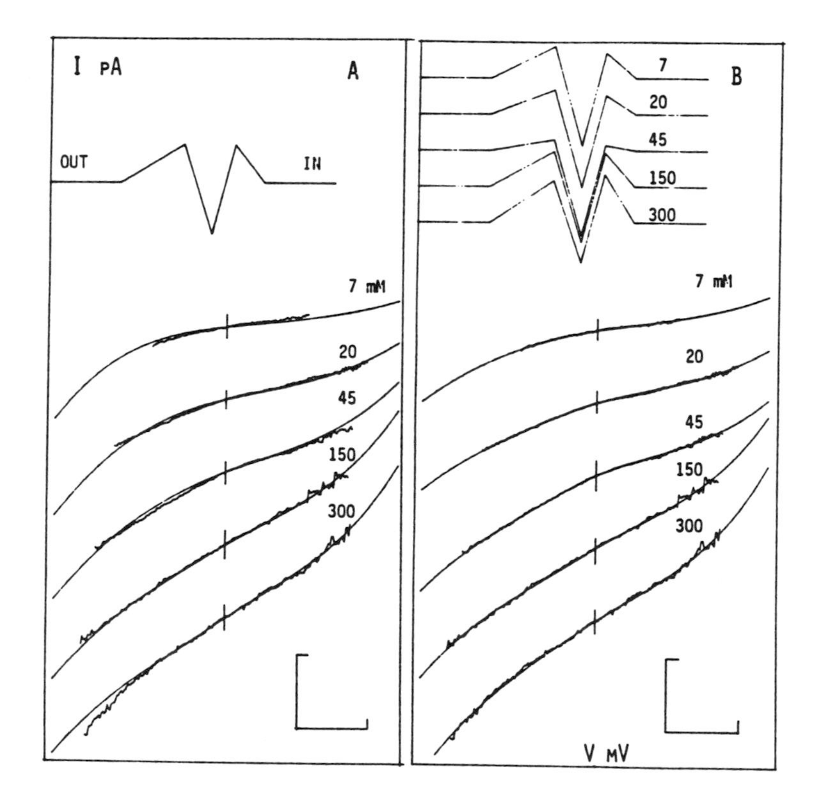

FIGURE 3. Comparison of experimental and theoretical I-V relations for the static 2B1S model for the AChR channel in symmetrical solutions having the indicated concentrations of Cs.

responds to a theoretical curve, with the bottom profile corresponding to the lowest concentration and the uppermost profile to the highest concentration). Although such steady-state conformational changes that alter the static energy profile as a function of concentration are not inconceivable (43,44), a more attractive alternative is to allow the energy profiles to fluctuate on the time scale of ion hopping.

B. Behavior of the Fluctuating 2B1S Model

The presently available I-V data for the AChR channel are not sufficiently detailed to allow a unique fit to the fluctuating 2B1S model with all 15 independent parameters allowed to be variable. To reduce the number of free parameters, some or all of the energies or locations can be fixed to gain a sense of the general behavior of the model. Fixing either the 6 energies of the 6 locations for ion translocation reduces the number of free parameters from 15 to 9. Fixing the

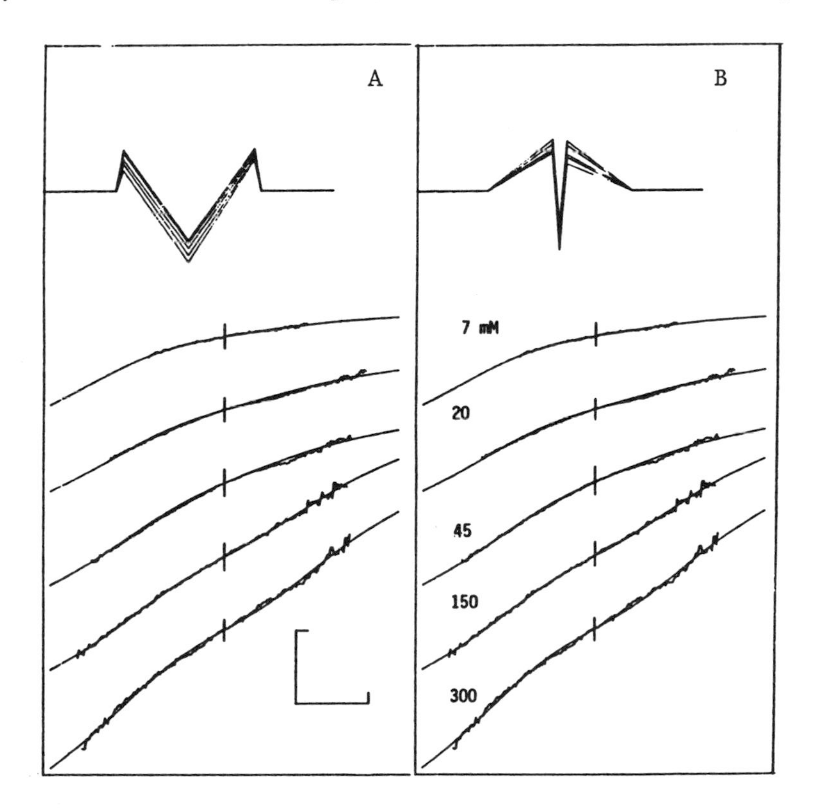

FIGURE 4. Fit of I-V data to static energy profiles for a centrally located site and very different barrier shapes.

locations is a particularly attractive strategy in that the zero-voltage properties are totally independent of these.

In preliminary explorations of the static 3B2S model we found (17,18) that the site location could be displaced as much as 15% from the center with little deterioration of fit quality. We also found that equally good fits were obtained for a centrally located site and 2 very different asymmetric barrier shapes, as shown in Fig. 4, which compares theoretical curves drawn for a static 2B1S model having the indicated energy profiles with the experimentally measured I-V behavior (the lowest profile corresponds to the lowest I-V curve). It therefore seems apparent that, in exploring the effects of fluctuations, it will be reasonable to locate the site in the middle of the field and the barriers anywhere between their extremes in Fig. 4.

Accordingly, the effects of fluctuating energy profiles

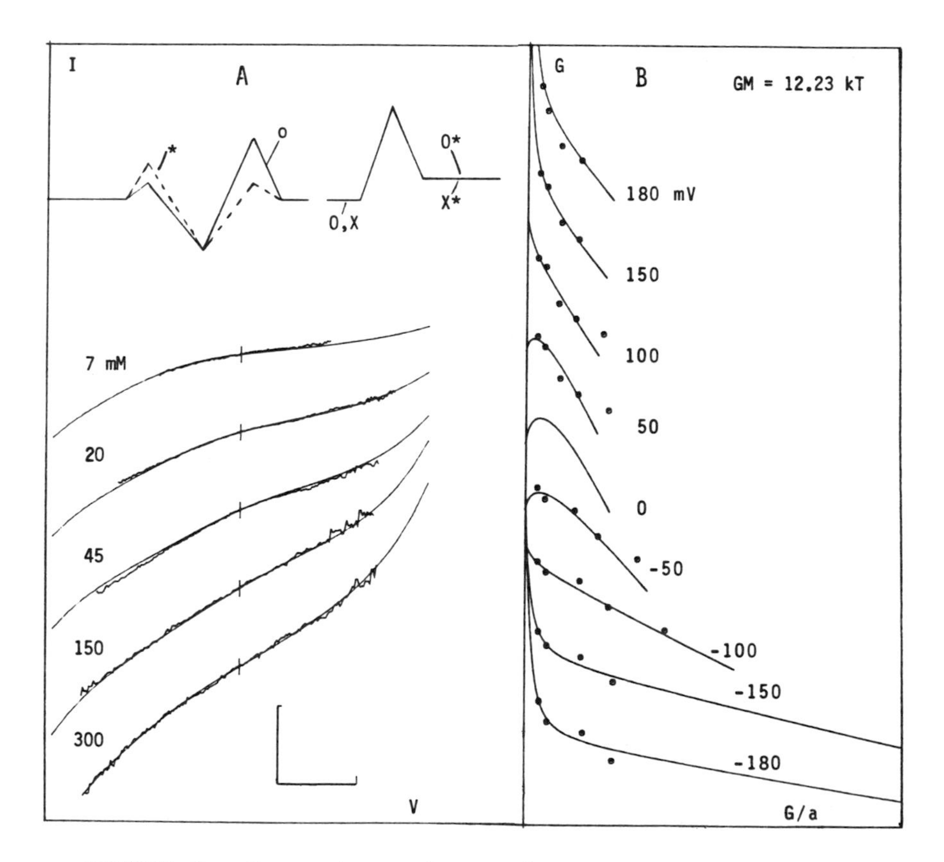

FIGURE 5. Comparison of experimental I-V data with theoretical expectations of the fluctuating 2B1S model.

are explored by allowing only the energy levels, but not the locations, of the barriers and site to vary. Fixing the site at the same locus in conformations * and o also removes any voltage dependence due to fluctuations in the occupied conformation. Figure 5 shows that the fluctuations 2B1S model can describe the I-V behavior quite well (R-factor = 0.048) with only the energy levels variable when the well is fixed in the middle and the barriers are fixed between the extremes of Fig. 4. Figure 5A presents the I-V expectations; while Fig. 5B presents Eadie-Hofstee plots of conductance (G) vs. conductance divided by activity (G/a), which is a particularly convenient way to describe conductance vs. concentration behavior, which will be discussed in a later section. For such plots the horizontal scale is 2000 pS/M and the vertical scale is 50 pS. The solid profile for ion translocation is for conformation o; the dashed profile, for conformation *. (Note that the 2 profiles for the fluctuations are almost indistinguishable in this figure). Note that the separation between the energy levels for states 0 and X (cf. Fig. 1, lower) has not been indicated in the right hand diagram in this and subsequent figures. This separation can be read directly from the depth of the well of the translocation profile for conformation o (solid profile) in A since it is given directly by G_2 (recall Fig. 1). The X-X* profile is lower than the 0-0* profile by this amount. (It also happens in Fig. 5 that G^*_2 nearly equals G_2).

Figure 6 illustrates that adequate fits are also obtained with different locations of the well and barriers. Figure 6A shows fits when the barriers are symmetrical (all d's = 0.25), and 6B shows fits when the site is moved away from a central location. Notice from the comparable heights of the energy barriers for ion translocation and structural fluctuation that the ion hopping and fluctuations in structure occur on the same time scale.

A feature common to the profiles in Figs. 5 and 6 is the pronounced asymmetry of the energy profile for ion premeation for each of the channel conformations. In one conformation a larger barrier is found to face one side and a smaller barrier to face the other in all cases. In the other conformation the situation reverses. Such an asymmetric energy profile has been recognized (23,29,34,35) as a requirement for a 1-ion channel to produce a multi-ion type of behavior. This is because an asymmetric profile, fluctuating at a frequency comparable to that of ion hopping, makes it possible for an ion to enter the site, and also to leave it, by a lower barrier than is possible with a static profile. For example, an ion moving from left to right can wait for the left hand barrier to be at its lowest height before jumping into the site. It can then wait for the barrier to the right to fluctuate to a lower level before jumping out. In addition, the level of the well can also fluctuate so that an ion could enter a strongly binding site (a deep well) and then wait for the kinetic energy of a structural fluctuation to raise the level of the well before it leaves. Of course, the ion must reside in the site long enough to take advantage of the fluctuations in barrier height and well depth. This is why ion hopping and structural fluctuation must occur in the same frequency range for the coupling between the fluctuating energy profile and ion hopping to occur.

IV. THE FREQUENCY RANGE WHERE COUPLING OCCURS BETWEEN CHANNEL FLUCTUATIONS AND ION TRANSLOCATIONS

Let us now examine the behavior when we deliberately force the fluctuation rates into ranges where, although the

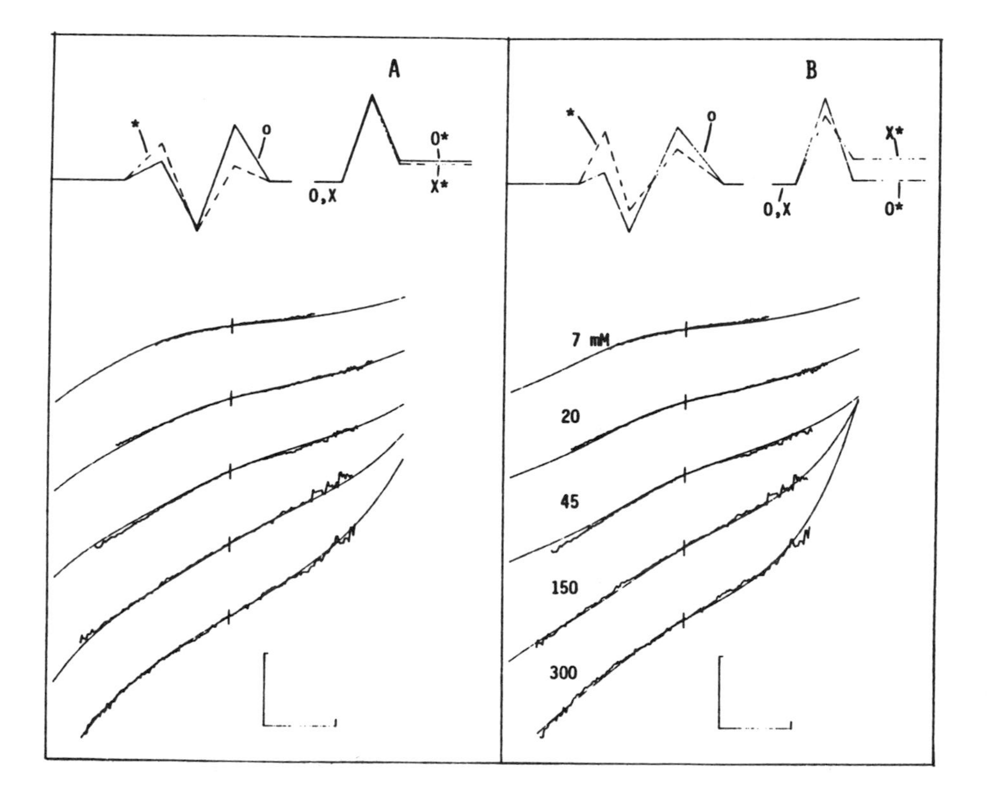

FIGURE 6. Insensitivity of the model to precise location of the site and barriers.

very best fit is not obtained, the experimental data still constrain the model to produce the best fit it can. By so doing we hope to define a range of frequencies in which the channel protein must fluctuate if, with the present model, it is to produce the kinds of experimental behavior described below.

Figures 7-10 illustrate the effect of varying the fluctuation rate constants by setting the activation energy GM for the fluctuation from the o to the * conformation at various fixed values and then finding the values for the 8 remaining energies that give the best fit to the experimental I-V data. Figure 7 plots the adequacy of fit (indicated by the log of the standard deviation for the I-V data) so obtained as a function of GM ranging from 0.0 to 18.9 kT (recall that Eq. 8 relates these energies to the rate constants). Figures 8 and 9 present the corresponding I-V and G-C behavior, respectively, for representative values of GM; and Fig. 10 summarizes graphically the variation in all the other rate constants that occur as GM is varied.

It is apparent in Fig. 7 that the best fit to the I-V data occurs for GM = 12.23 kT (corresponding to 30 megahertz; the manner of weighting the I-V data and statistical criteria for the fit are discussed in (18). This is the fit whose details were presented in Fig. 5. The sharp minimum in Fig. 7 shows that the fit deteriorates rapidly for values only 1.5 kT different from this.

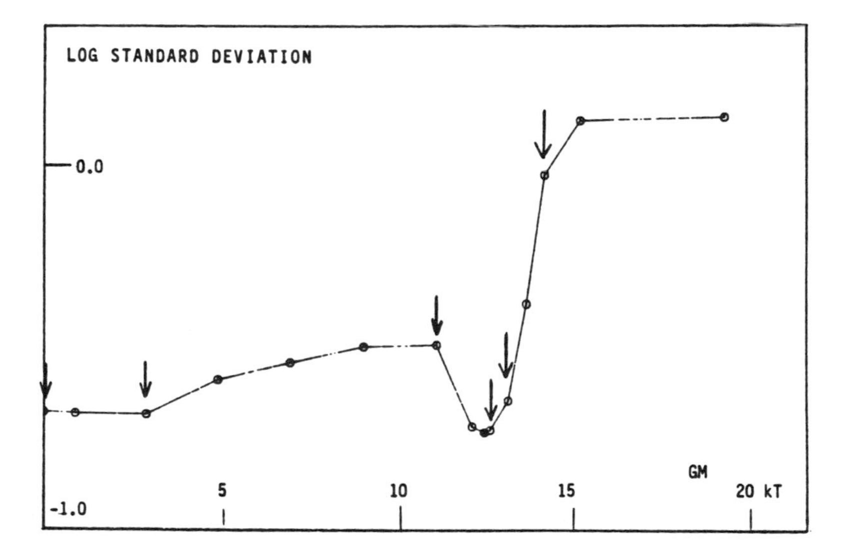

FIGURE 7. Adequacy of fit as a function of the energy barrier GM.

The deterioration can be seen on careful scrutiny of Fig. 8, which compares I-V data with theoretical expectations for the six representative values of GM indicated by the arrows in Fig. 7. The upper subfigures are for values larger than the optimum; while the lower subfigures are for values smaller than the optimum. Note that the fit becomes much worse for slower (increases in GM) than for faster (decreases in GM) fluctuations. Indeed, when the fluctuation rate is at its maximum (GM = 0.0 kT) the fits on the I-V plots are only slightly worse than the optimal. Indeed, at fluctuation fre-

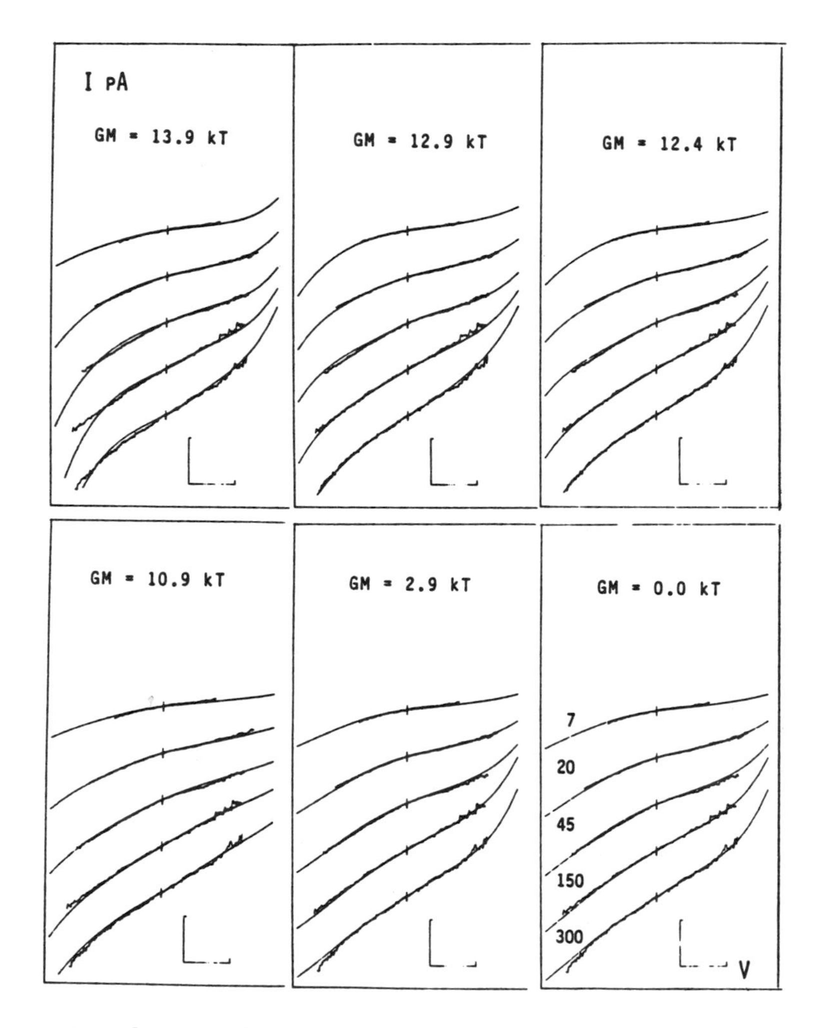

FIGURE 8. I-V behavior expected for 6 representative values of the fluctuation rate M from state 0 to 0*.

quencies higher than optimal the differences in theoretically expected behavior only show up clearly when the concentration dependence of the conductance is plotted, as can be seen from the Eadie-Hofstee plots in Fig. 9. (For the present, look only at the theoretical curves and ignore the data points which will be discussed later). At sufficiently high frequencies (GM < 2.9 kT) the curves approach the expected (see below) straight line behavior of a static 1-ion model.

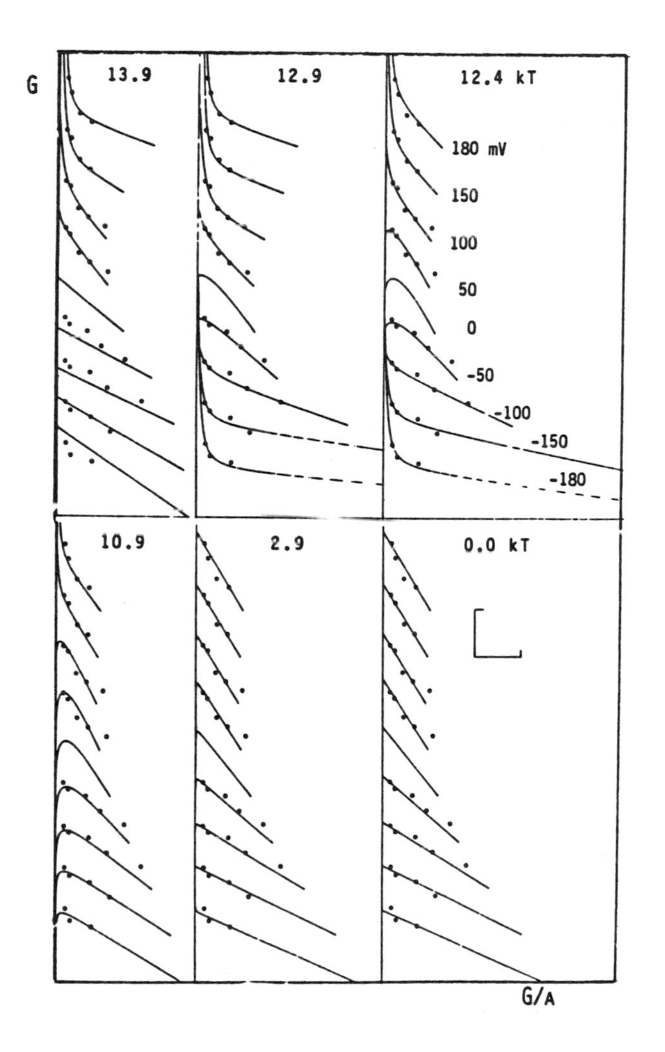

FIGURE 9. Eadie-Hofstee (or Scatchard)-like plot of chord conductance (G) vs. specific conductance (G/a) to illustrate the complex non-monotonic behavior characteristic of the fluctuating barrier model.

So far we have discussed the effect of varying the fluctuation frequency solely in terms of GM, the activation energy determining the rate constant M, for the fluctuation of the empty channel from the o to * conformation. It is also useful to scrutinize what values were found for the other rate constants for the variations of M examined above. This can be done with the aid of Fig. 10, which summarizes, in the compact form of energy profile plots, the values of all the rate constants computed for the indicated values of GM. Figure 10 shows the relative rates of structural fluctuations (right) and ion translocations (left). Recall that comparable energy barrier heights imply comparable rate constants, with rate constants being inversely related to the exponential of the barrier height through Eq. 8. First examine the activation energies for fluctuation at the right of Fig. 10 and notice that while the barrier becomes smaller for the empty channel as GM is decreased, the barrier remains roughly

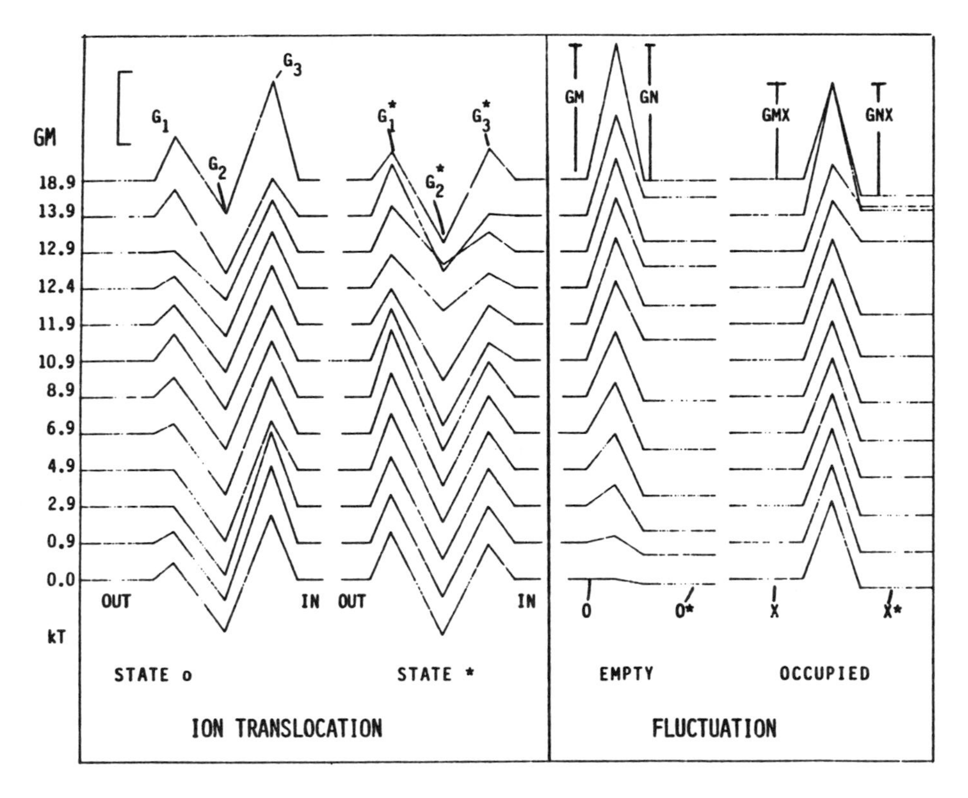

FIGURE 10. Energy diagrams summarizing the rate constants for the best fits to the data for the GM values indicated at the left.

constant in height for the occupied channel, being always comparable in height to the barriers for ion translocation at the left. This implies that, with decreasing GM, the fluctuation rates for the occupied channel do not become rapid relative to the ion translocation rates. This contrasts with the fluctuation behavior for the empty channel where, as the forward fluctuation rate for the empty channel is increased (with decreasing GM), the backward rate also increases.

Note that associated with the optimum in the fluctuation frequency M around 16-30 megahertz are rate constants for N in the range 62-500 megahertz, for MX from 38-43 megahertz, and for NX from 780-1800 megahertz, consistent with the energy profiles in Fig. 10. This is reflected in the finding that, when a parallel exploration was undertaken with MX (instead of M) as the fluctuation rate constant that was deliberately varied, only a rather narrow range of values (GMX between 10 and 12.2 kT) were acceptable.

These findings suggest that, whereas the empty channel can fluctuate rapidly between the o and * conformations, the occupied channel is stabilized in one of these. That is, the presence of an ion in the channel stabilizes the conformation of the channel. Although there is no intuitively satisfying explanation yet for such behavior, it is presumably a consequence of the particular current-voltage-concentration dependence imposed by the experimental data for the AChR channel.

Further details can be gleaned by comparing the energy levels for X and X* (or 0 and 0*) states at the right of Fig. 10. From such a comparison it is apparent that the * state is favored at the highest fluctuation frequencies, whereas the o state is favored at frequencies nearer to the optimal. Clearly the dependence of occupancy on fluctuation and hopping frequencies is very complex, as it must be, given the multiple paths available (see Fig. 2) by which a site can be occupied in a channel with a fluctuating energy profile.

V. DISCUSSION

A. A Complex, Multi-Ion Like, Non-Monotonicity of Conductance as a Function of Concentration is Expected Theoretically

One of the most interesting aspects of the behavior of fluctuating profile models is their ability to produce a multi-ion like behavior in a 1-ion channel. This manifests itself as flux coupling (29) or as a non-monotonic dependence of conductance on concentration (23,34). In marked contrast to this is the behavior expected for any 1-ion model with a static profile, which must possess a monotonic Michaelis-Menten type of behavior for the conductance measured at a given (fixed) voltage. This can be easily proven to be true regardless of the number of barriers or the complexity of profile (cf. Ref. 28). On an Eadie-Hofstee (or Scatchard) type plot, where chord conductance (G) at a given voltage is plotted vs. specific conductance (G/a), a monotonic behavior appears as a simple straight line (45), which is strictly true, of course, only in the absence of surface potential effects. A non-monotonic behavior, on the other hand, appears as a curve which may be concave or convex.

A markedly curvilinear behavior for the theoretical expectations is seen in Fig. 9 around the optimal fluctuation frequency (GM = 10.9 to 12.9 kT) and indicates the kind of multi-ion like behavior expected for the single-occupancy fluctuating 2B1S model in the region where structural fluctuations and ion hopping are coupled. This behavior is most pronounced when the fluctuation frequencies are in a range near the ion translocation rate (recall Fig. 5B). Looking at any of the sets of curves in Fig. 9 at this frequency (e.g., for GM = 12.4 kT) as a function of voltage, it is clear that the E-H plots change shape from voltage to voltage, exhibiting a conductance maximum around zero voltage and a nonlinear "foot" at high voltage. This kind of behavior, with its unprecedented voltage sensitivity, is quite different from what would be expected from a surface potential effect, which should produce a foot but no maximum at any voltage. Such behavior, indeed, provides an experimentally testable feature of the fluctuating 2B1S model, enabling it to be distinguished from static 1-ion models.

Examining the behavior of Fig. 9 in more detail, we note that at GM = 12.4 kT a maximum is clearly shown in the theoretical curves for +50 mV, 0 mV, and -50mV. However, on increasing GM by only 0.5 kT to 12.9 kT, the maximum disappears. When GM is increased 1 kT more (to 13.9 kT) there is no longer any sign even of the saturation and for negative voltages a monotonic (straight line) behavior is expected. The fit is so poor at this and higher values of GM, particularly at negative voltages, that one can confidently exclude fluctuation rates slower than 5.7 megahertz for the present model. On the other hand, from the lower portion of Fig. 9 it can be seen that, as the fluctuation rate of the empty channel becomes faster than optimal, the maximum becomes more prominent and the foot is lost. For GM = 10.9 kT a maximum is expected for all voltages more negative than +100 mV, and the foot is seen only above +150 mV. Such behavior also seems excluded by present data (see next section); so that a fluctuation frequency corresponding to GM of 12.6 kT + or -0.5 kT is suggested for the model. It is apparent thus that for fluctuations slower than optimal one sees a foot but no maximum; whereas for fluctuations faster than optimal one sees a maximum but no foot.

B. Experimentally Observed Chord Conductances as a Function of Concentration and Voltage

Although the I-V measurements of Dani and Eisenman were not designed for characterizing the low voltage conductances with the precision that is required for testing the above theoretical expectations, particularly around zero voltage, it is still instructive to compute chord conductances at various voltages from their I-V data to see to what extent the concentration dependence of the conductances are in accord with theoretical expectations. Figure 11 presents experimental I-V and chord conductance data as jagged curves; and it is clear that the experimental chord conductances "blow up" around zero voltage. A variety of filtering procedures were tried in order to circumvent this problem. The most satisfactory was found to be the use of a 6th degree polynomial (with a first coefficient of 0) to describe the data. Such a

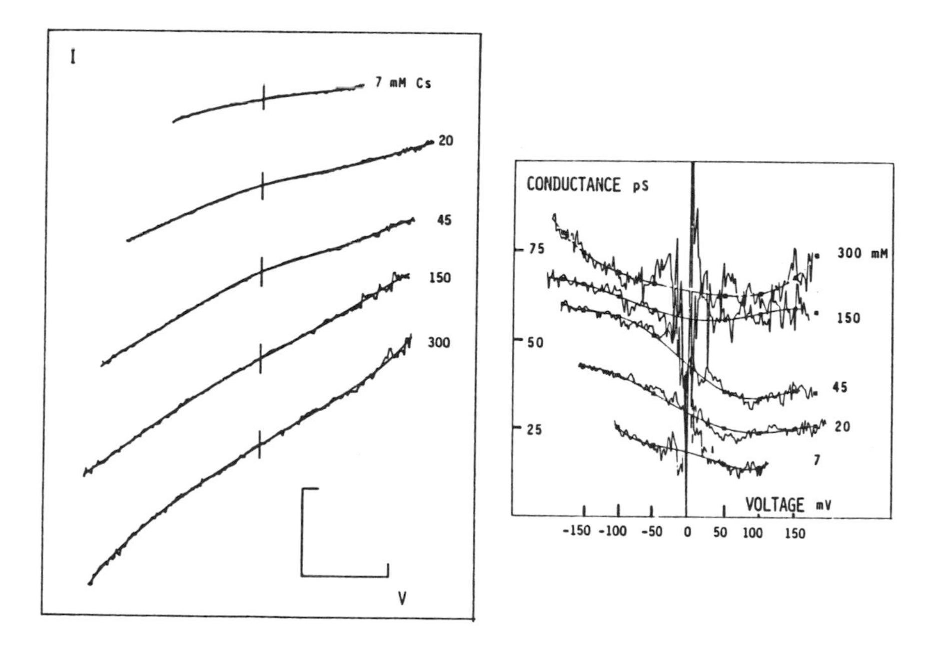

FIGURE 11. Polynomial representation of the I-V and chord conductance data.

representation is shown by the smooth curves in Fig. 11. The polynomial neither loses any important detail nor appears to produce any spurious inflections. Polynomially generated "experimental" chord conductances at selected voltages are indicated by dots in Fig. 11; and these have been used as the "experimental" data points in all plots of G vs. G/a. Chord conductances generated in this way are the most reliable that can be extracted at low voltage from existing data; and none will be used for voltages smaller in magnitude than 50 mV.

C. Comparison of Theoretical and Experimental Chord Conductances

The experimental conductances (data points) are compared with the theoretical expectations (smooth curves) for various locations of site and barriers in Figs. 5B, 9 and 12B. The best agreement between theory and data for the conductance-

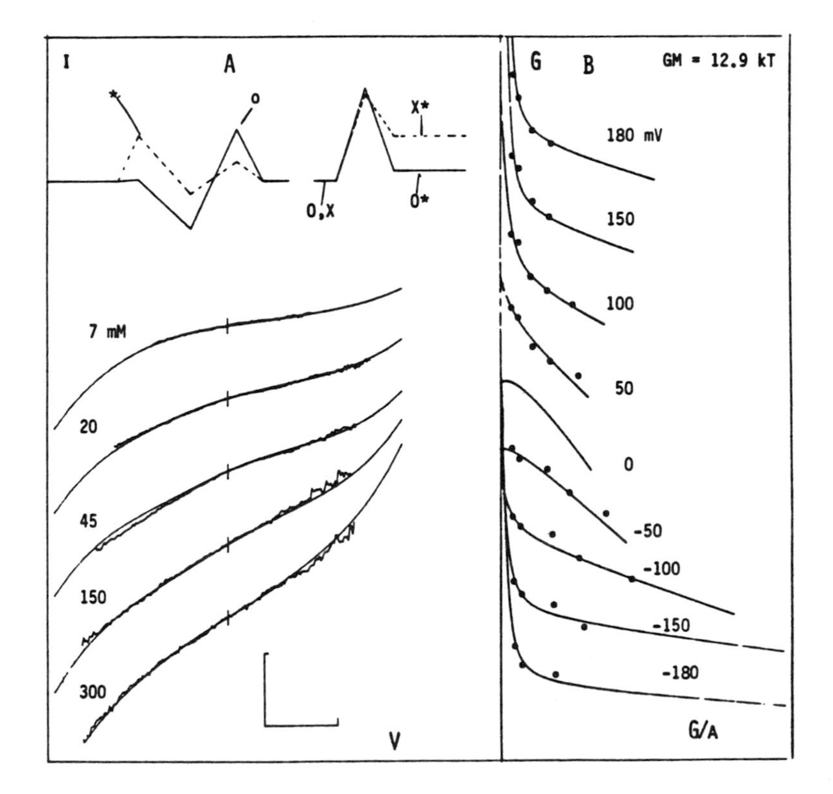

FIGURE 12. I-V and Eadie-Hofstee plots for GM = 12.9 kT.

concentration behavior occurs in Fig. 9 for GM = 12.9 kT; and the I-V and G vs. G/a, as well as the energy profiles corresponding to this fit, are shown in Fig. 12 for comparison with Fig. 5. Figure 9 confronts the experimental data with the theoretical expectations, already noted to be quite sensitive to small variations in the fluctuation rate. Comparing Figs. 5 and 12 it is seen that the I-V's are better fit in Fig. 5, while the conductances are better fit in Fig. 12. This difference reflects the fact that the parameters of Fig. 12 represent the lower voltage data better than those of the fit of Fig. 5, and vice versa (see Ref. 18).

It is important to note that we reach the same general conclusions about the effect of fluctuations regardless of whether we consider Fig. 5 or Fig. 12 to be more correct. However, notice how extremely sensitive the G vs. G/a behavior is to the fluctuation rate constant M. The two figures differ chiefly in that the value of GM of 12.23 kT in Fig. 5 implies that M = 30 megahertz, while the value for GM of 12.9 kT in Fig. 12 implies that M = 16 megahertz.

The maximum predicted at high concentrations near zero voltage in Fig. 9 (at GM = 12.4 kT or 10.9 kT; also recall Fig. 5) has not yet been reported for the AChR channel; although a foot at low concentrations, like that predicted at sufficiently high positive and negative voltages at GM = 12.4 kT, has been observed (20,39). Such a maximum should certainly be searched for experimentally; but its absence would not exclude the fluctuating 2B1S model since a good fit to the conductance data is seen at GM = 12.9 kT and shows almost no maximum. The really crucial test of this model is the exquisite sensitivity to voltage near V = 0 of the theoretically expected changes in G vs. G/a shape. It should be noted, however, that for sufficiently fast fluctuations the fit to straight line Eadie-Hofstee plots in Fig. 9 is not bad (see curves for GM at 0 and 2.9 kT). Nevertheless, the curvilinear plots for GM ranging between 12.4 kT and 12.9 kT seem to be a significantly better representation of the data points than are these straight lines.

D. Physical Implications of the Fluctuation Frequency for the Best Fit

The best fit of the present model occurs with GM in the range 12.23 - 12.9 kT (corresponding to a frequency of fluctuation between 16 megahertz and 30 megahertz). The fluctuation rate constants M, N, MX, NX corresponding to the energy profiles in Figs. 5 and 12 all lie in the range 10-1000 megahertz. Fluctuations at such frequencies occur commonly in proteins, helix-coil transitions typically having time constants around 1 nanosecond (31) and "hinge-bending" motion of different globular regions occurring in the 0.1 picosecond to 10 microsecond range (30).

A particularly tempting speculation would be to identify the 2 conformations of the model with 2 slightly different tilts of the 5 amphiphilic putative channel-forming subunits suggested by Guy (2) and by Young et al. (3). The units lining the channel are postulated to be in an arrangement like that proposed for the gap junction channel (33); and the gating of the channel is suggested to occur by its opening and closing when the subunits are tilted between two extremes. A rocking (lifetime 10-100 nanoseconds) between 2 less extreme tilts, intermediate between fully open and fully closed, to produce the fluctuations inferred here is attractive for the following two reasons. First, because it might occur at the appropriate frequency since the concerted motions of relatively large regions of the protein seem analogous to "hinge-bending". Second, because, for an appropriate pivot point, such a rocking would be compatible with the observed shifts in translocation energy profiles common to all the different situations examined here where in one conformation the larger barrier faces the inside solution whereas in the other conformation the larger barrier faces the outside solution.

In the case examined in this paper with a fixed, centrally located site, the postulated rocking can only alter the depth of the site and the free energy profiles for access to it. An alternative, suggested by the "tilting subunit" picture, would be for a site to be capable of being moved from one side to the other of the channel as the subunits rocked, much as in Lauger's model for a carrier (34). In the early exploration of the model this case actually gave a slightly better fit to the data. However, since the conclusions relating to fluctuation frequency are no different from those presented here, further examination of this interesting case will be put off to the future.

What I have in mind as underlying the postulated 16-30 megahertz fluctuation is a partial opening or closing of the channel. This could occur between several relatively stable intermediate open states (for example, the first substate for closing or the last substate for opening) of the normal gating process. Such substates are likely to exist on the reaction path for gating between the fully open and fully closed situations; and would serve as the "lubricant" to the full action of gating.

As mentioned in the Introduction, an attractive feature of a structural fluctuation which occurs on the same time scale as ion hopping is that such fluctuations can facilitate transport by enabling a higher transport rate to be manifested for a given level of selectivity. Thus, we can imagine a channel-lining subunit fluctuating between slightly different conformations and in this way enabling the affinity of the site to be propagated across the membrane without paying the full price of a low exit rate required from a deep energy well in a static channel.

Although a fluctuation frequency of 16-30 megahertz is too rapid to be detected by present electrical measurements (but see Ref. 32,46, its existence is not without possibility of detection in a channel like that of the AChR which has been shown to be capable of being significantly occupied by divalent cations (39,47). For the postulated fluctuations occur within the so-called "Mossbauer Window". Thus, divalent copper ion might be introduced into the channel as a Mossbauer probe or, alternatively, divalent manganese ion might be used as an electron spin resonance probe for such fluctuations.

ACKNOWLEDGMENTS

I thank Peter Lauger, Sergio Ciani, John Sandblom, and Dick Horn for valuable discussion, Chris Clausen and Ken Lange for their algorithms for nonlinear regression analysis, and Avi Ring for help with writing Fortran programs.

REFERENCES

1. Noda, M., Shimizu, S., Tanabe, T., Takai, T., Kayano, T., Ikeda, T., Takahashi, H., Nakayama, H., Kanaoka, Y., Minamino, Y., Kangawa, K., Matsuo, H., Raftery, M. A., Hirose, T., Inayama, S., Hayashida, H., Miyata, T., and Numa, S., Nature 312:121 (1984).
2. Guy, H. R., Biophys. J. 45:249 (1984).
3. Young, E. F., Ralston, E., Blake, J., Ramachandran, J., Hall, Z. W., and Stroud, R. M., Proc. Natl. Acad. Sci. USA 82:626 (1985).
4. Eisenman, G., and Horn, R., J. Membr. Biol. 76:197 (1983).
5. Hille, B., and Schwarz, W., J. Gen. Physiol. 72:409 (1978).
6. Etchabest, C., Ranganathan, S., and Pullman, A., Fed. Eur. Biochem. Soc. 173:301 (1984).
7. Matthew, J. B., Ann. Rev. Biophys. Biophys. Chem. 14:387 (1985).
8. Polymeropoulos, E. E., and Brickmann, J., Ann. Rev. Biophys. Biophys. Chem. 14:315(1985).

9. Schroder, H., Eur. Biophys. J. 12:129 (1985).
10. Adams, P. R., Biophys. J. 25:70a (1979).
11. Lewis, C. A., and Stevens, C. F., in "Membrane Transport Processes" (C. F. Stevens and R. W. Tsien, ed.) Vol. 3, p. 133. Raven Press, New York, 1979.
12. Horn, R., and Brodwick, M., J. Gen. Physiol. 75:297 (1980).
13. Urban, B. W., Hladky, S. B., and Haydon, D. A., Biochim. Biophys. Acta 602:331 (1980).
14. Sandblom, J., Eisenman, G., and Hagglund, J., J. Membr. Biol. 71:61 (1983).
15. Eisenman, G., and Sandblom, J., in "Physical Chemistry of Transmembrance Ion Motions" (G. Spach, ed.), p. 329. Elsevier, Amsterdam, 1983.
16. Hagglund, J. V., Eisenman, G., and Sandblom, J. P., Bull. Math. Biol. 46:41 (1984).
17. Eisenman, G., and Dani, J. A., in "Proceedings of 3rd International Conference on Water and Ions in Biological Systems", p.437. Union of Societies for Medical Sciences, Bucharest, Romania, 1985.
18. Eisenman, G., and Dani, J. A., in "Proceedings of International School on Ionic Channels. Santiago, Chile" (R. Latorre, ed.), p. 53. Plenum, New York, 1986.
19. Latorre, R., Alvarez, O., Cecchi, X., and Vergara, C., Ann. Rev. Biophys. Biophys. Chem. 14:79 (1985).
20. Dani, J. A., and Eisenman, G., Biophys. J. 45:10 (1984).
21. Moczydlowski, E., Alvarez, O., Vergara, C., and Latorre, R., J. Memb. Biol. 83:273 (1985).
22. Dani, J. A., and Eisenman, G., J. Gen. Physiol. Submitted, (1986).
23. Lauger, P., Stephan, W., and Frehland, E., Biochim. Biophys. Acta 602:167 (1980).
24. Eisenman, G., Sandblom, J., and Walker, J. L., Jr., Science 155:965 (1967).
25. Almers, W., and McClesky, E. W., J. Physiol. 353:608 (1984).
26. Hess, P., and Tsien, R. W., Nature 309:453 (1984).
27. Eisenman, G., Latorre, R., and Miller, C., Biophys. J. in press (1986).
28. Lauger, P., Biochim. Biophys. Acta 311:423 (1973).
29. Ciani, S., Biophys. J. 46:249 (1984).
30. McCammon, J. A., Rep. Prog. Phys. 47:1 (1984).
31. Karplus, M., and McCammon, J. A., Ann. Rev. Biochem. 53:263 (1983).
32. Sigworth, F. J., Biophys. J. 47:709 (1985).
33. Unwin, P. N. T., and Zamphighi, G., Nature 283:545 (1980).
34. Lauger, P., Current Topics in Membranes and Transport 21:309 (1984).

35. Lauger, P., Biophys. J. 48:369 (1985).
36. Schrodinger, E., in "What is Life?", Cambridge University Press, Cambridge 1967.
37. Marchais, D., and Marty, A., J. Physiol. 297:9 (1979).
38. Eisenman, G., Dani, J. A., and Sandblom, J., in "Ion Measurements in Physiology and Medicine" (M. Kessler, D. K. Harrison, and J. Hoper, ed.), p. 54. Springer-Verlag, Berlin, Heidelberg, 1985.
39. Dani, J. A., and Eisenman, G., Biophys. J. 47:43a (1985).
40. Powell, M. J. D., in "Numerical Analysis. Dundee 1977" (G. A. Watson, ed.), Lecture Notes in Mathematics No. 630, Springer-Verlag, Berlin, 1977.
41. Eyring, H., Lumry, R., and Woodbury, J. W., Rec. Chem. Prog. 10:100 (1949).
42. Hamilton, W. C., in "Statistics in Physical Science" Ronald Press, New York, 1964.
43. Von Hippel, P. H., and Schleich, T., in "Biological Macromolecules, Vol. 2; Structure and Stability of biological Macromolecules" (S. N. Timasheff and G. Fasman, eds.), p. 417. Marcel Dekker, New York, 1969.
44. Record, M. T., Jr., Anderson, C. F., and Lohma, T. M., Q. Rev. Biophys. 11:103 (1978).
45. Eisenman, G., Sandblom, J., and Neher, E., Biophys. J. 22:307 (1978).
46. Yellen, G., J. Gen. Phys. 84:157 (1984).
47. Adams, D. J., Dwyer, T. M., and Hille, B., J. Gen. Physiol. 75:493 (1980).

α-HELICAL ION CHANNELS RECONSTITUTED INTO PLANAR BILAYERS

Günther Boheim
Sabine Gelfert

Department of Cell Physiology
Ruhr-Universität Bochum
Bochum, W-Germany

Günther Jung

Institute of Organic Chemistry
Universität Tübingen
Tübingen, W-Germany

Gianfranco Menestrina

Department of Physics
Universita degli Studi di Trento
Povo, Italy

I. INTRODUCTION

Information processing by way of biological membranes is carried out on the basis of ion movements through various specific transmembrane protein channels. Two representative channels are the acetylcholine receptor from the postsynaptic membrane at the neuromuscular junction (1) and the Na-channel from excitable cells (2). Ion translocation through these channels is modulated by discrete movements of the so-called

---

This work was supported by the Deutsche Forschungsgemeinschaft (SFB 114).

gates which are controlled either by ligand binding, i.e. chemical transmitters as in case of the acetylcholine receptor, or by the membrane electric field as in case of the Na-channel or by both, ligand binding and membrane voltage, which is observed with the Ca-dependent K-channel (3). The acetylcholine receptor from the electric organ of the electric ray (Torpedo) consists of five subunits (four homologous proteins, $\alpha_2, \beta, \gamma, \delta$) the amino acid sequences of which had been determined by applying the recombinant DNA technique (4). Similarly the amino acid sequence of the tetrodotoxin-sensitive Na-channel protein from the electric organ of the electric eel (Electrophorus electricus) was determined which revealed four homologous domains (5). The size of these domains is comparable to that of the acetylcholine receptor subunits.

Membrane spanning parts of integral proteins were found to consist of mainly hydrophobic amino acids and to adopt α-helical structure as revealed by 3-dimensional electron microscopy and subsequent image reconstruction (6). It is assumed that the hydropathy profile (7) of an amino acid sequence reveals its α-helical membrane spanning parts and thus may give an indication of the actual channel structure. On the basis of this assumption the respective ion channels are formed by a bunch of circularly arranged α-helical rods of varying number, e.g. five in case of the acetylcholine receptor (8) and four in case of the Na-channel (5). Since α-helices bear an intrinsic dipole moment (9), it is expected that its transmembrane orientation will be influenced by the strength and direction of the membrane electric field. Thus it is of significant importance to investigate this problem by using helical model peptides.

A well-known ion channel forming molecule of helical structure is the polypeptide antibiotic alamethicin (10,11). Channel properties are unique in two points : 1. the channel formation rate is strongly dependent on membrane voltage and alamethicin concentration, 2. a fixed sequence of usually 5 - 7 stable non-integral conductance states are adopted after channel activation. Sometimes we resolved up to 11 discrete current levels within a single activity burst. In the following we will present a short review of our studies on the structure/function relationship of relatively short-chained (up to 20 amino acids) helical polypeptides. It is concluded that the voltage dependence directly results from the dipole moment of the extended helix and that channel state stabilization requires the presence of particular amino acids as well as end groups at the C- and N-terminal parts.

II. ALAMETHICIN SEQUENCE ANALOGS

The amino acid sequences of alamethicin AL30, some natural (suzukacillin SK40; trichotoxin TT50) and chemically modified (AL-OBzl, Boc-AL) analogs as well as of some related synthetic polypeptides (P5, P10, P15, P20, P16, P19) are listed in

TABLE 1. Amino acid sequences of alamethicin, some natural and chemically modified analogs, as well as of some related synthetic polypeptides (12,13).

Alamethicin (AL30, main component)
Ac-Aib-Pro-Aib-Ala-Aib-Ala-Gln-Aib-Val-Aib-Gly-Leu-Aib-
1 2 6 7 11
-Pro-Val-Aib-Aib-Glu-Gln-Pheol
14 18 19 20

Suzukacillin (SK40, main component)
Ac-Aib-Ala-Aib-Ala-Aib-Ala-Gln-Aib-Aib-Aib-Gly-Leu-Aib-
-Pro-Val-Aib-Aib-Gln-Gln-Pheol

Trichotoxin (TT50, main component)
Ac-Aib-Gly-Aib-Leu-Aib- - -Gln-Aib-Aib-Aib-Ala-Ala-Aib-
-Pro-Leu-Aib-Iva- - -Gln-Valol

Alamethicin-O-benzyl (AL-OBzl)
Ac-Aib-Pro-Aib-.... ....-Aib-Aib-Glu-Gln-Pheol
OBzl

t-Butyloxycarbonyl-alamethicin (Boc-AL)
Boc-Pro-Aib-.... ....-Aib-Aib-Glu-Gln-Pheol

P5 - P20 : Boc-(Ala-Aib-Ala-Aib-Ala)$_n$-OMe, n = 1 - 4

P16 : Ac-Aib-(Ala-Aib-Ala-Aib-Ala)$_3$-OMe

P19 : Ac-Aib-(Ala-Aib-Ala-Aib-Ala)$_3$-Glu-Gln-Pheol

Abbreviations :
Aib = α-aminoisobutyric acid (2-methylalanine),
Iva = isovaline (2-ethylalanine), Pheol = phenylalaninol,
Ac = acetyl, OBzl = O-benzyl, OMe = O-methyl,
Boc = t-butyloxycarbonyl
All chiral constituents are of the L-configuration with the exception of D-Iva .

Table 1 (12,13). The crystal structure of alamethicin (14) is an α-helix which extends from the N-terminus up to the C-terminal end except for one 3/10-helical turn after Pro(14). The crystal structure of two fragments, the N-terminal undecapeptide and the C-terminal nonapeptide, just reveal the same helical structures (15, 16). Consequently single helical rods are formed in unpolar environment of approximately 3.2 nm (alamethicin) and 3.0 nm (P20) length, respectively. Since the equivalent charges are +1/2 elementary charge at the N-terminal end and -1/2 elementary charge at the C-terminal part (9) these helical rods bear a large intrinsic dipole moment of 70 - 75 D (17). By comparing polypeptide behavior with that of biological channel proteins one has to take into consideration that the small polypeptide molecules ($M_r \approx 2000$) are very flexible in comparison for instance to the significantly larger protein subunits of the acetylcholine receptor channel (M_r = 39000 - 65000). Note that the helices of the synthetic polypeptides P5 to P20 are composed exclusively of hydrophobic amino acids.

A. Multi channel I/V-characteristics

Examples of current/voltage (I/V)-characteristics of alamethicin-type polypeptides are shown in Fig. 1 : AL30 (A), AL-OBzl (B) and Boc-AL (C). There is no qualitative difference in the strongly voltage dependent current increase. If the corresponding conductivity λ is described in terms of

$$\lambda \quad \exp\{a \cdot FV/RT\} ,$$

the following approximate values for the gating charges a ± 5% are calculated : a(AL30) = 5.0, a(AL-OBzl) = 2.9 and a(Boc-AL) = 3.8 . Similar I/V-curves are obtained in the presence of other natural analogs, e.g. SK40 and TT50. This indicates that the voltage dependence inducing principle constitutes a common feature of the molecular structure.

In addition to the strongly voltage dependent conductance a weakly voltage dependent conductance exists which increases with antibiotic concentration.

B. Single channel current fluctuations

In contrast to the similarities of the I/V-characteristics striking differences are seen in the single channel properties (Fig. 2 A-C). The OBzl-group near the C-terminal end shortens the lifetimes of the channel as well as those of the channel states by at least one order of magnitude (B) as compared to corresponding lifetimes found with alamethicin (A). Whereas the conductance values of the various channel states

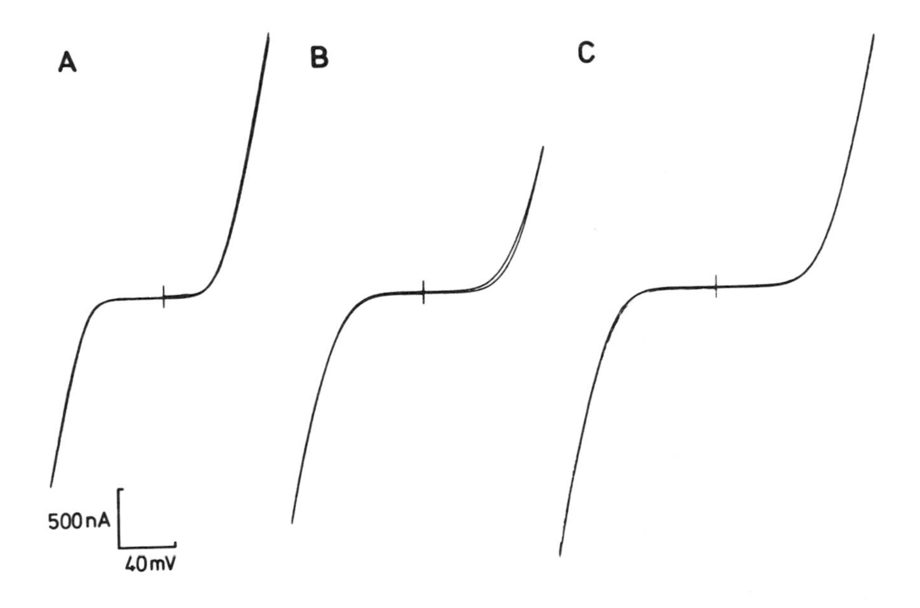

FIGURE 1. Current/voltage (I/V)-characteristics of alamethicin AL30 (A) and the two chemically modified derivatives AL-OBzl (B) and Boc-AL (C) showing the strongly nonlinear voltage dependence in current increase.

Experimental conditions :
Salt solution : 1 M KCl, 10 mM Hepes, pH 7.2;
Polypeptide concentration on cis side : A. AL30 o.25 μg/ml; B. Al-OBzl 0.50 μg/ml; C. Boc-Al 0.25 μg/ml;
Lipid solution : 1,2-POPC / 1,2-DOPE (9:1, molar ratio) in hexane/ethanol (99:1 by vol.);
Membrane forming procedure : Montal-Mueller technique (18);
Bilayer area : $2.0 - 2.5 \cdot 10^{-4}$ cm^2;
Temperature : 20 °C; Period duration of voltage change : 200 s;
I/V scaling bars are the same for all three curves.
Positive voltage sign refers to the cis side to which polypeptides were added, i.e. trans side is virtual ground. Positive current direction is that from cis to trans side. For further experimental details see (19).

Abbreviations :
AL 30, AL-OBzl and Boc-AL see : Table 1.
1,2-POPC : 1-palmitoyl-2-oleoyl-glycero-3-phosphocholine
1,2-DOPE : 1-oleoyl-2-oleoyl-glycero-3-phosphoethanolamine

remain unchanged, the size of the most probable channel state is decreased significantly. According to Boheim and Kolb (19) such a reduction in the mean channel state is reflected in a lower steepness of the I/V-curve consistent with the smaller value of a(AL-OBzl) found.

In case of the N-terminal Boc-group modification discrete conductance levels could not be resolved at the time resolution of our electronic recording device of 1 kHz (C). In order to demonstrate voltage dependent channel formation at the level of small Boc-AL-induced currents we applied voltage jumps and recorded the subsequent current relaxations. Note that the width of the current noise in (C) is comparable to the current difference of two higher neighbouring channel states of alamethicin in (A) under similar experimental conditions. It is evident that the exchange of Ac-Aib- by the Boc-group leads to the destabilization of alamethicin channel states. The strongly voltage dependent I/V-characteristic indicates, however, that alamethicin-type channel formation occurs.

Further single channel experiments with other natural AL30 analogs (see Table 1) gave the following results : 1. Alamethicin AL50 differs from AL30 in the exchange of Glu(18) by Gln. This has no effect on single channel properties (20). 2. The exchange positions of suzukacillin SK40 in reference to AL50 are - Ala in place of Pro(2) and Aib in place of Val(9). This has no effect on single channel properties (21). 3. Paracelsin A, B, C and D have been highly purified and sequenced (12). Exchange positions in reference to AL50 are - Ala in

FIGURE 2. Single channel current fluctuation pattern at low concentrations of alamethicin AL30 (A), AL-OBzl (B) and Boc-AL (C). Single channel kinetics have changed significantly at positive and negative voltages due to the small chemical modifications near the C- and N-terminal parts of the alamethicin molecule.

Experimental conditions :
Salt solution : 1 M KCl, 10 mM Hepes, pH 7.2;
Nominal concentration on cis side of all the three polypeptides (A-C) : 50 ng/ml;
Lipid solution : 1,2 - POPC / 1,2-DOPE / cholesterol (87.5:10:2.5, molar ratio) in hexane/ethanol (99:1 by vol.);
Temperature : 18 °C;
Applied voltages (upper trace / lower trace) : A. +107 mV / -140 mV ; B. +85 mV / -110 mV ; C. +115 mV / -115 mV (final voltage after voltage-jump from 0 mV).
Lengths of ordinate scaling bars correspond to 200 pA in each case (A-C), whereas abscissa time scale in C is enlarged by a factor of two.

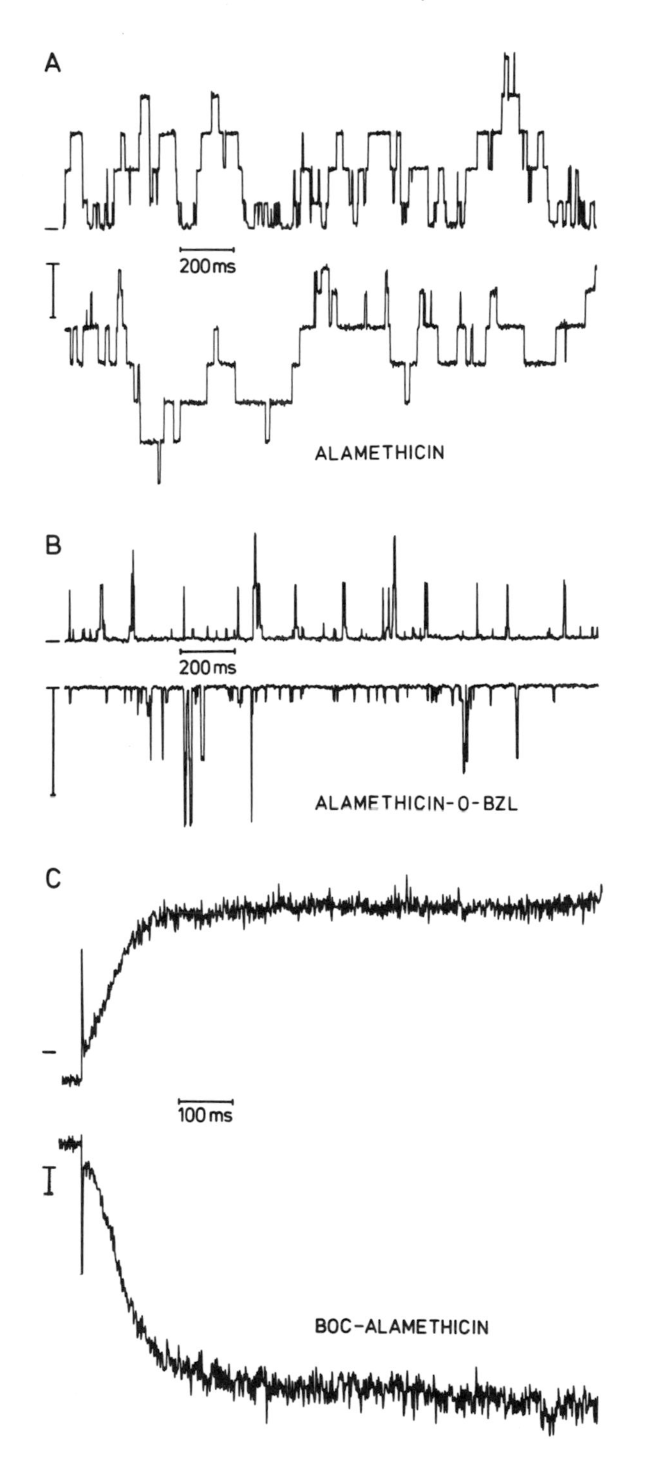
A
200ms
ALAMETHICIN
B
200ms
ALAMETHICIN-O-BZL
C
100ms
BOC-ALAMETHICIN

place of Pro(2), Leu in place of Val(9) and Aib in place of Leu(12). Our experimental data did not indicate qualitative changes in single channel behavior. 4. In contrast, properties of the natural analog trichotoxin TT50 are similar to those of Boc-AL. TT50 induces an I/V-curve of comparable steepness and discrete current levels from stable channel states could not be resolved as well (21). Besides amino acid exchanges comparable to those in SK40 and the paracelsins there are two positions in TT50 significantly different from the AL30 sequence - Ala(6) and Glu(18) are missing, i.e. the trichotoxin family (12) is shorter by 2 amino acids. Whereas Ala(6) is placed in a region of large natural variety due to microheterogeneity, Glu(18) means the same position which had been chemically modified in AL-OBzl leading to the strong decline in channel state lifetimes.

III. HYDROPHOBIC HELICAL POLYPEPTIDES

Alamethicin secondary structure in crystals is completely helical (14-16). However, it does not form a straight elongated rod, because a bend in the helix axis is imposed by the amino acid Pro(14). Pro is found with all natural alamethicin analogs at this peculiar position, also with trichotoxin. A variety of polypeptides were synthesized consisting exclusively of hydrophobic amino acids and forming straight elongated helical rods with a large intrinsic dipole moment (15,17). On the basis of the SK40 pentapeptide fragment - Ala(2) to Ala(6) - the polypeptides P5 to P20 were designed (see Table 1). Then the N-terminal Boc-group was exchanged **for** Ac-Aib-(P16) and finally the C-terminal -OMe group for -Glu-Gln-Pheol (P19), i.e. by the hydrophilic C-terminal segment of the AL30 sequence (positions 18 to 20). Details of the experimental results obtained in the presence of P5 to P20 have been presented elsewhere (13,22).

A. Multi channel I/V-characteristics

The strongly voltage dependent I/V-characteristic typical of alamethicin is observed with each of the synthetic helical polypeptides, whereby helix length varied between 5 and 20 predominantly hydrophobic amino acids (13,22). An example of a planar lipid bilayer modified by the pentadecapeptide P15 is given in Fig. 3 . The shape of the I/V-curve turns out to be nearly independent of the helix length of the straight elongated molecules P5 to P20, i.e. a formal gating charge a = 4.7 is calculated in each case. In contrast, the actual polypep-

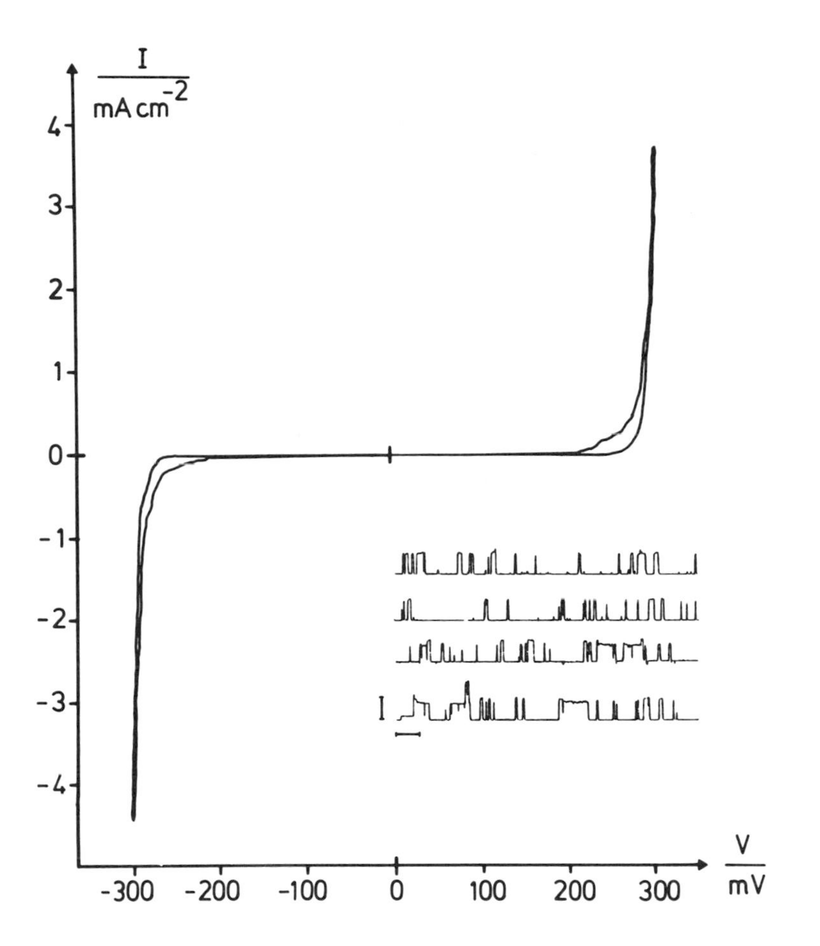

Figure 3. I/V-characteristic and single channel current fluctuations (inset) of the synthetic polypeptide P15 Boc-(L-Ala-Aib-L-Ala-Aib-L-Ala)$_3$-OMe.

Experimental conditions : I/V-curve -
Salt solution : 1 M KCl, pH = 6 unbuffered;
Polypeptide concentration on both sides : 2 µg/ml;
Lipid solution : 1,2-OPPC in hexane/ethanol (9:1 by vol.);
Temperature : 22 - 23 °C;
Period duration of voltage change : 50 s .

Inset : single channel currents - same conditions except for
Polypeptide concentration on both sides : 40 µg/ml;
Temperature : 5 °C; applied voltage : +193 mV;
Scaling bars : vertical 7.5 nS , horizontal 500 ms .
Abbreviation :
1,2-OPPC = 1-oleoyl-2-palmitoyl-glycero-3-phosphocholine

tide concentrations which are needed to induce a reference conductivity of $100\ \mu S/cm^2$ at a characteristic voltage of V_c = 200 mV increase strongly with decreasing helix length. In the presence of AL30 V_c is reached at about 0.1 µg/ml. Concentrations of 10 µg/ml and 20 µg/ml are needed in case of P20 and P15, respectively, which are two orders of magnitude larger than that with AL30. Even higher are the activity inducing concentrations of P10 (100 µg/ml) and P5 (2 mg/ml). The large differences in these reference concentrations seem to reflect the variability of the corresponding distribution coefficients for the polypeptides between the aqueous phase and the membrane phase. In each case V_c depends on the 8th - 9th power of the polypeptide concentration, i.e. V_c decreases strongly by increasing the concentration of a given polypeptide.

The synthetic polypeptides P16 and P19 behave similarly to P5 to P20. The predominant feature which is common to all these natural and synthetic polypeptides constitutes the helical molecule structure. Thus it is conceivable that the large intrinsic dipole moment of the α-helix causes the strong voltage dependence in channel formation by its interaction with the membrane electric field. A weakly voltage dependent conductance is observed in the presence of P5 to P20, too.

B. Voltage-jump current-relaxation experiments

Alamethicin-type antibiotics and the synthetic polypeptides possess a significant structural difference which results from the helix bend inducing property of Pro(14) in alamethicin. Here the dipole moment has a vector component vertical to the main helix axis. If the membrane electric field and the main helix vector are oriented antiparallel, the torque from the vector product of the vertical dipole moment component and the electric field would tend to drive the helix into the stable parallel orientation. This is different with the straight elongated helices of the synthetic polypeptides. The dipole moment vector and the direction of the helix axis coincide. No torque arises in a position antiparallel to the electric field because of the metastable situation.

The results of voltage sign reversal experiments relating to the considerations given above are shown in Fig. 4 A-C . AL30 channels are immediately destroyed after voltage sign reversal and they reform in an exponential time course, until the new steady state is reached (A). In contrast P10-channels which are representative for the other polypeptide channels are not much affected by the same voltage-jump procedure (B). However, if the voltage-free state (0 mV) is adopted for a short time, P10-channels sense the abrupt change in voltage sign and disappear. Thereafter current-relaxation proceeds to

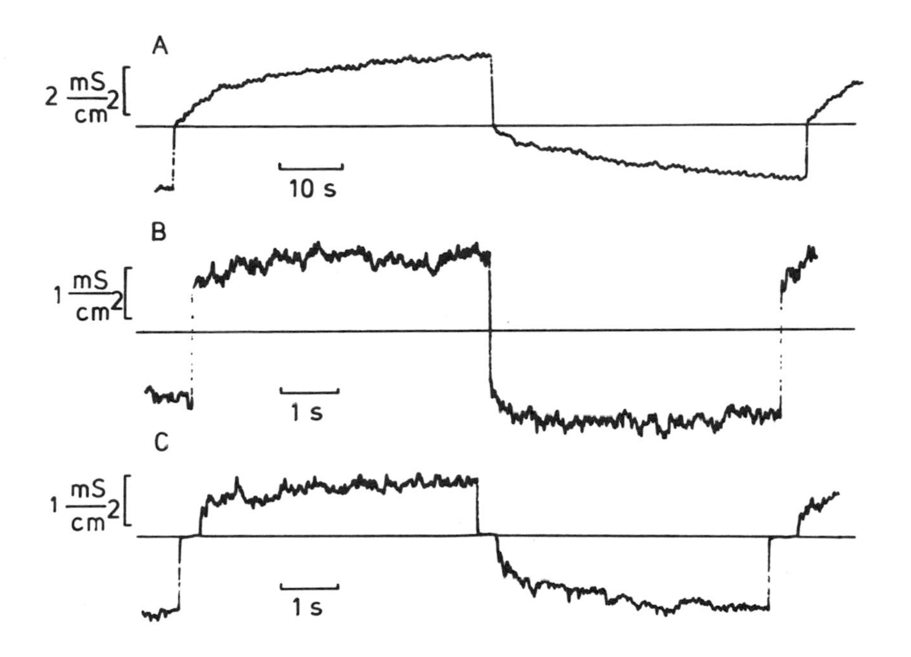

FIGURE 4. Voltage-jump current-relaxation experiments in the presence of alamethicin (A) and the polypeptide P10 Boc-(L-Ala-Aib-L-Ala-Aib-L-Ala)$_2$-OME (B, C).
A. Membrane voltage was changed from +190 mV to -190 mV and back to +190 mV. In each case the P10-induced membrane current which had been established at the preceding voltage vanished immediately after the voltage jump. While other channels are formed, current relaxation proceeds exponentially to the new steady state.
B. Membrane voltage was changed from +265 mV to -265 mV and back to +265 mV. With P10 more than 50% of the preceding current is left after voltage sign reversal. Consequently only a small relaxation amplitude is recorded. Furthermore the relaxation process is more than one order of magnitude faster than that in case of alamethicin (A).
C. Membrane voltage was changed from +260 mV to -260 mV in a two step process via 0 mV and then back to +260 mV in the same way. Now the preceding current is nearly vanished and new channels are formed comparable to A.

Experimental conditions are the same as in Fig. 3 (I/V-curve) except of :
Polypeptide concentration on both sides : A. AL30 0.06 µg/ml;
B. and C. P10 100 µg/ml.

the new steady state characterized by the final constant voltage (C). We interpret this observation by the action of thermal disordering forces at 0 mV which tend to drive the helices out of the direction vertical to the membrane plane. As a consequence a component of the dipole moment within the membrane plane builds up which imposes a torque upon the helical molecule comparable to the situation with AL30.

It is interesting to notice that the voltage sign reversal experiments represent reliable tests in order to discriminate between the straight elongated helices of the polypeptides (P5 to P20) and the AL30 helix which possesses a bend at Pro(14) (14,16).

C. Single channel current fluctuations

In contrast to the similarities with respect to the I/V-characteristics of alamethicin and synthetic polyeptide channels, current fluctuations on the single channel level reveal considerable differences. The recording of single channel current fluctuations is the method of choice in order to obtain detailed information on the "fine structure" of conformational changes of channel proteins, i.e. on the gating processes. This information cannot be obtained by multi-channel experiments, e.g. by vesicle flux measurements (23). The inset of Fig. 3 demonstrates that the current fluctuation pattern of single P15 channels differs significantly from that of alamethicin channels (Fig. 2A). The peculiar sequence of discrete non-integral current levels within bursts of single channel activity is not observed. Instead, the P15 channel opens up to a quite large hole of 7.6 nS conductance corresponding to a channel lumen diameter of about 1.8 - 2.0 nm and to the conductance level of the 8. - 9. state of the alamethicin channel. Generally current levels of open P15 channels are much more noisy than those of alamethicin channel states. Occasionally step-like current changes of the open P15 channel level of about 1 nS are recorded. This is on the order of the conductance difference of two higher neighbouring states in case of the alamethicin channel.

The mean conductance value of these large size channels is higher the shorter the length of the polypeptide helix. In 1 M KCl the following conductances are obtained : P20, 6.2 nS; P15, 7.6 nS; P10, 11.2 nS ; P5, 10.7 nS. The product of conductance value times helix length is nearly constant except for the P5 channel. This gives evidence for P5 dimer formation in order to bridge the lipid phase. Crystal structure data indicate head-to-tail dimerization (24). It is interesting to note that at the high voltages applied (>180 mV) Boc-Al forms the same type of large size channels of several nS conductance (25) as the N-terminal Boc-polypeptides.

The exchange of the Boc-group in P15 by Ac-Aib- in P16 does not change the general feature of channel formation properties. However, the mean lifetimes of the large size channel events are shorter by nearly one order of magnitude. Resolved bursts of transitions between two open state levels of about 1 nS conductance difference are observed quite often. This indicates a channel substructure similar to that of alamethicin.

In addition to the Pro(14)-induced helix bend the synthetic polypeptides P5 to P20 lack the hydrophilic C-terminal amino acid sequence -Glu-Gln-Pheol of alamethicin. In exchange for the -OMe group this partial sequence has been added to P16 to yield P19, the Pro-lacking analog of alamethicin. Surprisingly P19 did not form the non-integral sequence of single channel current levels. Short living current steps in the ms time range as a result of forming channels of large conductance were observed. Thus the hydrophilic C-terminus is not sufficient for channel state stabilization. Because of sterical hindrance the voluminous hydrophilic part has to be removed more distantly from the middle axis of the channel. Apparently, this is achieved by the helix bend at Pro(14) in alamethicin.

IV. CONCLUSIONS

The purpose of our investigations presented above was twice - first to find the main structural prerequisite for voltage dependent channel formation of rod-like helical molecules and second to elucidate the structure/function relationship of single amino acids with respect to channel state stabilization by way of amino acid exchange at distinct sequence positions. Our experimental results unequivocally demonstrate that the large intrinsic dipole moment of an α- or 3/10-helical molecule causes a strong voltage dependence in ion channel formation. The molecular mechanism of channel formation is most simply understood on the basis of the dipole flip-flop gating model (11). We do not want to list the large number of evidences for this model and refer to the original paper. Since all the helical synthetic polypeptides lack Gln(7) of alamethicin in their sequence, there is no way that this amino acid can provide the strong voltage dependence of alamethicin channel formation as proposed by Fox and Richards (14).

Our analyses of the effects of single amino acid exchange on ion channel properties give evidence for three positions within the alamethicin sequence which are critical for channel state stabilization : Ac-Aib(1), Pro(14), and Glu(18) [or Gln (19), respectively]. Whereas the exchange of Pro to Ala in position 2 did not affect channel gating, channel state destabilization occurred after substituting Ac-Aib- by the Boc-group in position 1. The Boc-group which has freedom for rotational movements is more voluminous than the Ac-group and does not fit into the slim shape of the helix. This sterical hindrance for molecule association may prevent the establishment of smaller aggregates which vary statistically in size according to the barrel stave model (11).

Amino acid exchanges between positions 1 and 14 seem not to influence channel properties. However, the lack of Pro(14) destabilizes channel states as seen with the P19 analog. Again this may be caused by a sterical hindrance, i.e. the hydrophilic C-terminal sequence -Glu-Gln-Pheol seems to be too voluminous in order to allow helices to arrange in form of small stable circular arrays. Only if Pro(14) provides the helix bend, the sterical prerequisites are met. However, a further crucial point for the long-lived stabilization of the various alamethicin channel states seems to be the Glu-Gln (Gln-Gln) cluster. Since each of the two Glx provides two functional arms to both nearest neighbours, they may form hydrogen bonds and thus build up a stable closed circle of hydrogen bridges at the C-terminal side (19). If there is only one Glx available as in case of the trichotoxins, no stabilization occurs and no channel states are resolved. The chemical modification of Glu(18) to Glu(OBzl) leads to a significant destabilization of the AL-OBzl channel states as well (Fig. 2B). All the alamethicin channel properties mentioned above are in agreement with channel formation by helical rods, i.e. by those molecular structures which are observed in the crystallized state. There is no indication for a ß-barrel structure (26) of the C-terminal nonapeptide in alamethicin. Especially the voltage dependence of channel state distributions (19) is exactly described only by the helix dipole flip-flop gating model (11).

REFERENCES

1. Popot, J. L., and Changeux, J.-P., Physiol. Rev. 64:1162 (1984).
2. French, R. J., and Horn, R., Ann. Rev. Biophys. Bioeng. 12:319 (1983).
3. Methfessel, C., and Boheim, G., Biophys. Struct. Mech. 9:35 (1982).
4. Noda, M., et al., Nature 302:528 (1983).
5. Noda, M., et al., Nature 312:121 (1984).
6. Hendersen, R., and Unwin, P. N. T., Nature 257:28 (1975).
7. Kyte, J., and Doolittle, R. F. J., J. Molec. Biol. 157: 105 (1982).
8. Guy, H. R., Biophys. J. 45:249 (1984).

9. Hol. W. G. J., Prog. Biophys. Molec. Biol. 45:149 (1985).
10. Jung, G., Brückner, H., Schmitt, H., in "Structure and Activity of Natural Peptides" (Voelter, W., and Weitzel, G., eds.), p. 75. de Gruyter, Berlin, 1981.
11. Boheim, G., Hanke, W., and Jung, G., Biophys. Struct. Mech. 9:181 (1983).
12. Brückner, H., Graf, H., and Bokel, M., Experientia 40: 1189 (1984).
13. Jung, G., Katz, E., Schmitt, H., Voges, K.-P., Menestrina, G., and Boheim, G., in "Physical Chemistry of Transmembrane Ion Motions" (Spach, G., ed.), p. 349. Elsevier Publ., Amsterdam, 1983.
14. Fox, R. O., Richards, F. M., Nature 300:325 (1982).
15. Bosch, R., Jung, G., Schmitt, H., and Winter, W., Biopolymers 24:961 (1985).
16. Bosch, R., Jung, G., Schmitt, H., and Winter, W., Biopolymers 24:979 (1985).
17. Rizzo, V., Schwarz, G., Voges, K.-P., and Jung, G., Eur. Biophys. J. 12:67 (1985).
18. Montal, M., and Mueller, P., Proc. Natl. Acad. Sci. USA 69:3561 (1972).
19. Boheim, G., and Kolb, H.-A., J. Membrane Biol. 38:99 (1978).
20. Boheim, G., J. Membrane Biol. 19:277 (1974).
21. Hanke, W., Methfessel, C., Wilmsen, H.-U., Katz, E., Jung, G., and Boheim, G., Biochim. Biophys. Acta 727:108 (1983).
22. Menestrina, G., Voges, K.-P., Jung, G., and Boheim, G., J. Membrane Biol. submitted.
23. Mathew, M. K., and Balaram, P., Molec. Cell. Biochem. 50: 47 (1983).
24. Bosch, R., Jung, G., Schmitt, H., Sheldrick, G. M., and Winter, W., Angew. Chem. Int. Ed. Engl. 23:450 (1984).
25. Jung, G., Becker, G., Schmitt, H., Voges, K.-P., Boheim, G., and Griesbach, S., in "Peptides, Structure and Function" (Hruby, V. J., and Rich, D. H., eds.), p. 491. Pierce Chem. Comp., Rockford, Ill., 1983.
26. Hall, J. E., Vodyanoy, I., Balasubramanian, T. M., and Marshall, G. R., Biophys. J. 45:233 (1984).

ATP SYNTHASE: GENETICS AND MECHANISM HUMAN AND THERMOPHILIC F_1

Yasuo Kagawa
Hajime Hirata
Shigeo Ohta

Department of Biochemistry
Jichi Medical School
Minamikawachi, Tochigi-ken

Morio Ishizuka

Department of Industrial Chemistry
Chuo University
Bunkyo-ku, Tokyo

Yukio Karube

Tokyo Institute of Technology
Nagatsuda, Yokohama

I. INTRODUCTION

ATP synthase is the complex enzyme (1) for ATP synthesis in mitochondria, chloroplasts and bacterial plasma membranes. This enzyme utilizes energy of proton flux through these biomembranes (2, 3). It consists of a catalytic portion, called F_1 (4) and a proton channel portion, called Fo (5). F_1 is an oligomeric protein that contains five different kinds of polypeptides, labeled alpha to epsilon in order of decreasing molecular weight, and its beta subunit is the catalytic site (3). An operon coding for FoF_1 of Escherichia coli (EF_1)(6, 7) and genes for some subunits of chloroplasts

(CF_1)(8,9), yeast mitochondria (a partial sequence of the beta subunit)(10) etc. have been sequenced. However, these F_1's are unstable in the absence of ATP-Mg, and can not be reconstituted from the primary structure of the subunits. Moreover, none of the genes of F_1 has ever been cloned from animal cells.

To elucidate the mechanisms of proton driven ATP synthesis, we sequenced genes for F_1's obtained from both thermophilic bacterium PS3 (TF_1)(11) and a human cell line (HF_1) (12). TF_1 can be reconstituted from its five subunits after its higher structure has been completely destroyed (13). Thus, TF_1 is suitable for enzyme level studies on its nucleotide binding, proton translocation and genetics. Human mitochondrial DNA is the samllest among those reported, and some of the human cell lines can be cultured without serum for controlling the energetic and ionic conditions (14). Thus, HF_1 is suitable for cell level studies on coordination of mitochondria, nucleus and cytosol.

II. SEQUENCES OF THERMOPHILIC AND HUMAN F_1's

A. Gene for Thermophilic F_1

Hybrid plasmids pFT 1503 containing E. coli F_1 gene and pMCR 533 containing E. coli Fo gene were obtained from Dr. Futai (7) and the inserted DNA's were used as probes. Mixed oligonucleotides, 14mer corresponding to a part of TFo subunit (MFIGV)(15), and 13mer corresponding to a part of TF_1 the beta subunit (YHEM)(16) were synthesized with a DNA synthesizer and also used as probes. Since the probes used were either genes of E. coli or similar DNA, it was essential to use M13 phage containing $TFoF_1$ gene to separate contaminating E. coli DNA. The replicative form M13 phage prepared from the positive single stranded recombinant used for dot blot hybridization (17) with the probes was digested with Sal I. The 755 base pair Sal I fragment (inserted TF_1 gene) was isolated. The Pst 9.3 kilo base pair fragment hybridized with probes of both Fo and F_1. DNA was sequenced by both the method of Maxam and Gilbert (18) as modified for thermophilic DNA (19), and the method of Sanger (20). The data of sequence were analyzed with a software (GENETYX, Tokyo).

Despite the high homology in amino acid sequences of F_1's, the codon usage for TF_1 beta is quite unique. Except for Gln, Lys and Glu, amino acids are coded by codons ending with G or C. Even these exceptions were not found in the codon usage of an extreme thermophile (19), perhaps to stabilize nucleic acid structure at high temperature. However, these thermophilic genes contained Shine-Dalgarno sequence and were well expressed in E. coli. The thermophilic beta gene was first ligated to the Hind III-Eco RI site of pUC18 to orient the gene, and then ligated to the same restriction site of an expression vector pKK223-3. Then the E. coli harboring this expression vector was induced with IPTG. Since large amounts of thermophilic gene products were obtained, site directed mutagenesis of the gene is under investigation. The physico-chemical analysis of TF_1 subunits is possible, though it is very difficult in mesophilic F_1's. There was a terminator at the 3' end of the epsilon gene, which was stabilized by a long stem (13 base pairs).

B. Gene for Human F_1 Beta Subunit

Our attempts to isolate genes for human F_1 by the DNA hybridization method with probes obtained from the genes for EF_1 and CF_1 have been unsuccessful. Thus, a human HeLa cell cDNA library established in an expression vector λgt11 (21), kindly provided by Dr. P. Nielsen of Biocenter, Basel, was used. The library was screened with antiserum against the yeast F_1 beta subunit. A positive plaque contained a full-length cCNA for the beta subunit. The insert was cut out with the restriction enzyme EcoRI, and fragmented with several restriction enzymes or by sonication after circularization with T4 DNA ligase. The small fragments were cloned in M13 mp18 or mp19 vector by the method of Messing et al.(22). The nucleotide sequence of the insert was determined by the dideoxy chain termination method (20). The genomic DNA for the beta subunit was cloned by hybridization method with the cDNA of this subunit.

The structure of the human F_1 beta gene was quite different from that of other F_1 beta genes in the following respects: 1. It is very large (9 kilo base pairs) and separated into at least 6 exons. 2. It codes for a signal prepeptide (mol. wt. 51,568.3) for crossing the mitochondrial membrane after the translation. 3. It has no operon structure like genes for EF_1 and TF_1. 4. It contains three stop codons and a Shine-Dalgarno-like sequence in the 5' upstream region of its initiation codon. Thus, no fusion protein was detected when it was expressed with λgt11. 5. The first exon codes the signal peptide, and the fifth exon starts from the middle portion of the 8-azido ATP-binding site (23).

C. Amino Acid Sequences of TF_1 and HF_1 Beta Subunits

```
        1                   20                  40                  60                  80
a)TF1   ------------------MTRGRVIQVMGPVVDVKFENGHLPAIYNALKIQHKARNENEVDIDLTLEVALHLGDDTVRTIA
b)CF1   MRINPTTSDPGVSTLEKKNLGRIAQIIGPVLNVAFPPGKMPNIYNALIVKGRDTAGQPMN--VTCEVQQLLGNNRVRAVA
c)HF1   ---------TSPSPKAGAATGRIVAVIGAVVDVQFDEGLPPIL-NALEVQGRETR-------LVLEVAQHLGESTVRTIA
d)EF1   ------------------MATGKIVQVIGAVVDVEFPQDAVPRVYDALEVQNGNER-------LVLEVQQQLGGGIVRTIA

                            100                 120                 140                 160
a)      MASTDGLIRGMEVIDTGAPISVPVGQVTLGRVFNVLGEPIDLEGDIPADARRDPIHRPAPKFEELATEVEILETGIKVVD
b)      MSATDGLTRGMEVIDTGAPLSVPVGGPTLGRIFNVLGEPVDNLRPVDIRTT-SPIHRSAPAFTQLDTKSLIFETGIKVVN
c)      MDGTEGLVRGQKVLDSGAPIKIPVGPETLGRIMNVIGEPIDERGPIKTKQF-APIHAEAPEFMEMSVEQEILVTGIKVVD
d)      MGSSDGLRRGLDVKDLEHPIEVPVGKATLGRIMNVLGEPVDMKGEIGEEER-WAIHRAAPSYEELSNSQELLETGIKVID

                            180                 200                 220                 240
a)      LLAPYIKGGKIGLFGGAGDGKTVLIQELIHNIAQEHGGISVFAGVGERTREGNDLYHEMKDSGVIS-------KTAMVFG
b)      LLAPYRRGGKIGLFGGAGVGKTVLIMELINNIAKAHGGVSVFGGVGERTREGNDLYMEMKESGVINEQNIAESKVALVYG
c)      LLAPYAKGGKIGLFGGAGVGKTVLIMELINNVAKAHGGYSVFAGVGERTREGNDLYHEMIESGVINLKDAT-SKVALVYG
d)      LMCPFAKGGKVGLFGGAGVGKTVNMMELIRNIAIEHSGYSVFAGVGERTREGNDFYHEMTDSNVID-------KVSLVYG
```

FIGURE 1. Alignment of the protein sequences of the F_1 beta subunits obtained from a) thermophilic bacterium PS3 (TF_1)(11), b) spinach chloroplasts (CF_1)(8), c) human mitochondria, HeLa cell (HF_1)(12) and d) Escherichia coli (EF_1)(6,7). Identical residues are boxed. The signal presequence of HF_1 beta has been removed.

```
                         260                 280                 300                 320
a)      QMNEPPGARMRVALTGLTMAEYFRDEQGQDVLLFIDNIFRFTQAGSEVSALLGRMPSAIGYQPTLATEMGQLQERITSTA
b)      QMNEPPGARMRVGLTALTMAEYFRDVNEQDVLLFIDNIFRFVQAGSEVSALLGRMPSAVGVQPTLSTEMGSLQERITSTK
c)      QMNQPPGARARVALTGLTVAEYFRDQEGQDVLLFIDNIFRFTQAGSEVSALLGRIPSAVGYQPTLATDMGTMQERITTTK
d)      QMNEPPGNRLRVALTGLTMAEKFRD-EGRDVLLFVDNIYRYTLAGTEVSALLGRMPSAVGYQPTLAEEMGVLQERITSTK

                         340                 360                 380                 400
a)      KGSITSIQAIYVPADDYTDPAPATTFSHLDATTNLERKLAEMGIYPAVDPLVSTSRALAPEIVGEEHYQVARKVQQTLER
b)      EGSITSIQAVYVPADDLTNPAPATTFAHLDATTVLSRGLAAKGIYPAVDPLDSTSTMLQPRIVGEEHYEIAQRVKETLQR
c)      KGSITSVQAIYVPADDLTDPAPATTFAHLDATTVLSRAIAELGIYPAVDPLDSTSRIANPNIVGSEHYDVARGVQKILQD
d)      TGSITSVQAVYVPADDLTDPSPATTFAHLDATVVLSRQIASLGIYPAVDPLDSTSRQLDPEVVGQEHYDTARGVQSILQR

                         420                 440                 460                 480
a)      YKELQDIIAILGMDELSDEDKLVVHRARRIQFFLSQNFHVAEQFTGQPGSYVPVKETVRGFKEILEGKYDHLPEDRFRLV
b)      YKELQDIIAILGLDELSEEDRLTVARARKIERFLSQPFFVAEVFTGSPGKYVGLAETIRGPQLILSGELDSLPEQAFYLV
c)      YKSLQDIIAILGMDELSEEDKLTVSRARKIQRFLSQPFQVAEVFTGHMGKLVPLKETIKGFQQILAGEYDHLPEQAFYMV
d)      YQELKDIIAILGMDELSEEDKLVVARARKIQRFLSQPFFVAEVFTGSPGKYVSLKDTIRGFKGIMEGEYDHLPEQAFYMV

                         500
a)      GRIEEVVEKAKAMGVEV----
b)      GNIDEATAKAMNLEMESKLKK
c)      GPIEEAVAKADKLAEEHSS--
d)      GSIEEAVEKAKKL--------
```

FIGURE 1. Continued.

The amino acid sequences were deduced from the nucleotide sequences. The sequence of TF_1 beta was consistent with amino acid composition, partial peptide sequences and apparent molecular weight of the purified TF_1 beta subunit (51,995.6 from the sequence). The molecular weight of HF_1 beta subunit was determined to be 51,568.3 from the sequence and was identical with that purified from human heart mitochondria and also with that detected in HeLa cell mitochondria by radioimmunoelectrophoresis. Fig. 1 shows the aligned amino acid sequences of the F_1 beta subunits obtained from thermophilic bacterium, spinach chloroplasts (8), human mitochondria, and E. coli (6,7). Identical residues are boxed. The homologies of TF_1 with the beta subunits obtained from CF_1, HF_1 and EF_1 were 66.0 %, 67.7 % and 65.3 %, respectively. There are 270 common residues of the beta subunits in CF_1, HF_1 and EF_1, and 241 of these residues (89.3 %) were also common to TF_1 beta subunit. Harr plots of the amino acid sequences of the subunits in Fig. 1 were also linear. This high homology is probably a consequence of the location of the active site of F_1 on the beta subunit. As discussed in detail, especially for EF_1 (6, 7), the beta subunits show homologies with nucleotide binding proteins. The location of nucleotide binding site was confirmed by photoaffinity labelling of 8-azido ATP (23). A region of acid-base cluster (#390-405, TF_1) which may react with H^+-flux and change the conformation to release bound ATP (24) was also conserved in the beta subunits. A metal reactive site of EF_1 beta (25)(Ser^{174}) was conserved but the other ion reactive dicyclohexylcarbodiimide binding site of the TF_1 beta subunit (Glu^{190}) was distant from that of other beta subunits (16). In short, most reactive sites for ATP and H^+ in other F_1's were also found in HF_1 and TF_1.

D. SECONDARY STRUCTURES OF THE SUBUNITS OF HF_1 AND TF_1

The secondary structures of HF_1 and TF_1 were deduced by the Chou and Fasman's calculation (26)(Figs. 2 and 3). The helix and sheet contents of TF_1 beta determined by circular dichroism spectrometry were 34 % and 23 %, respectively (13), which were close to the calculated contents of 30.1 % and 22.3 %. Fourier transform infrared spectroscopy of TF_1 beta also revealed the presence of about 50 % of the secondary structure containing a large amount of antiparallel sheet (a sharp peak at 1640 cm^{-1})(27), which appears as a beta-meander of the ATP binding site (Fig. 3). Residues forming reverse turns are well conserved in the four beta subunits in Fig. 1 (Gly=92.3 %, Pro=82.0 %, Tyr=75.0 %) and contribute to keep

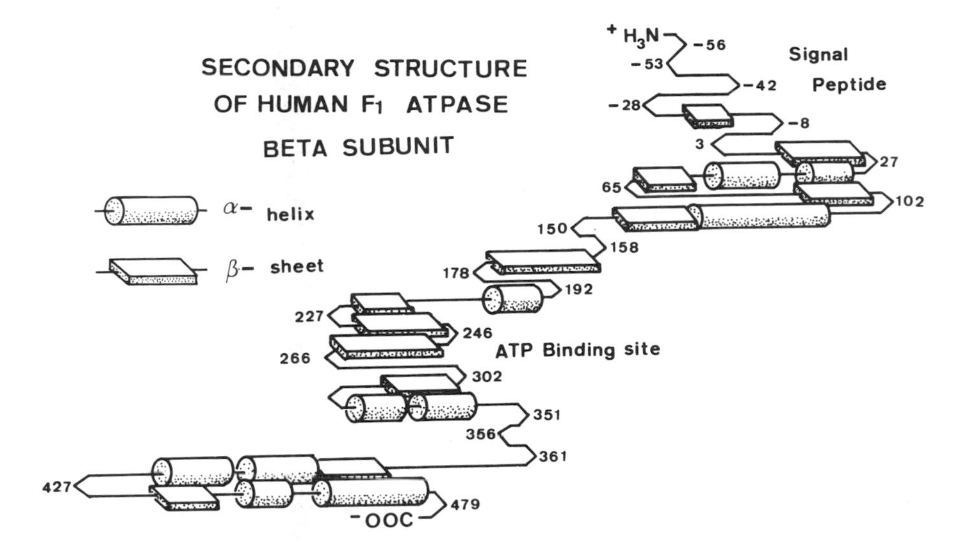

FIGURE 2. Secondary structure of the human F_1 beta subunit.

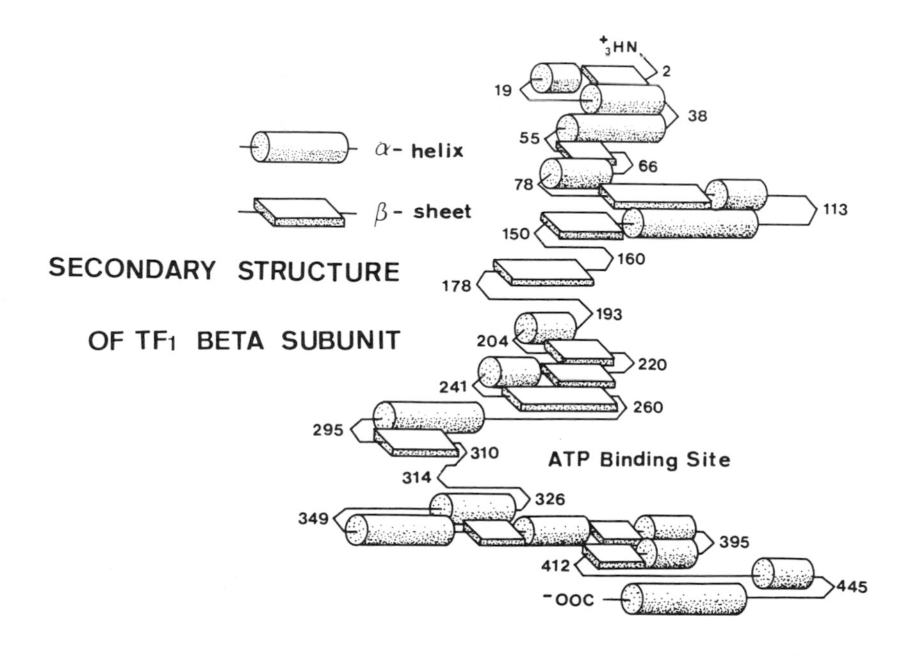

FIGURE 3. Secondary structure of the TF_1 beta subunit.

TABLE I. External Polarity in the Epsilon Subunits

		TF_1 epsilon			EF_1 epsilon		
N-terminus		#1	MKT---	2	#-1	MAMT--	1
Revers turn	no.1	#10	VTPDEG	1	#15	QMFSGL	0
	no.2	#26	KAKSGE	3	#25	QVTGSE	1
	no.3	#34	ILPGHI	1	#35	IYPGHA	1
	no.4	#52	LKKGGK	3	#53	IVKOHG	2
	no.5	#63	VSGGFL	0	#64	YLSGGI	0
	no.6	#70	VRPDNV	2	#84	AIRGQD	2
	no.7	#95	ARKSGR	3	#105	HISSSH	2
C-terminus		#132	---EMK	3	#130	---LSS	1
total external ionic residues				18			10

the tertiary structure of the subunits homologous to each other. The Rossmann fold in EF_1 beta (6) was not typical.

The secondary structure of TF_1 was not lost until about 90°C. The substituted residues in the helices and sheets were mostly formers of the structures with higher propensities (26) than those of mesophilic beta subunits. A role of ionic bonds in thermophilic proteins has been pointed out, and in fact, there are many substitutions of polar residues around revers turns of TF_1. Since the epsilon subunit is the smallest one in F_1, and in this case the reverse turns are exposed externally, the residues of TF_1 and EF_1 at the termini and reverse turns are compared (Table I).

These external ionic groups (18 in thermophilic epsilon, while 10 in E. coli epsilon) may intensify the subunit-subunit interaction through ionic bonds. The stability and reconstitutability may be achieved by many small changes throughout the thermophilic subunits without change in the homologous functioning sites. In fact, these subunits of TF_1 and EF_1 are interchangeable in making active hybrid F_1's (28), yet the external polarity, internal hydrophobicity, propensities forming helices and sheets, and internal packing (29) are higher in subunits of TF_1 than in those of mesophilic F_1's. Thus, the exact comparison of homologous proteins obtained from mesophiles and a thermophile, and different ion translocating biomembranes was achieved.

III. INTERACTION OF ATP WITH F_1

A. ATP Binding Sites of HF_1 and TF_1

F_1's show the following homologies with nucleotide binding proteins (numbers are amino acid residues from the N-terminus): a region (#144-165, TF_1) homologous to regions of oncogene ras p21 (#5-36), adenylate kinase AMP binding pocket (#15-21) and rabbit myosin S_1 head (#165-193); a region (#193-206, TF_1) homologous to Rec A protein (#266-282); and a region (#235-263, TF_1) homologous to ATP/ADP antiporter (#283-291) and phosphofructokinase (#93-111)(6,7,12). One of the 8-azido ATP binding site (23)(Lys^{297} of TF_1) was not present in EF_1 beta. It is interesting that the exon coding for another 8-azido ATP binding sites (Tyr^{307} of TF_1) close to that Lys is separated by an intron in HF_1, and the fact may indicate nonspecific binding of the azido compound to Lys.

The crosslinking of the alpha and beta subunits of mesophilic F_1 by the divalent 3'aryl-8-azido ATP strongly suggests that the catlytic site of F_1 is located at the interface of the two subunits. Since TF_1 contains no firmly bound ATP, treatment of TF_1 with the same divalent ATP analogue resulted in the crosslinking of the alpha-alpha-beta subunits (30). Thus, the tight ATP binding site on the alpha subunit may also be interfacial.

B. Synthesis of F_1 Bound ATP without H^+-Transport

Although the formation of an anhydride bond between ADP and Pi requires a large free energy change in aqueous solution, this catalytic step may not be the energy requiring step in ATP synthesis on ATP synthetase. Direct measurement of enzyme bound nucleotides with ^{31}P-NMR confirmed that the equilibrium constant for the enzyme-bound nucleotide system,

K = [enzyme-ADP-Mg-product]/[enzyme-ATP-Mg-substrate]

is 1 (31). This "equalization of the internal thermodynamics" has not been confirmed by NMR study on F_1's, but the tight binding of AT(D)P to TF_1 was confirmed by NMR study (Yokoyama and Kagawa, to be published). The true substrate of ATP synthase is ATP-Mg complex rather than the free nucleotide, as shown by the divalent cation dependent diastereoisomer preference of TF_1 (32). The stereochemistry of the ATPase reaction was shown to invert the structure of $[^{16}O, ^{17}O, ^{18}O]$ thiophosphate using TF_1 which does not contain endogenous ATP-Mg (32). This experiment supports the in-line S_{N2} reaction, and rules out the presence of X-P intermediate, metaphosphate, and transition state of pseudorotation during the reaction.

Mitchell adovocated the hypothesis that $2H^+$ driven by the

TABLE II. H^+-Transport into FoF_1 Liposomes and Synthesis of TF_1-Bound ATP without H^+-Flux in the Presence of Various Divalent Cations

Divalent cation	ATPase activity (% of Mg-ATPase)	H -transport (% of Mg system)	Bound ATP (ATP/TF_1)
Zn^{2+}	107	15	0.02
Mn^{2+}	104	77	0.18
Co^{2+}	102	77	0.22
Cd^{2+}	102	29	0.02
Mg^{2+}	100	100	0.20
Ca^{2+}	7	7	0.02

(M. Yohda, M. Yoshida and Y. Kagawa, to be published)

proton motive force directly attack the O^- of PO^{4-} in a complex with $ADPO^-$ and Mg^{-2} (33). However, ATP synthesis takes place on purified F_1 in the absence of proton flux (34)(TABLE II). The energy of ATP formation from ADP and Pi may be derived from the energy of conformational change for the ATP binding. If ATP-metal complex directly release H^+, we can expect a parallel relationship between metal-dependent ATPase and H^+-transport. However, as shown in TABLE II, both reactions proceeded differently. On the other hand, H^+-transport paralleled synthesis of F_1-ATP-metal complex. Mitchell's original idea involving the formation of metaphosphate was also denied by the results of stereochemistry (33). The binding constant of ATP and F_1 is very high (35). Thus, the energy requiring step in ATP synthesis is the ATP-release step of the F_1-bound ATP as discussed by recent reviews (3,36,37).

IV. DIRECT MEASUREMENT OF H^+-TRANSPORT OF F_1

A. H^+-Current of ATP Synthase in a Lipid Membrane

Since ATP-driven H^+-transport was demonstrated in the FoF_1-liposomes (38), many systems measuring the ATP-coupled ion transport have been reported. Net ATP-synthesis was demonstrated in FoF_1-liposomes driven by H^+-gradient (39) and an external electric field (40). These results supported Mitchell's hypothesis (2) at physiological level. The direct electrical measurement of H^+-pump activity in a planar lipid bilayer is not so easy as that of ion-channel activity be-

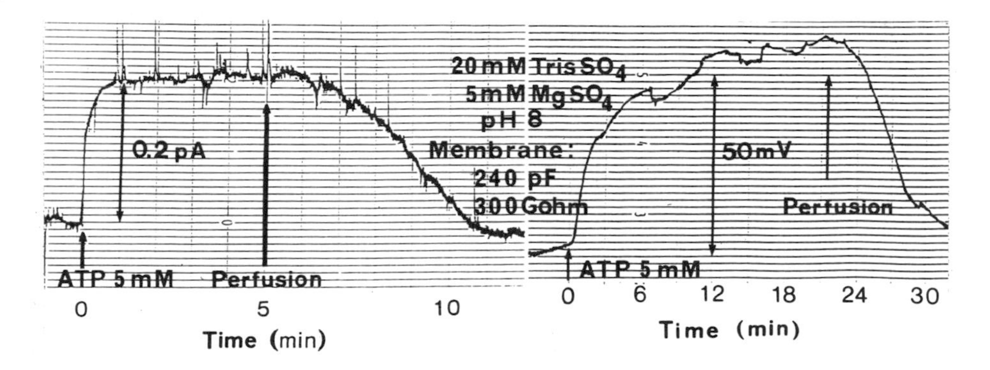

FIGURE 4. Generation of electric current (left) and membrane potential (right) by ATP synthase on a planar lipid membrane. Measurement of the current was carried out with a picoammeter under voltage clamp. The trans chamber was held at virtual ground and the signal was filtered at 5 Hz with an RC low-pass filter. The membrane potential was measured with an electrometer (Keithley 617). The cis chamber was grounded and the potential of the trans chamber was monitored. The addition and perfusion of ATP were performed on the cis side. Soybean phospholipids and thermophilic ATP synthase were purified and proteoliposomes formed were fused with a preformed planar lipid bilayer from the cis side in the presence of 20 mM $CaCl_2$.

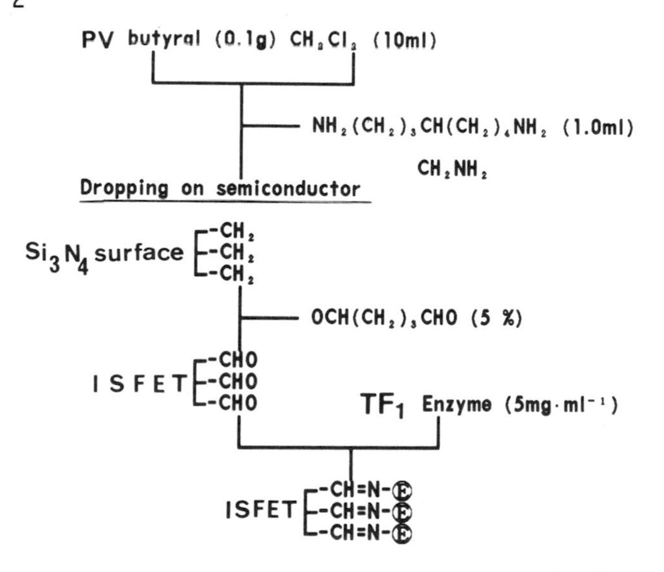

FIGURE 5. Immobilization of TF_1 on ISFE Transistor with polyvinylbutyral and glutaraldehyde.

cause of the very few ion flux per molecule. Turnover rate of FoF_1 is only 20/sec, compared with 10^5 ions/sec per channel protein. However, recent refined method by Hamamoto et al.(41) enabled us the direct measuremnet of H^+-pump activity of FoF_1. Hirata incorporated $TFoF_1$ into a planar membrane and obtained a current of 0.2-1.0 pA, corresponding to $2x10^4$-10^5 FoF_1 per hole of 0.2 mm diameter (Fig. 4). The highest voltage attained by FoF_1 during ATP hydrolysis was about 100 mV, depending on the stability of the membrane. The H^+-current was specifically abolished by azide, AMPPNP, etc. The ATP-induced H^+-current of $TFoF_1$ in a planar lipid bilayer was stopped by an applied potential of -180 mV. Free energy change of ATP hydrolysis in that condition was -554 mV ([ADP][Pi]/[ATP] = 10^{-4}, and the standard free energy change of ATP hydrolysis is -30.5 kJ). Thus, g ions of H^+ transported per mole of ATP hydrolyzed is -554/-180 = 3.08. This result is inconsistent with Mitchell's hypothesis at molecular level which assumes $2H^+$/ATP (33).

B. Direct Measurement of H^+ Production by F_1 with Transistor

The unsolved problem is how H^+ is liberated from ATP synthase during ATP hydrolysis. Although H^+ can be released from Fo, the acid-base cluster hypothesis assumes the liberation of H^+ from F_1 by a conformational change (25). To show this, in collaboration with Prof. Y. Karube, TF_1 was directly connected on the surface of a transistor (Fig. 5). The transistor is called ISFET (ion selective field effect transistor), the surface of which is covered with Si_3N_4, and rapidly responds to H^+. As shown in Fig. 6, the circuit was devised to substract the change in bulk phase pH. This ATP biosensor responded to the ATP concentration as shown in Fig. 7. The highest response of the biosensor was obtained at pH 9, which is the optimal pH for TF_1.

V. Epilogue

Proton driven ATP synthesis in FoF_1 liposomes (38) and passive proton translocation through Fo (5) strongly supported Mitchell's chemiosmotic theory (2). The molecular mechanism of proton translocation has been elucidated by sequencing the primary structures of the catalytic site, analyzing stereo-specific ATP binding to F_1, and measuring the proton current of FoF_1 and F_1. The primary structure of FoF_1 is not homologous with that of other E_1E_2 type ATPases including Na,K-ATPase (42), Ca-ATPase, K-ATPase etc, but rather similar to that of myosin ATPase. This difference was

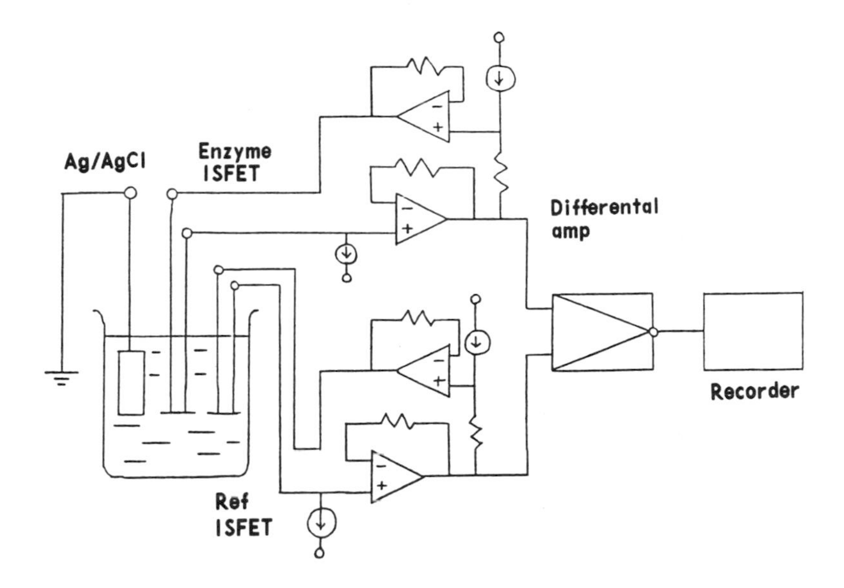

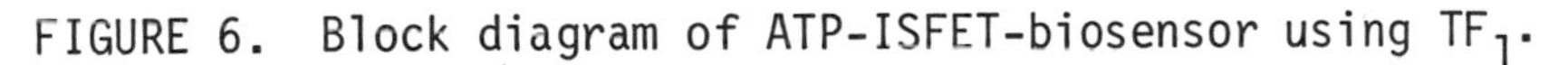
FIGURE 6. Block diagram of ATP-ISFET-biosensor using TF_1.

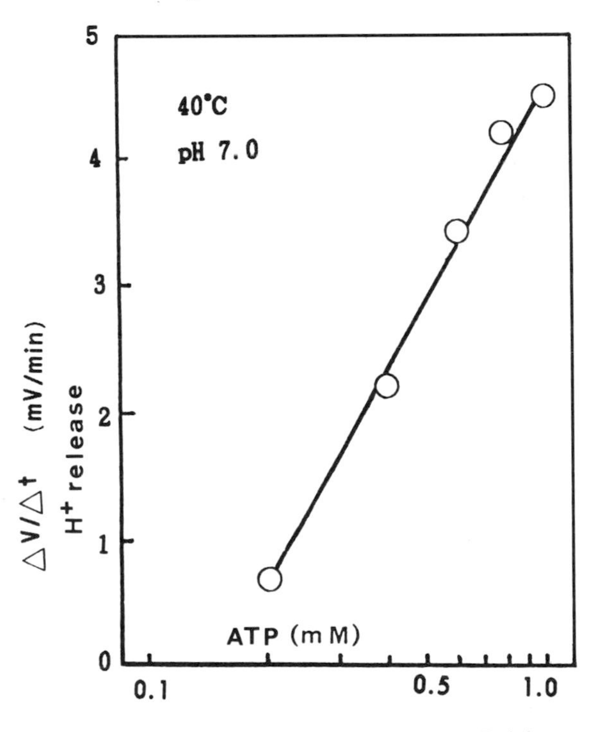

FIGURE 7. Calibration curve of ATP-ISFET-biosensor using TF_1.

also shown by the stereochemical studies (32). During the ATPase reaction, E_1E_2 type retained the configuration of [^{16}O,^{17}O,^{18}O] thiophosphate, while FoF_1 type inverted it.

The direct participation of translocated protons in anhydride bond formation during ATP synthesis was disproved by the net synthesis of F_1-bound ATP without H^+-flux, and $3H^+$/ATP stoichiometry. The conformational change of FoF_1 is the energy requiring step for ATP synthesis. Detachment of ATP from the ATP-F_1-complex is not caused by the direct action of the membrane potential, because the true substrate is the negatively charged Δ, β, γ, ATP-Mg complex. How is the energy of proton flux converted into conformational energy of F_1 to release ATP? Mitchell summarized current mechanistic hypotheses, including his rolling well hypothesis (43). There must be an energy converting device for proton-reactive conformational change. This device may be an acid-base cluster of F_1, like that in hemoglobin which changes affinity for the ligand by protonation. The acid-base cluster hypothesis is thus proposed (Fig. 8) (3, 24).

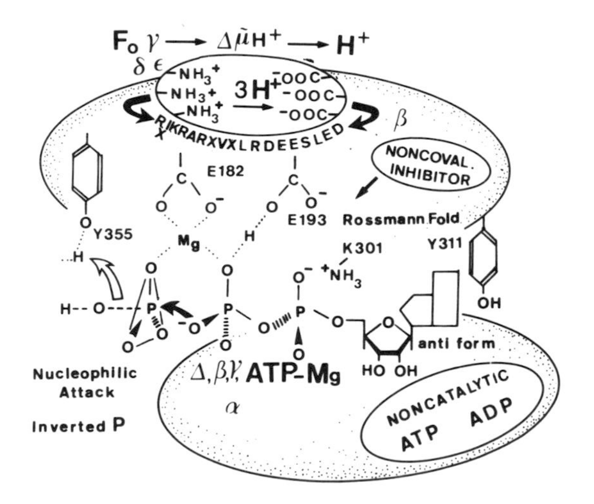

FIGURE 8. Acid-base cluster hypothesis. The cluster of acidic and basic residues causes conformational change in response to the flux of $3H^+$ and release the Δ, β, γ, ATP-Mg complex. ATP was formed at the Rossmann fold (or β-meander shown in Figs. 2 and 3) on the α, β-interface without H^+ flux via nucleophilic attack.

REFERENCES

1. Hatefi, Y., Ann. Rev. Biochem. 54:1015 (1985).
2. Mitchell, P., J. Biochem. 97: 1 (1985).
3. Kagawa, Y., in "Bioenergetics" (Ernster, L. ed), p.149. Elsevier, Amsterdam, 1984.
4. Pullman, M. E., Penefsky, H. S., Datta, A., and Racker, J., Biol. Chem. 235: 3322 (1960).
5. Okamoto, H., Sone, N., Hirata, H., Yoshida, M., and Kagawa, Y., J. Biol. Chem. 252: 6125 (1977).
6. Kanazawa, H., and Futai, M., Ann. N. Y. Acad. Sci. 402: 45 (1982).
7. Walker, J. E., Eberle, A., Gay, N. J., and Saraste, M. O., Biochem. Soc. Trans. 10: 203 (1982).
8. Zurawski, G., Bottomely, W., and Whitfeld, P. R., Proc. Natl. Acad. Sci. USA. 79: 6260 (1982).
9. Shinozaki, K., and Sugiura, M., Nuc. Acid. Res. 10: 4923 (1982).
10. Saltzgaber-Muller, J., Kunapuli, S. P., and Douglas, M.G., J. Biol. Chem. 258: 11465 (1983).
11. Kagawa, Y., Ishizuka, M., Saishu, T., and Nakao, T., in "International Conference on Energy Transduction in ATPase",p.84, Yamada Conference, Kobe, 1985.
12. Ohta, S., and Kagawa, Y., J. Biochem. (Tokyo) 99: in press (1986).
13. Yoshida, M., Sone, N., Hirata, H., and Kagawa, Y., J. Biol. Chem. 252: 3480 (1977).
14. Saishu, T., Hamamoto, T., Kagawa, Y., Ohta, T., and Takaoka, T., J. Biochem. (Tokyo) 97:1079 (1985).
15. Hoppe, J., and Sebald, W., Biochim. Biophys. Acta. 768: 1 (1984).
16. Yoshida, M., Poser, J.W., Allison, W.S., and Esch, F.S., J. Biol. Chem. 256:148-153 (1981).
17. Bagchi, K. M., Banerjee, C. A., Roy, R., Chakrabarty, I., and Gupta, K. N., Nuc. Acid. Res. 10: 6487 (1982).
18. Maxam, A., and Gilbert, W., Methods. Enzymol. 65: 499 (1980).
19. Kagawa, Y., Nojima, H., Nukiwa, N., Ishizuka, M., Nakajima, T., Yasuhara, T., Tanaka, T., and Oshima, T., J. Biol. Chem. 259: 2956 (1984).
20. Sanger, F., Nickelen, S., and Coulson, A. R., Proc. Natl. Acad. Sci. USA. 74: 5463 (1977).
21. Young, R. A., and Davis, R. W., Proc. Natl. Acad. Sci. USA 80: 1194 (1983).
22. Messing, J., Crea, R., and Seeburg, P. H., Nuc. Acid Res. 9: 309 (1900).
23. Hollemans, M., Runswick, M. J., Fearnley, I. M., and Walker, J. E., J. Biol. Chem. 258: 9307 (1983)

24. Kagawa, Y., J. Biochem. 95: 295 (1984).
25. Noumi, T., Mosher, M. E., Natori, S., Futai, M. and Kanazawa, H., J. Biol. Chem. 259: 10071 (1984).
26. Chou, P. T., and Fasman, G. D. Adv. Enzymol. 47: 45 (1978).
27. Ohta, S., Tsuboi, M., Yoshida, M., and Kagawa, Y., Biochemistry 19: 2160 (1980).
Takeda, K., Hirano, M., Kanazawa, H., Nukiwa, N., Kagawa, Y., and Futai, M., J. Biochem. 91: 695 (1982).
29. Argos, P., Rossmann, G. M., Grau, M. U., Zuber, H., Frank, G., and Tratschin, D., Biochemistry 18: 5698 (1979).
30. Schafer, J. H., Rathgeber, G., Dose, K., Masafumi, Y., and Kagawa, Y., FEBS Lett. 186: 275 (1985).
31. Nageswara Rao, B. D., Kayne, F. K., and Cohn, M., J., Biol. Chem. 254: 2689 (1979).
32. Senter, P., Eckstein, F., and Kagawa, Y., Biochemistry 22: 5514 (1983).
33. Mitchell, P., and Moyle, J., Eur. J. Biochem. 4: 530 (1986).
34. Feldman, R. I., and Sigman, D. S., J. Biol. Chem. 257: 1676 (1982).
35. Grubmeyer, C., Cross, R. L., and Penefsky, H. S., J. Biol. Chem. 257: 12092 (1982).
36. Cross, R. L., Ann. Rev. Biochem. 50: 681 (1981).
37. Amzel, L. M., and Pedersen, P. L., Ann. Rev. Biochem. 52: 801 (1983).
38. Kagawa, Y., Biochim. Biophys. Acta 265: 297 (1972).
39. Sone, N., Yoshida, M., Hirata, H., and Kagawa, Y., J. Biol. Chem. 252: 2956 (1977).
40. Kagawa, Y., Curr. Top. Membr. Transport 16: 195 (1982).
41. Hamamoto, T., Carrasco, K., Matsushita, H., Kaback, R., and Montal, M., Proc. Natl. Acad. Sci. USA. 82: 2570 (1985).
42. Kawakami, K., Noguchi, S., Noda, M., Takahashi, H., Ohta, T. Kawamura, M., Nojima, H., Nagano, K., Hirose, T., Inayama, S., Hayashida, H., Miyata, T., and Numa, S. Nature 316: 733 (1985).
43. P, Mitchell., FEBS. Lett. 182: 1 (1985).

MOLECULAR PROFILE OF A COMPLEX OF MITOCHONDRIAL ELECTRON-TRANSPORT CHAIN AND H^+ PUMP ATPase

Takayuki Ozawa
Morimitsu Nishikimi
Hiroshi Suzuki
Masashi Tanaka
Yoshiharu Shimomura

Department of Biomedical Chemistry
Faculty of Medicine
University of Nagoya
Nagoya, Japan

I. INTRODUCTION

Since fragmentation of mitochondrial electron-transport chain into four complexes by Hatefi (1,2) and discovery of the tripartite particles of the inner membrane by Fernandez--Moran (3), a large number of experiments have been undertaken to understand the molecular architecture of the mitochondrial energy transducing unit. MacLennan and Tzagoloff isolated the oligomycin-sensitive ATPase and established this to be the seat of ATP hydrolysis and ATP synthesis (4-6). At the same time, Racker in a series of studies identified the head-piece of the tripartite unit as oligomycin-insensitive ATPase (F_1) and the base piece or membrane sector of the tripartite unit as the system which, together with the head-piece, carried out coupled ATP hydrolysis and synthesis (7-10). So, all the units that participate in energy coupling could be isolated and defined, namely, the four electron-transfer complexes and the F_0-F_1 tripartite system. However, details of the molecular architecture of each unit and the three dimensional array of the units and F_0-F_1 system have not been fully clarified. As Mitchell has proposed the chemiosmotic coupling hypothesis and nominated the F_0-F_1 system as H^+ pump ATPase, the reality of three dimensional interaction of the electron-transfer units and H^+ pump ATPase is especially important to assess his hypothesis (11-13).

Therefore, we have intended to clarify the detailed molecular architecture and three dimensional array of mitochondrial energy transducing system.

II. CONSTITUENTS OF ENERGY TRANSDUCING SYSTEM

A. Complex I (NADH-CoQ Oxidoreductase)

Complex I transfers electrons and protons from NADH to coenzyme Q (CoQ), and is the energy coupling site I. However, information on Complex I is still tentative because of its high molecular weight, 7×10^5, and many polypeptide subunits, 16 ∿ 18 (14). We are attempting to isolate individual polypeptides and subfragments of Complex I.

We have attempted to isolate and to characterize individual polypeptides. At the first step, Complex I was fractionated into three subfragments — hydrophobic proteins, the soluble iron-sulfur protein (IP), and the soluble NADH dehydrogenase of flavoprotein (FP) — by use of deoxycholate and cholate according to the method of Galante and Hatefi (15). They were further purified by hydrophobic affinity chromatography and gel filtration. We found that iron-sulfur clusters were associated with polypeptides of 51 and 23 kDa in FP subfraction (16). Further identification of individual polypeptides such as CoQ-binding protein is proceeding as described in the next paragraph.

B. CoQ-Binding Proteins

CoQ is an essential constituent of the energy-transducing respiratory chain of mitochondria. Many hypotheses have been postulated concerning the role of CoQ in electron transfer and H^+ pumping activity. However, no definite number of mobile or bound CoQ in the chain has been estimated.

We have established an isolation method for CoQ-binding protein (QP) from either mitochondria or from electron transfer particles (17). By this method, at least 29% (0.87 nmol/mg of protein) of the total CoQ in the mitochondrial inner membrane was found to be bound to QP (18). Furthermore, QP was isolated from Complex I or from Complex III in the form of a protein-CoQ complex. As shown in Fig. 1, QP from Complex I (QP-I) was identified as a 13-kDa polypeptide, a constituent of IP, and QP from Complex III (QP-III) as a

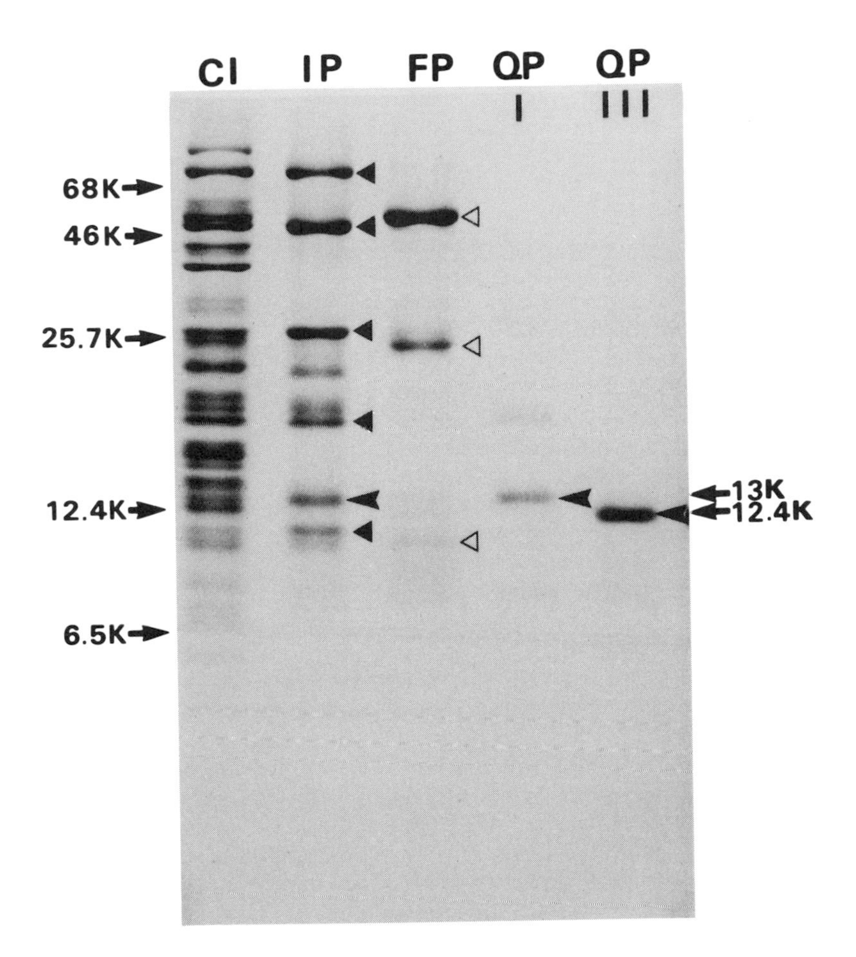

FIGURE 1. Polypeptide subunits in Complex I. Beef heart mitochondrial Complex I (CI) and its iron-sulfur (IP) and flavoprotein (FP) subfractions were subjected to SDS-urea gel electrophoresis. A couple of iron-sulfur clusters were found to be associated with polypeptides of 51 and 23 kDa in the FP subfraction (16). CoQ-binding protein purified from Complex I (QP-I) and that from Complex III (QP-III) are shown by arrows. QP-I was identified as one of polypeptides in IP subfraction of 13 kDa (19). Solid arrowheads indicate the subunits of IP, and open arrowheads those of FP.

12.4-kDa polypeptide (19). It is now recognized that QP-I and QP-III are new constituents of Complex I and III.

C. Complex III (ubiquinol-cytochrome *c* oxidoreductase)

Complex III transfers electrons from ubiquinol to cytochrome *c* and involves a close interaction with phospholipids (20). It was postulated that the complex plays a central role in H^+ pumping from inside to outside of the mitochondrial inner membrane by the electron-transfer chain (13).

It was found that iron-sulfur protein in the complex (FeS-III) could be isolated from the complex with depletion of boundary phospholipids, especially cardiolipin, by washing the complex with detergent on a hydrophobic interaction affinity column of phenyl-Sepharose (21-23). This presented the notion that boundary cardiolipin plays a role as a hoop in the complex, keeping the subunits as a single unit (24). In fact, after removal of boundary cardiolipin, all subunits of a single source of the complex could be separated from each other by detergent-exchange chromatography, as shown in Fig. 2 (25). After the purification, cytochrome c_1 was demonstrated to be coupled to the iron-sulfur protein, resulting in reconstitution of the antimycin-insensitive pathway of electrons in the complex (26). Cytochrome *b* was very labile when it was purified to a single polypeptide, but was stable when it formed a complex with subunit IX, suggesting that subunit IX stablilizes the cytochrome (27).

Antibody against each subunit, such as anti-iron-sulfur protein antibody, could be prepared after the purification (28). The content of the iron-sulfur protein in submitochondrial particles was determined by methods of radioimmunoassay and immunoblotting. The results showed that the mitochondrial inner membrane contains two times as much of the immunoreactive iron-sulfur protein as is expected for Complex III (29).

D. Complex IV (cytochrome *c*-oxygen oxidoreductase)

Cytochrome oxidase (Complex IV), located at the terminus of the mitochondrial electron-transfer chain, transduces the redox energy into the driving force for ATP production. The oxidase consists of seven major subunits. We have succeeded in crystallizing the oxidase in a form of complex with a ligand, cytochrome *c* (30). By using a detergent as another ligand, crystals of the oxidase itself were obtained (31). The electron microscopic observations and electron diffrac-

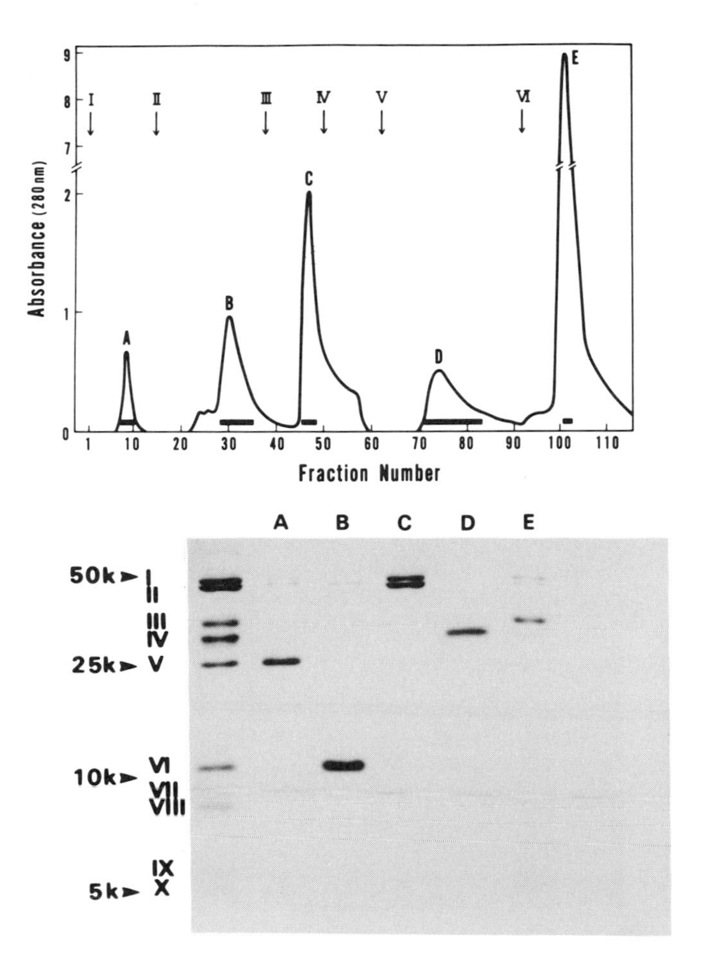

FIGURE 2. Isolation of polypeptide subunits by detergent-exchange chromatography of Complex III. [Upper] After removal of phospholipids from Complex III on a phenyl-Sepharose CL-4B column, an iron-sulfur protein fraction (peak A) was eluted with buffer containing 1% deoxycholate (I). Subsequently, two fractions of subunit VI (peak B) and core proteins (peak C) were eluted with 1.5 M (II) and 3 M (III) guanidine hydrochloride, respectively. After equilibration of the column with the buffer (IV), a cytochrome c_1 fraction (peak D) was eluted with the buffer containing 1% $C_{12}E_8$ (V). Finally, a cytochrome *b*-rich fraction (peak E) was eluted with the buffer containing 2% SDS (VI). Arrows indicate the change of buffers. [Lower] SDS-polyacrylamide gel electrophoretic patterns of the fractions (A-E) from the phenyl-Sepharose CL-4B column are shown.

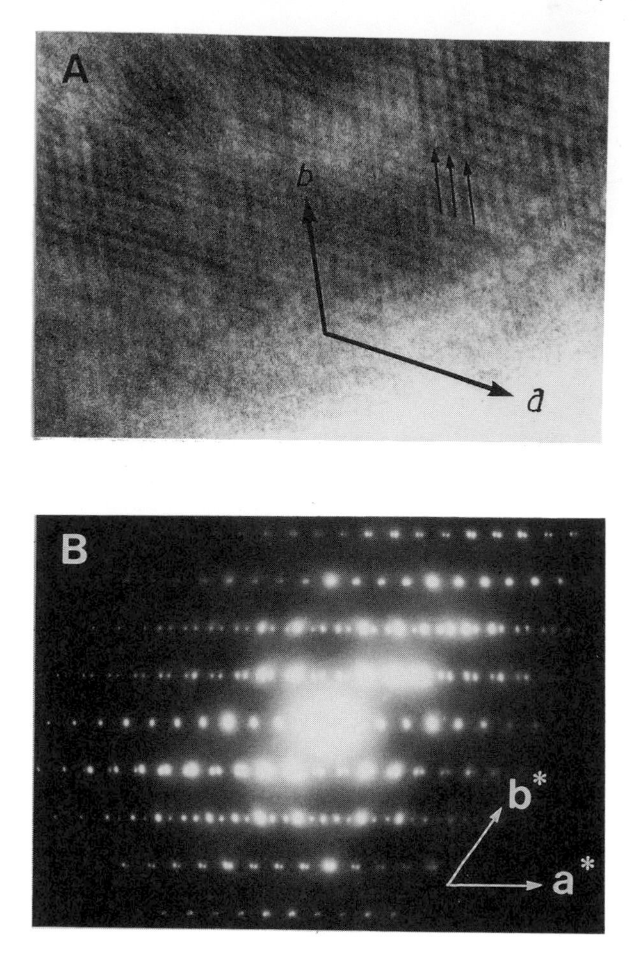

FIGURE 3. Crystal of cytochrome oxidase and its electron diffraction patterns. [A] Crystals of the oxidase were fixed with 1% glutaraldehyde overnight, then negatively stained with 2% ammonium molybdate, pH 7.4. A stained single crystal on a carbon-coated grid was observed with 1,000-kV electrons in a Hitachi HU-1000D ultra-high-voltage electron microscope. Arrows indicate individual oxidase molecules. [B] An electron diffraction pattern from the (hk0) plane of the crystal was photographed under an illumination level of ≃0.25 electrons/$Å^2$ per sec. 1 cm = 0.43 $Å^{-1}$. The crystal is monoclinic in the space group $P2_1$, with unit cell dimensions a = 92 Å, b = 84 Å, and c = 103 Å, and $\alpha = \beta = 90°$, $\gamma = 126°$.

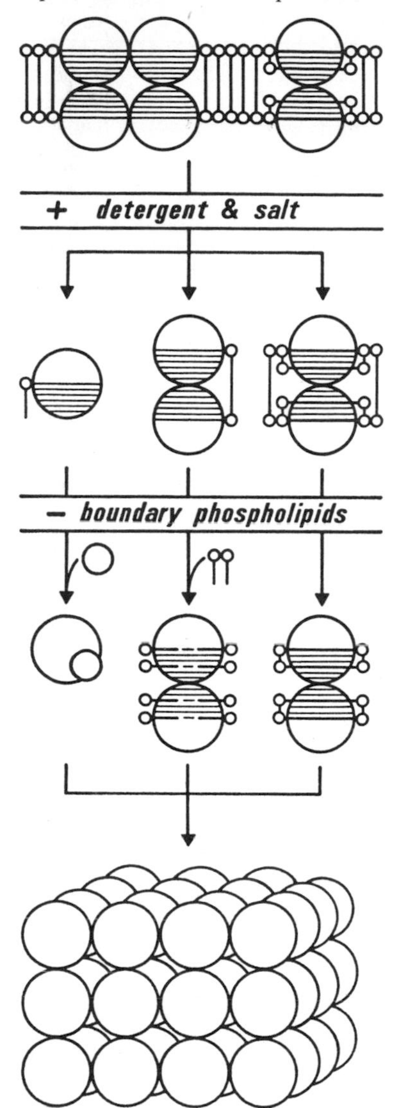

FIGURE 4. Principle for crystallization of membrane proteins. The scheme outlines the process of crystallization of the membrane proteins. As Green (38) noted, membranes are composed of bimodal proteins (spheres) and phospholipids (dumbbells). After removal of boundary lipids, complex formation of the proteins with the specific polar ligand (small spheres), amphiphilic detergent (half-dumbbells), or with cardiolipin (double-dumbbells) promotes formation of the unimodal protein-ligand complex and the three-dimensional crystal.

tion studies revealed the dimensions and angles of the unit cell in the crystal (Fig. 3). The three-dimensional structure of the oxidase, which we expect to elucidate by X-ray diffraction studies, will give us fundamental information on the mechanism of energy transduction.

The principle for the crystallization of these membrane proteins is schematically presented in Fig. 4 (32). Other examples of membrane protein crystals obtained by use of this principle are complex III (33,34) and the calcium-binding protein in sarcoplasmic reticulum (35).

III. SUPRAMOLECULE

A. Separation of the Fundamental Assembly of the Electron-Tranfer Complexes and H^+ Pump ATPase

It has been established that both the electron-transfer chain and oligomycin-sensitive ATPase are located in the mitochondrial inner membrane. However, an understanding of the nature of their assembly and interaction has been elusive, in spite of its importance for unraveling the energy coupling mechanism.

We have intended to isolate the fundamental assembly of the electron-transfer complexes and H^+ pump ATPase. The mitochondrial inner membrane was sonicated in the presence of 2 mM EDTA essentially following 'solubilization' of the inner membrane by the method of Tzagoloff *et al.* (36). The 'solubilized' inner membrane was subjected to sucrose density-gradient centrifugation in the presence of a relatively low concentration (0.3%) of a non-ionic detergent, $C_{12}E_8$, to prevent reconstitution of the membrane.

The distribution of protein and enzymic activities among the fractions is shown in Fig. 5. Two major peaks of protein were obtained. One peak showed neither NADH- nor succinate-oxidase activity and its ATPase activity was relatively oligomycin-insensitive. The other showed the full oxidase activity and its ATPase was completely oligomycin-sensitive.

Subunit polypeptides patterns of both peaks detected by SDS-urea electrophoresis, however, were essentially the same, as shown in Fig. 6. On the other hand, electron microscopic images of both fractions were quite different. The fraction having no oxidase activity showed small aggregates of particles arranged in short chain-like structures, while the fraction having full oxidase activity had depolymerized units of tripartite nature, namely, particles of 190 to 210 Å attached to a satellite particle with the diameter of F_1 (96 Å) via a cylindrical stalk, as shown in Fig. 7.

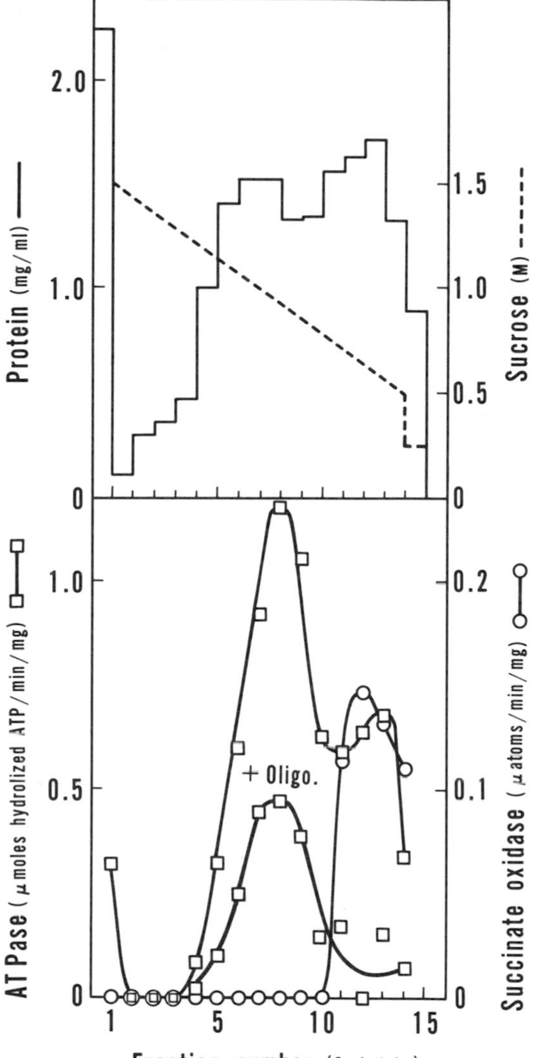

FIGURE 5. Isolation of 'supramolecule' by sucrose density-gradient centrifugation. 'Solubilized' mitochondrial inner membrane obtained by the method of Tzagoloff *et al.* (36) was subjected to density-gradient centrifugation at 79,000 x *g* for 180 min. Protein content, oxidase activity, and ATPase activity with and without oligomycin in each fraction are illustrated.

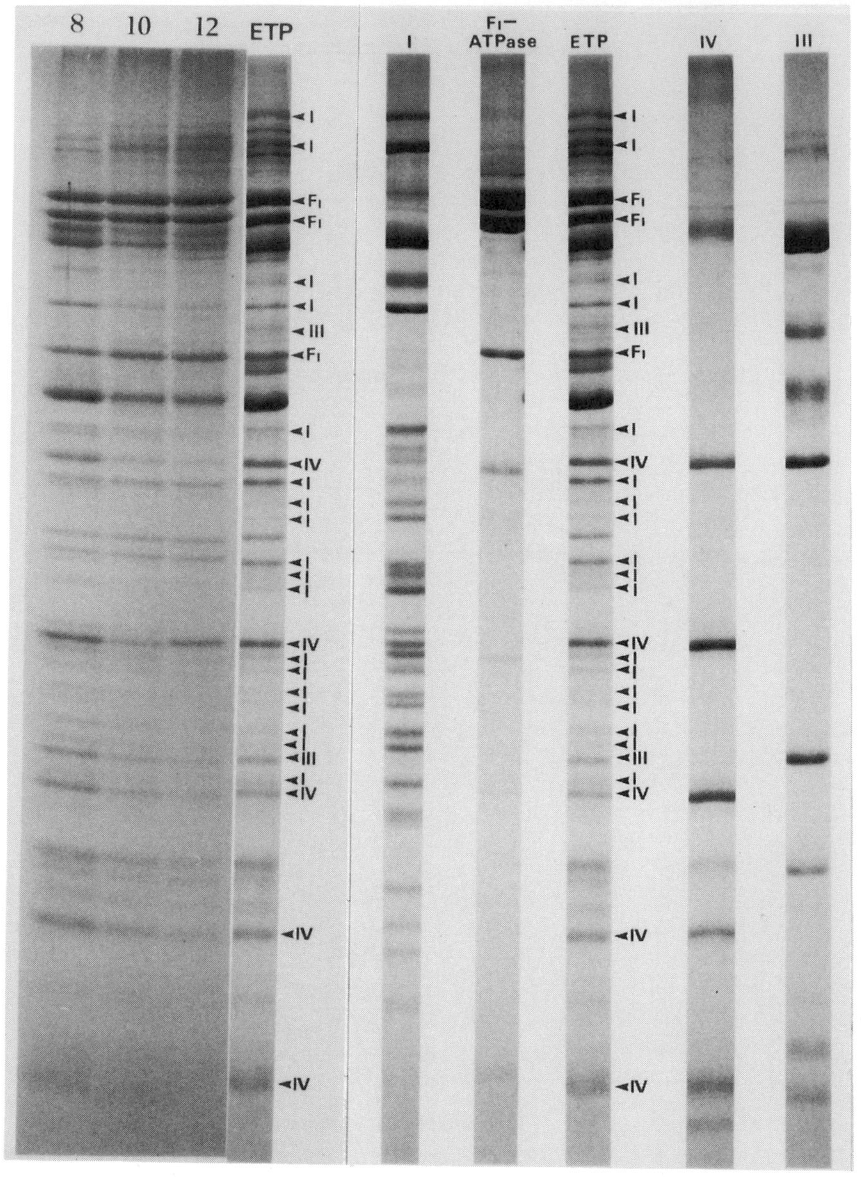

FIGURE 6. SDS-urea electrophoresis patterns of the fractions separated by sucrose density-gradient centrifugation. [Left] Fractions separated as shown in Fig. 5 were subjected to SDS-urea electrophoresis. Fraction numbers are on the top of each column. [Right] Identification of polypeptide subunits of the mitochondrial inner membrane (ETP) by comparison with those of Complex I-IV (I-IV) and F_1-ATPase.

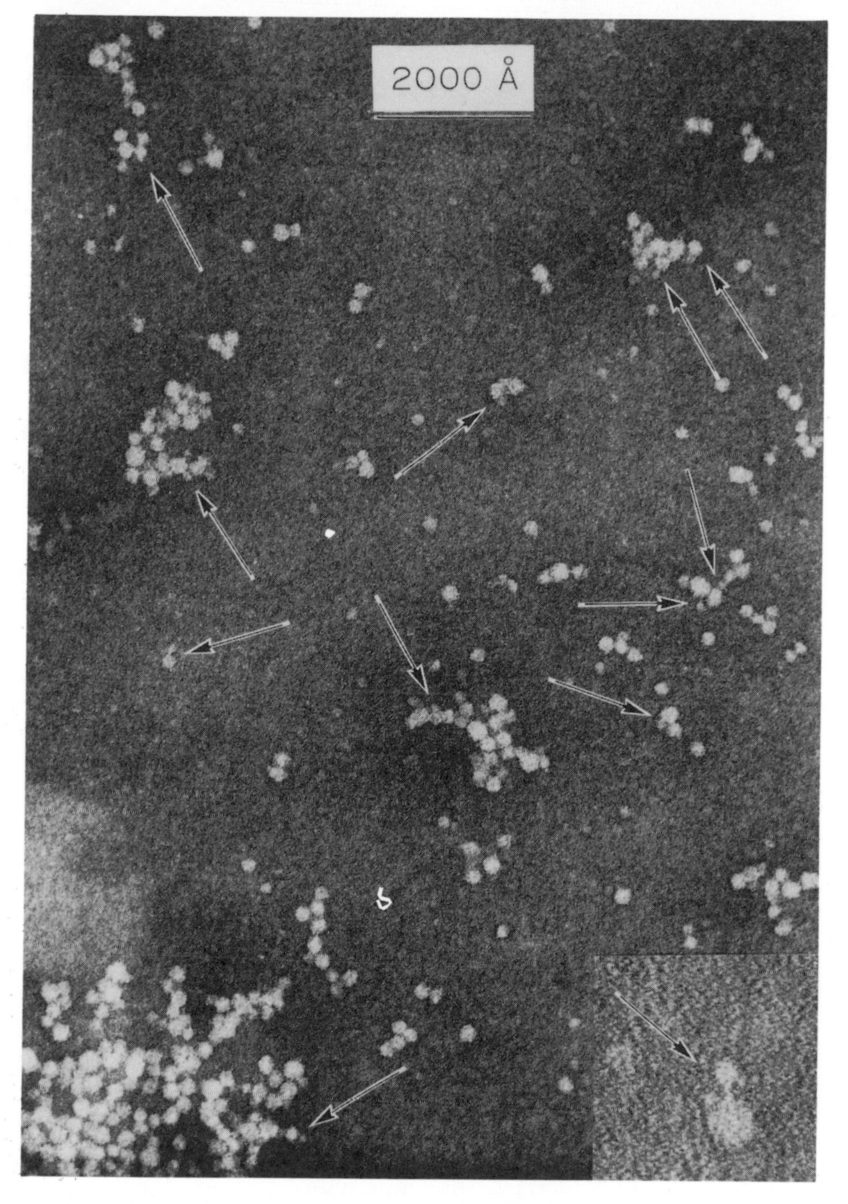

FIGURE 7. Negatively stained image of 'supramolecule'. A sample of fraction No. 12 in Fig. 4 which had both the oxidase and oligomycin-sensitive ATPase activities was negatively stained by 2% ammonium molybdate. Tripartite particles with an asymmetric dumbbell-like shape are indicated by arrows. [Insert] An individual particle of typical tripartite structure is indicated.

B. Assembly of the Supramolecule

From the fact that the composition of the tripartite particle corresponds closely to that of the original inner membrane in terms of polypeptide subunits (Fig. 6) and in terms of the percentage of the total protein attributable to F_1 (18 versus 16%), and since, by dialysis with phospholipids, the particles could reconstitute vesicular membrane with the characteristic headpiece-stalk-basepiece structure of the original, we conclude that the particle is the fundamental assembly of the electron-transfer complexes and H^+ pump ATPase. Thus we have designated it 'supramolecule'.

By proteinase treatment of the membrane, Ozawa and Asai (37) estimated the percentage of the total protein attributable to F_1 in the mitochondrial inner membrane to be 16%. They concluded that one unit of the inner membrane, which includes one head-piece, stalk, and the corresponding membrane sector, should have a 'molecular' weight of 2×10^6, and implied that the unit of the inner membrane which combines with one head-piece consists of $6 \sim 8$ complexes of average molecular weight $200 \sim 300 \times 10^3$. From the electron microscopic image of the supramolecule and its content of the complexes, we postulate the fundamental assembly of the supramolecule as illustrated in Fig. 8; namely, one molecule each of Complex I (700 kDa), Complex II (200 kDa), Complex III (250 kDa), transphosphatase (250 kDa) and two molecules of Complex IV (150 kDa), together with the F_0 part of the H^+ pump ATPase (150 kDa), are closely assembled into one unit (200 x 70 Å) having a 'molecular' weight of $1.8 \sim 1.9 \times 10^6$ and F_1 projecting from it.

Our assumption could explain the negatively stained image of the inner membrane originally reported by Fernandez-Moran *et al.* (3); namely, the presence of two to three 'molecules' in the thickness of specimen (average, 600 Å), which makes the distance between F_1 molecules *ca.* 10 Å on the electron micrograph, as schematically presented in Fig. 9. At present, we do not have enough information on the other protein fraction, representing *ca.* 50% of the protein and having no oxidase activity, to conclude whether that fraction represents "immature" supramolecules or decomposition products of them. The presence of two times as much of the immunoreactive iron-sulfur protein as is expected for Complex III (29) may be related to this problem.

From these observations, it could be concluded that about half of the protein in the inner membrane exists in the form of tightly assembled complexes forming 'supramolecules' which have both the oxidase and ATPase activities. The close

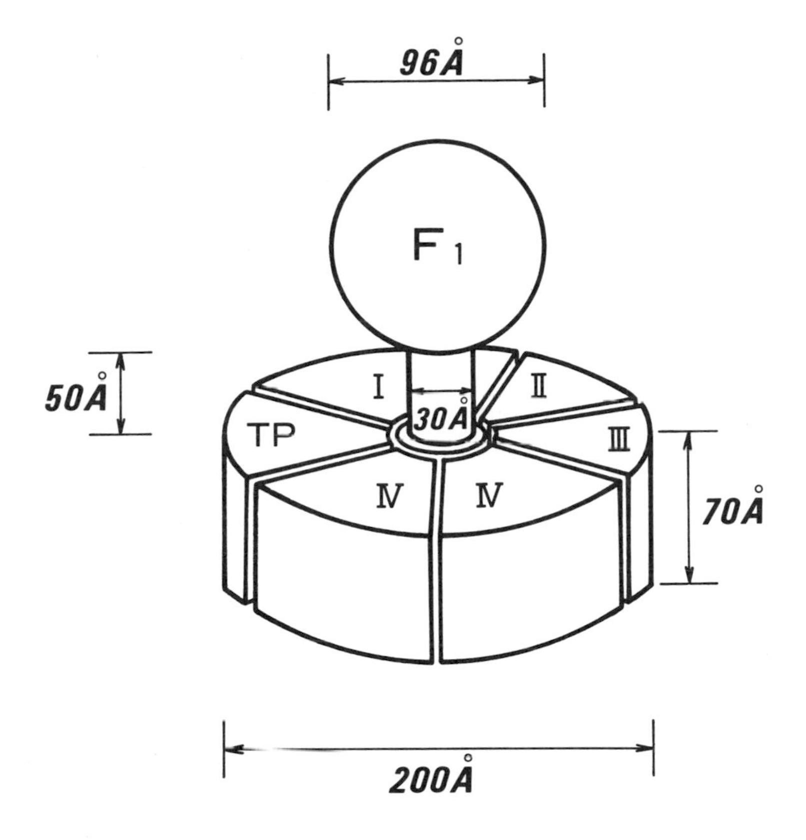

FIGURE 8. Schematic presentation of 'supramolecule'. Based on the electron micrographic image and the cytochrome content of the tripartite particles of an asymmetric dumbbell-like structure, the 'supramolecule', the fundamental assembly of the electron-transfer complexes and H^+ pump ATPase, is schematically presented. One molecule each of Complex I (700 kDa), Complex II (200 kDa), Complex III (250 kDa), transphosphatase (250 kDa), and two molecules of Complex IV (2 x 150 kDa), together with F_0 (150 kDa), are closely assembled into one unit of 200 x 70 Å (1.8 ∿ 1.9 x 10^6 Da) and F_1 projects from the unit. Complex I ~ IV : I ~ IV. Transphosphatase: TP.

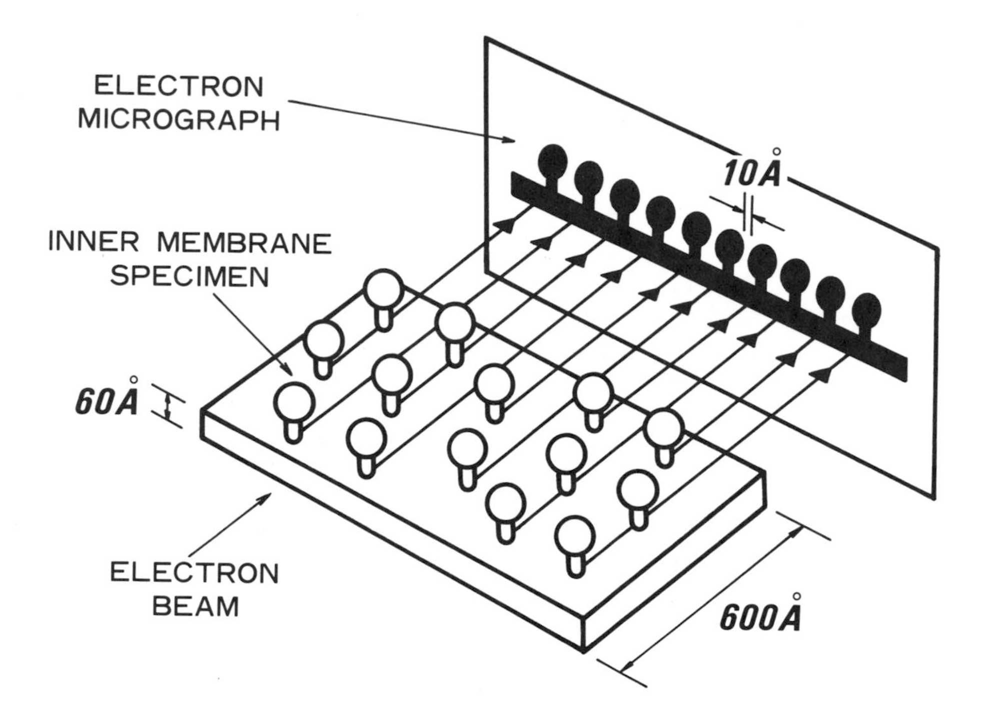

FIGURE 9. Rationalization of negatively-stained image of the mitochondrial inner membrane in terms of the 'supramolecule' concept. Assuming that two to three 'supramolecules' exist in a specimen of the mitochondrial inner membrane used for the negative stainning (average thickness of 600 Å), the average distance of 10 Å between F_1 projections on the electron micrograph originally observed by Fernandez-Moran *et al.* (3) can be explained.

conformational interaction between H^+ pump ATPase and the electron-transfer complexes in the supramolecule would play a crucial role in energy transduction leading to ATP formation.

IV. RETROSPECT AND PERSPECTIVE

The separation and reconstitution of the electron -transfer complexes by Hatefi *et al.* (1,2), the discovery of the head-piece and stalk structure by Fernandez-Moran *et al.* (3), and the isolation of oligomycin-sensitive ATPase by Tzagoloff *et al.* (4,5) with the identification of the head-piece as F_1 by Racker *et al.* (7,8) gave us the basic outline of the mitochondrial energy transducing machinery.

However, the close interaction of the membrane-protein with phospholipid, especially with the boundary type, and the presence of numerous polypeptide subunits in the machinery have complicated our further understanding of the mechanism of energy transduction, despite the presentation of the chemiosmotic hypothesis by Mitchell (11-13).

Advances in affinity chromatography, in the chemistry of detergents, and in immunochemical techniques made it possible to identify and to quantify the essential subunits comprising the electron-transfer chain. At present, we can schematically present the chain as shown in Fig. 10.

The isolation and characterization of the essential assembly of the electron-transfer chain and H^+ pump ATPase, the 'supramolecule', will provide us with crucial information to understand the mechanism of energy transduction.

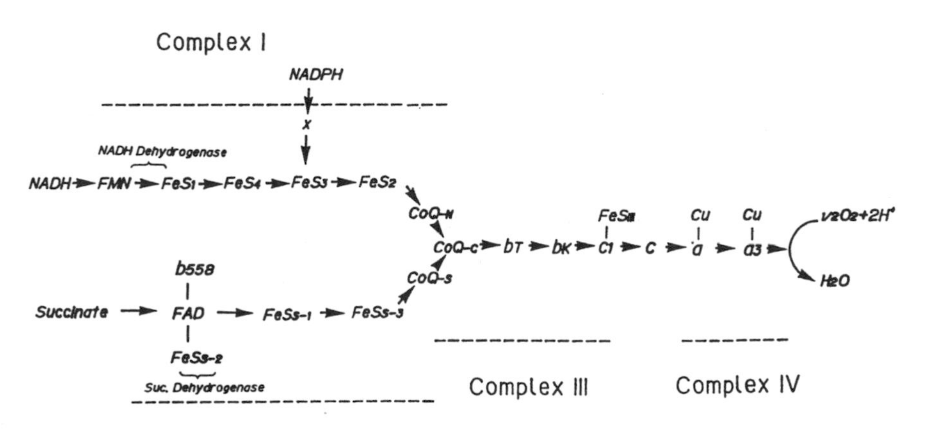

FIGURE 10. Schematic presentation of mitochondrial electron-transfer chain.

REFERENCES

1. Hatefi, Y., Haavik, A. G., Fowler, L. R., and Griffiths, D. E., J. Biol. Chem. 237:2661 (1962).
2. Hatefi, Y., Haavik, A. G., and Griffiths, D. E., J. Biol. Chem. 237:1676 (1962).
3. Fernandez-Moran, H., Oda, T., Blair, P. V., and Green, D. E., J. Cell Biol. 22:63 (1964).
4. Tzagoloff, A., Byington, K. H., and MacLennan, D. H., J. Biol. Chem. 243:2405 (1968).
5. MacLennan, D. H., Smoly, J. M., and Tzagoloff, A., J. Biol. Chem. 243:1589 (1968).
6. Tzagoloff, A., MacLennan, D. H., and Byington, K. H., Biochemistry 7:1596 (1968).
7. Pullman, M. E., Penefsky, H. S., Datta, A., and Racker, E., J. Biol. Chem. 235:3322 (1960).
8. Kagawa, Y., and Racker, E., J. Biol. Chem. 246:5477 (1971).
9. Kagawa, Y., and Racker, E., J. Biol. Chem. 241:2461 (1966).
10. Racker, E., and Horstman, L. L., J. Biol. Chem. 242:2547 (1967).
11. Mitchell, P., Nature (London) 191:144 (1961).
12. Mitchell, P., in "Chemiosmotic Coupling in Oxidative and Photosynthetic Phosphorylation", Glynn Research Ltd, Bodmin, England, 1966.
13. Mitchell, P., FEBS Lett. 43:189 (1974).
14. Hatefi. Y., Galante, Y. M., Frigeri, L., and Stiggall, D. L., in "Structure and Function of Biomembranes" (Yagi, K. ed.), p.167. Japan Sci. Soc. Press, Tokyo, 1979.
15. Galante, Y. M., and Hatefi, Y., in "Methods in Enzymology" (Fleischer, S. and Packer, L. ed.), Vol. 53, p.15. Academic Press, New York, 1978.
16. Nishikimi, M., Shimomura, Y., Yamada, H., and Ozawa, T., Biochem. Biophys. Res. Commun. 120:237 (1984).
17. Suzuki, H., and Ozawa, T., Biochem. Int. 9:563 (1984).
18. Suzuki, H., and Ozawa, T., Biochem. Biophys. Res. Commun. 124:889 (1984).
19. Suzuki, H., and Ozawa, T., Biochem. Biophys. Res. Commun. (submitted) (1986).
20. Shimomura, Y., and Ozawa, T., Biochem. Int. 5:1 (1982).
21. Shimomura, Y., and Ozawa, T., Biochem. Int. 8:187 (1984).
22. Shimomura, Y., Nishikimi, M., and Ozawa, T., Biochem. Int. 8:19 (1984).
23. Shimomura, Y., Nishikimi, M., and Ozawa, T., J. Biol. Chem. 259:14059 (1984).
24. Ozawa, T., in "Transport and Bioenergetics in Biomembranes" (Sato, R. and Kagawa, Y. ed.), p.1. Japan

Sci. Soc. Press, Tokyo, 1982.
25. Shimomura, Y., Nishikimi, M., and Ozawa, T., Anal. Biochem. in press (1985).
26. Shimomura, Y., Nishikimi, M., and Ozawa, T., J. Biol. Chem. 260:15075 (1985).
27. Nakahara, H., Shimomura, T., and Ozawa, T., Biochem. Biophys. Res. Commun. 132:1166 (1985).
28. Nishikimi, M., Shimomura, Y., and Ozawa, T., Biochem. Int. 7:793 (1983).
29. Nishikimi, M., Shimomura, Y., and Ozawa, T., J. Biol. Chem. 260:10398 (1985).
30. Ozawa, T., Suzuki, H., and Tanaka, M., Proc. Natl. Acad. Sci. USA 77:928 (1980).
31. Ozawa, T., Tanaka, M., and Wakabayashi, Y., Proc. Natl. Acad. Sci. USA 79:7175 (1982).
32. Ozawa, T., J. Bioenerg. Biomembr. 16:321 (1984).
33. Ozawa, T., Tanaka, M., and Shimomura, Y., Proc. Natl. Acad. Sci. USA 77:5084 (1980).
34. Ozawa, T., Tanaka, M., and Shimomura, Y., Proc. Natl. Acad. Sci. USA 80:921 (1983).
35. Maurer, A., Tanaka, M., Ozawa, T., and Fleischer, S., Proc. Natl. Acad. Sci. USA 82:4036 (1985).
36. Tzagoloff, A., McConnell, D. G., and MacLennan, D. H., J. Biol. Chem. 243:4117 (1968).
37. Ozawa, T., and Asai, J., J. Bioenerg. 4:507 (1973).
38. Green, D. E., Ann. N. Y. Acad. Sci. 195:150 (1972).

PROPAEDEUTICS[1] OF IONIC TRANSPORT ACROSS BIOMEMBRANES

John F. Nagle

Departments of Physics
and Biological Sciences
Carnegie-Mellon University
Pittsburgh, Pennsylvania

I. INTRODUCTION

For over half a century it has been clear that, for linear transport of ionic currents, the driving potential is the total electrochemical potential,

$$\delta = \delta_{el} + \delta_{chem}, \quad [1]$$

where

$$\delta_{el} = q\,\Delta V \quad [2]$$

where q is the charge on the ion and ΔV is the voltage drop across the membrane, and

$$\delta_{chem} = kT \ln\,(c'/c'') \quad [3]$$

where, in the usual approximation of non-interacting ions, c' and c" are the ionic concentrations on either side of the membrane and kT is the thermal energy. In the phenomenological linear transport theory, the cornerstone of which is the

[1] For a fuller appreciation of this title the reader is encouraged to see pages 225-228 in the delightful book, The Wine of Life and other Essays on Societies, Energy & Living Things, by Harold Morowitz, St. Martins Press, New York, 1979.

Onsager reciprocal relations (1), it is always the total electrochemical potential δ that appears in the equations, not the separate electrical or chemical potentials. Because of the success of this theory, it is quite understandable that in current research areas, such as bioenergetics, it is often automatically assumed that rates of transport and ATP synthesis will depend only upon the magnitude of δ and not upon whether δ comes predominantly from electrical or chemical sources.

A few years ago, when working on the kinetics of proton transport for models of hydrogen bonded chains, we derived the result that the rate of transport need not be the same when the driving potential δ_{el} is electrical, as it is when the driving potential δ_{chem}, is chemical, even though the two driving potentials are chosen to have equivalent magnitudes, $\delta_{el} = \delta_{chem}$ (2). Since our particular results followed from some complicated calculations for rather specialized models of proton transport, we were at first concerned that there was something wrong with our calculations, in view of the general results of transport theory mentioned in the first paragraph. In fact, our results obey all the general theorems of linear transport theory, provided δ is small compared to kT, that is to say, provided that one is in the linear transport regime.

The remarkable general feature that I wish to elaborate upon today is that in the biophysics of membranes, often, one should not expect to be working in the linear transport regime because $kT_{ambient} \cong 25$ mV and typical electrochemical potentials are comparable to or larger than this. The theoretical approach that illustrates this is the one of solving the kinetic equations for specific models. This approach is incapable of deducing general theorems, such as the Onsager reciprocal relations, and it is theoretically and computationally somewhat pedestrian. However, this approach is fully rigorous for each kinetic model examined. Its particular utility is that it can be used to illustrate the complexity and variety of phenomena that may occur in nature, thereby providing us with counterexamples to conventions that may grow from too broad acceptance of concepts that may be limited to narrower ranges, such as the linear transport regime. In this regard, it may be helpful to consider a quote from Onsager (3) "The various types of selective transport through membranes present interesting problems of kinetics more or less closely related to thermodynamics. It would appear that the "irreversible thermodynamics" could be safely applied wherever the differentials of chemical potential are small (<<kT); but this is about the only quantitative result that does not depend on a reasonably detailed understanding of the kinetic mechanism." I may add that Onsager was well

aware that transmembrane potentials for nerve axons are well in excess of kT.

Concerning the correctness of our own kinetic calculations we were relieved to find that similar kinds of behavior followed from calculations that had been performed previously (4,5) or that appeared a little later (6) for ionic (non-protonic) transport. Indeed, as will be clear from the remainder of this talk, such nonlinear behavior has been mentioned, at least implicitly, many times before. However, its consequences have perhaps not been fully appreciated in the realm of membrane transport and bioenergetics. Therefore, it may be useful to review it on as elementary a level as possible, even though no claim to complete originality will be made.

The outline of this talk is first to review the simplest model of membrane transport, the free diffusion model. Even for this model the result when δ is comparable to kT requires some manipulations of the experimental variables, mentioned a long time ago by Kedem and Katchalsky (7), to make it look "normal", i.e., to make it fit the framework of conventional linear non-equilibrium thermodynamics. Next, the simplest thermally activated hopping model is reviewed. Again manipuplation of the experimental variables is required to make it look "normal", but the manipulations are different in this case from the free diffusion model case. This raises the question, which manipulation should an experimentalist use if the transport mechanism is unknown? Finally, we briefly review the results for an extended class of thermally activated hopping models which require a variety of different manipulations to give the "normal" result for $\delta/kT > 1$. At first, this may appear to be one of those messy situations that is best ignored. However, such an attitude would be a mistake, because the different kinds of nonlinear behavior predicted for different models for $\delta/kT > 1$ provides a way to discriminate experimentally between different models of membrane transport.

II. FREE DIFFUSION MODEL

Let us first consider the basic Nernst-Planck electrodiffusion equation for the rate of transport, J, in the x-direction of ions with charge q

$$J = - D\, dc/dx - (qD/kT)\, c\, dV/dx, \qquad [4]$$

where D is the coefficient of diffusion, qD/kT is electrical mobility, μe, c is the concentration of ions and V is the electrical potential; both c and V will be assumed to be functions of the x dimension only. This is the usual linear transport equation which embodies both Fick's law and Ohm's law. The breakdown of this linear transport occurs only for very large driving forces that are well outside the range that are relevant for typical lipid bilayer transport. It is also useful to divide Eq. [4] by c to obtain the mean ionic drift velocity v_D in terms of the electrical field force, F_{elec}, and the concentration or chemical force, F_{chem},

$$v_D = J/c = (D/kT)\ [-kT\ d(\ln c)/dx - q\ dV/dx] \qquad [5]$$

$$= \mu(F_{chem} + F_{elec})$$

where μ is the standard mobility which equals D/kT by the Nernst-Einstein theorem. Also, in this paper we will assume either that the counterions of opposite charge sign are immobile or that we can separate that current from the current of the ions of charge q; in the latter case, cross correlations between currents of charges of opposite sign will be assumed to be negligible.

Let us now apply these basic relations to the free diffusion model of membranes. In this model the membrane is simply a homogeneous slab of material of thickness d. The ion concentration on either surface of the membrane is equal to the ion concentration in the adjacent aqueous phase times a partition coefficient p. Inside the membrane the electrodiffusion equation is assumed to hold. Outside the membrane it is most convenient to assume well stirred aqueous phases, so the boundary conditions at the surface of the membrane are set by the bulk aqueous phase concentrations. We will also neglect any electrical double layer effects, so the full electrical potential is applied across the membrane. The partition coefficient p is not central to the results for transport; it simply changes J by a factor p and does not change v_D at all, so we will not explicitly display it in the following equations; alternately, the concentrations that appear in the equations pertain to just inside the membrane and not to the bulk phases.

If the driving force for transport is purely electrical, then one recovers Ohm's law from Eqs. [4] and [5], namely,

$$J = -c(qD/kT)\ dV/dx = c\ P\ \beta\ \delta \quad ;\ v_D = P\ \beta\ \delta \qquad [6]$$

where $\beta = 1/kT$, $P = D/d$ is an effective membrane permeability, and $\delta = -q\Delta V = -q(dV/dx)d$ is the electrical driving potential. On the other hand, if the driving force is purely chemical (concentration), then from Eq. [4]

$$J = P(c' - c'') \qquad [7]$$

where c' and c" are the two concentrations at either side of the membrane. Since Eq. [7] is just the exceedingly familiar Fick's law, one may find all this boringly trivial. But Fick's law is <u>not</u> in the usual form of linear transport equations. It does <u>not</u> express J in terms of the total electrochemical potential drop δ, which for the chemical case is

$$\delta = kT \ln(c'/c''). \qquad [8]$$

Indeed, Eq. [7] only becomes equivalent to Eq. [8] in the limit that c' and c" become equal. One may express J for the chemical case in terms of δ in a variety of ways, for example,

$$J = P\, c'(1 - e^{-\beta\delta}) = P\, c''(e^{\beta\delta} - 1) \qquad [9]$$

$$= P\, (c'c'')^{\frac{1}{2}}\, 2(\sinh(\beta\delta/2).$$

While the latter form is perhaps preferable because it treats both c' and c" symmetrically, the concentration dependence in any of the expressions for J is not very pretty. The equation for v_D is a little nicer in this regard if one divides J by the average concentration, $c = (c'+c'')/2$, to obtain

$$v_D = 2\, P \tanh(\beta\delta/2) \qquad [10]$$

but this expression is clearly not linear, but sublinear, in δ, and shows definite saturation for $\beta\delta > 1$. (See Fig. 1)

The preceding behavior of the free diffusion model can be made to appear more "normal" by the following manipulation of the experimental variables that appear in Eqs. [7-10]. One simply writes (7,8)

$$J = c_o\, P\, \beta\, \delta \qquad [11]$$

and chooses c_o such that Eq. [11] holds. For the free diffusion model in the general case of both electrical and chemical potentials this requires, using Goldman's (9) Eq.,

$$c_o = c''[(e^{\beta\delta} - 1)/\delta][q\Delta V/(e^{q\beta\Delta V} - 1)]. \qquad [12]$$

This rather contrived manipulation is certainly legitimate and for the free diffusion model it makes the transport linear in δ over a wider range of δ. (A similar kind of extension can be performed to allow the Onsager reciprocal relations to hold over a wider range as discussed by Sauer (10). It is this author's impression that most experimental results in the field of bioenergetics are not manipulated in this fashion. Irregardless, as we shall see shortly, the same

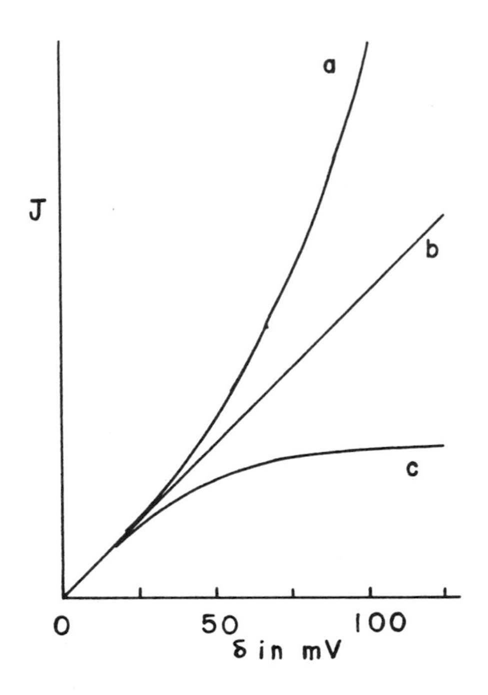

FIGURE 1. Current J versus δ in millivolts for kT = 25 mV. (a) $J = 2\sinh(\beta\delta/2)$, (superlinear); (b) $J = \beta\delta$, (linear); and (c) $J = 2\tanh(\beta\delta/2)$, (sublinear and saturating).

manipulation dose not work for other, equally or more plausible models of membrane transport.

III. SIMPLE THERMALLY ACTIVATED HOPPING MODEL

In the next section we will review the general case of activated hopping models across barriers of general, realistic potentials. For clarity, we deal in this section with the particularly simple model shown in Fig. 2, in which the membrane presents, in zero electric field, a very sharp symmetrical energy barrier to ion transport. The aqueous phase will be assumed to be well stirred with concentrations c' and c". In this section (but not in the next) we will ignore any complications from the ion being in any states other than the two aqueous solutions. In this case the steady state rate of transport is

$$J = k_o d(c' e^{q\beta\Delta V/2} - c'' e^{-q\beta\Delta V/2}) \qquad [13]$$

where k_o is the hopping rate with no electrical field.

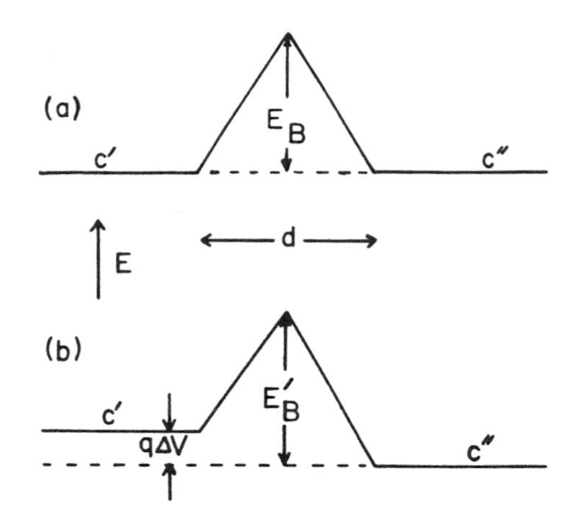

FIGURE 2. Energy E versus distance x across the membrane of thickness d and with concentrations c' and c" in the two bulk aqueous phases for the simple thermally activated hopping model. (a) With no electric field. Barrier height E_B. (b) With electric field driving ions from left to right. Barrier height $E_B' = E_B + q\Delta V/2$.

In the special case when the total electrochemical potential $\delta = q\Delta V$ is purely electrical, Eq. [13] yields

$$J = k_o d\ c \sinh(\beta\delta/2);\ v_D = k_o d \sinh(\beta\delta/2). \qquad [14]$$

In the other special case when δ is purely chemical,

$$J = k_o d(c' - c");\ v_D = k_o d \tanh(\beta\delta/2). \qquad [15]$$

For $\delta \ll 25$ mV, i.e. $\beta\delta \ll 1$, both Eqs. [14] and [15] reduce to the linear transport equation

$$v_D = k_o d\ \beta\delta/2. \qquad [16]$$

However, for larger δ one has the same problem for the chemical case that was discussed for the free diffusion model in the last section. For the electrical case there is a new feature, namely, the current given by Eq. [14] is clearly superlinear for large δ_{el}, as shown in Fig. 1, because $\sinh(\beta\delta/2)$ behaves like $(1/2)\exp(\beta\delta/2)$ for large $\beta\delta$.

For the general case of both electrical and chemical potentials one obtains

$$J = k_o d\ c" e^{-q\beta\Delta V/2}\ (e^{\beta\delta} - 1). \qquad [17]$$

In order to make this into a linear equation, $J = c_o L\beta\delta$, requires choosing

$$c_o = c'' \; [(e^{\beta\delta} - 1)/\delta] \; e^{-q\beta\Delta V/2}. \qquad [18]$$

For the purely chemical case with $\Delta V = 0$ this manipulation is the same as was discussed for the free diffusion model. For the purely electrical case this manipulation is even more contrived, since the effective concentration c_o used in the transport equation is only abstrusely related to the concentrations $c' = c''$.

IV. SOME OTHER MODELS OF TRANSMEMBRANE TRANSPORT

The preceding two examples, and their comparison, suffice for the purposes of this paper. However, it is worth mentioning briefly in this section that there is a spectrum of models that includes the preceding two models as special cases. The general model that includes all these cases was discussed by Johnson et al. many years ago (4). The potential barrier for this model is shown in Fig. 3. There are N-1 locations for the ions to reside inside the membrane as well as the two aqueous solutions with concentrations c' and c". (As originally presented each location is a pool capable of holding many ions. However, Lauger (5) has shown that single file pore models in which only one ion at a time can reside at each location within each pore have the same solution, Eq. [19], in the limit of low occupancy of ions in each pore). It is straightforward to show for this model that

$$J = k_o d \; (c' - c'' \, e^{-q\beta\Delta V})/\Sigma \;\; e^{\beta\Delta F_i} \qquad [19]$$

where ΔF_i is the difference in free energy, including the electrical potential, between the ith barrier and the aqueous solution with concentration c'.

From Eq. [19] it is easy to obtain the result for the simple hopping model in the last section by taking the special case of only one barrier in the middle of the membrane as shown in Fig. 2. As a field is applied, this term, and therefore the whole denominator varies as $\exp(-q\beta\delta V/2)$, so Eq. [19] becomes identical to Eq. [13]. To obtain the free diffusion case in the first section, first one lets N, the number of locations, become very large so that the sum in the denominator becomes an integral. Next, in the absence of an electric field ΔF_i is taken to be constant across the whole membrane, consistent with having the perfectly homogeneous membrane of the free diffusion model (ignoring image forces).

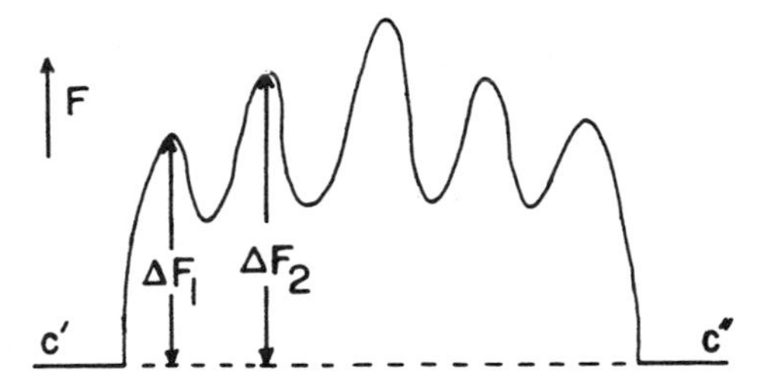

FIGURE 3. Schematic free energy F versus distance x across the membrane of thickness d for other models of membrane transport, with no electric field.

Performing the integral then yields Eq. [11] with c given by Eq. [12] and P given by $k_o d$.

It is perhaps worth emphasizing once again in this more general context that J does not have the same magnitude when δ is purely electrical as when it is purely chemical. This follows most easily from Eq. [19] when it is rewritten

$$J = k_o d\, c'' \, (e^{\beta\delta} - 1) \,/\, [e^{q\beta\Delta V} \, \Sigma_i \, e^{\beta\Delta F_i}]. \qquad [20]$$

The numerator of Eq. [20] is the same for a given δ whether it be electrical or chemical in origin. However, the denominator does not depend upon the concentrations of the aqueous phase but it does depend explicitly upon the electrical field, both in the ΔV and in the ΔF_i. Indeed, each different model of the membrane that has a different potential barrier profile (different ΔF_i) in zero electrical field yields a different J versus V (current/voltage) curve. The J/V curves are strongly superlinear for barriers with sharp maxima in the middle, e.g. Eq. [14], and also for trapezoidal barriers (11,12). For smoother quadratic barriers they are considerably less superlinear (13). For flat topped barriers they are linear (large number of locations) to slightly sublinear (finite number of locations).

V. DISCUSSION AND CONCLUSIONS

The preceding well-documented kinetic results for some very simple models suffice to show that, for quite modest transmembrane potentials of 25 mV or more, the rate of transport J, in addition to being non-linear, is also dependent upon whether the potential is electrical, δ_{el}, or whether it is chemical, δ_{chem}, not just on the sum, δ, identified as the total electrochemical potential. These particular results do not violate any of the theorems in linear transport theory, which can only be guaranteed for transmembrane potentials much less than 25 mV. Efforts to manipulate the equations to extend the results of linear transport theory, including the dependence of J only upon δ, to larger transmembrane potentials are shown to be not general, with different manipulations required for each different kinetic model; such efforts should not be encouraged.

It follows, therefore, that it should not be assumed in the study of membrane transport and in bioenergetics that rates should depend only upon δ and not upon δ_{el} and δ_{chem} separately. Of course, it is not precluded that rates might depend upon δ only. But it should be clearly understood that such an experimental result is not the consequence of general laws of transport thermodynamics. Rather, it is just one of many possibilities that may occur.

The logical structure of the area of membrane transport and bioenergetics is first, that there are many possible results regarding non-linearities in rates (already widely recognized) and also in the dependence of the rates upon δ_{chem} versus δ_{chem}. Second, there are many possible kinetic models for the transport/bioenergetic mechanisms. Clearly, careful measurements of the rates as a function of both δ_{el} and δ_{chem} combined with detailed kinetic calculations offer an oppotunity to discriminate between different models of transport and/or bioenergetic mechanisms.

The author has been engaged in two such studies. Earlier studies (14), already mentioned, elucidated the rate kinetics for models of permanent hydrogen bonded chains postulated to exist in membrane bound proteins that pump protons. (See also Brunger et al. (15).) One complication in such active transport systems, compared to passive transport, is that there are two essential steps, proton transport and chemical reactions. If one of the steps is always the rate limiting one, then no discrimination can be made for the mechanism of the other step. The more recent study concerns the passive permeability of lipid vesicles to protons. The remarkable result is that the conductance, $\sigma = J/\delta$, (not the permeability) is nearly independent of concentration of ions (i.e., pH) (16). The kinetic equations for most simple models (such as all the ones in this paper) are inconsistent with this. However, my analysis for the kinetics of transient hydrogen-bonded chain models, first suggested for liposomes by Nichols and Deamer (17), shows that several of them have this remarkable property. Now the dependence of the conductivity upon δ_{el} and δ_{chem} can be used to discriminate between the different models.

REFERENCES

1. Onsager, L., Phys. Rev. 37:405 and 38:2265 (1931).
2. Nagle, J. F., Mille, M., and Morowitz, H. J., J. Chem. Phys. 72:3959 (1980).
3. Onsager, L., in "The Neurosciences" (F.O. Schmitt, ed.), p. 75. Rockefeller University Press, New York, 1967.
4. Johnson, F. H., Eyring, H., and Polissar, M. J., in "The Kinetic Basis of Molecular Biology", Chap. 14, p. 754. Wiley, New York, 1954.
5. Lauger, P., Biochim. Biophys. Acta 311:423 1973.
6. Gradmann, D., Hansen, U.-P., and Slayman, C. L., in "Current Topics in Membranes and Transport" (C. L. Slayman, ed.), Vol. 16, p. 257. Academic Press, New York, 1982.
7. Kedem, O., and Katchalsky, A., Trans. Faraday Soc. 59: 1941 (1963).
8. Essig, A., and Caplan, S. R., Proc. Natl. Acad. Sci. USA 78:1647 (1981).
9. Goldman, D. E., J. Gen. Phys. 27:37 (1944).
10. Sauer, F., in "Handbook of Physiology" (J. Orloff and R. W. Berliner, ed.), appendix to Chap. 12, Section 8, p. 399. Renal Physiol. Amer. Phys. Soc. 1973.
11. Hall, J. E., Mead, C. A., and Szabo, G., J. Memb. Biol. 11:75 (1973).
12. Hladky, S. B., Biochim. Biophys. Acta 3622:71 (1974).
13. Neumcke, B., and Lauger, P., Biophys. J. 9:1160 (1969).
14. Nagle, J. F., and Tristram-Nagle, S., J. Memb. Biol. 74:1 (1983).
15. Brunger, A., Schulten, Z., and Schulten, K., Z. Phys. Chem. Neuer Folge 136:1 (1983).
16. Gutknecht, J., J. Memb. Biol. 822:105 (1984).
17. Nichols, J. W., and Deamer, D. W., Proc. Natl. Acad. Sci. USA 77:2038 (1980).

MEMBRANE PHOSPHOLIPID TURNOVER AND Ca^{2+} MOBILIZATION IN STIMULUS-SECRETION COUPLING

Yoshinori Nozawa

Department of Biochemistry
Gifu University School of Medicine
Gifu 500, Japan

I. INTRODUCTION

Physiological signals such as peptide hormones and neurotransmitters are recognized by and bind to specific receptors on the surface of their target cell membranes. These interactions then initiate biochemical and physico-chemical alterations in membranes which in turn allow stimulated cells to exert their various specific functions. It is now generally accepted that one of the earliest and crucial membrane changes in the stimulus-response coupling is the turnover of membrane phospholipids, especially inositol phospholipids. This original concept emerged from the experiments by Hokin and Hokin (1), who demonstrated that a considerable increase in the ^{32}P incorporation into the total phospholipids was associated with the amylase secretion in the pigeon pancreas slice stimulated with acetylcholine. The discovery of the "phospholipid effect" has consequently led to the idea of the "phosphatidyl-inositol (PI) response" defining the enhanced PI turnover coupled with cellular responses. In fact, rapid turnover of inositol phospholipids has been observed to occur in parallel with secretory response in other various cells or tissues, such as brain cortex, salivary glands, pituitary, adrenal medulla, platelets, neutrophils, mast cells, thyroid, smooth muscles, and nerve tissue.

The physiological significance of such agonist-stimulated breakdown of PI was suggested by Michell (2), who proposed the Ca^{2+}-gating theory. The Ca^{2+} mobilization within the cell is responsible for initiating the events involved in secretion, and it occurs as a result of increased permeability to Ca^{2+} of the plasma membrane and/or through release of Ca^{2+} from internal stores.

Further recent support has advanced the role of the receptor-activated hydrolysis of inositol phosholipids, particularly phosphatidylinositol 4,5-bisphosphate (PIP_2), by which 1,2-diacylglycerol (DG) and inositol trisphosphate (IP_3) are generated. The former neutral lipid is a potent activator of protein kinase C (3) and the latter water-soluble product serves as an effective releaser of Ca^{2+} from

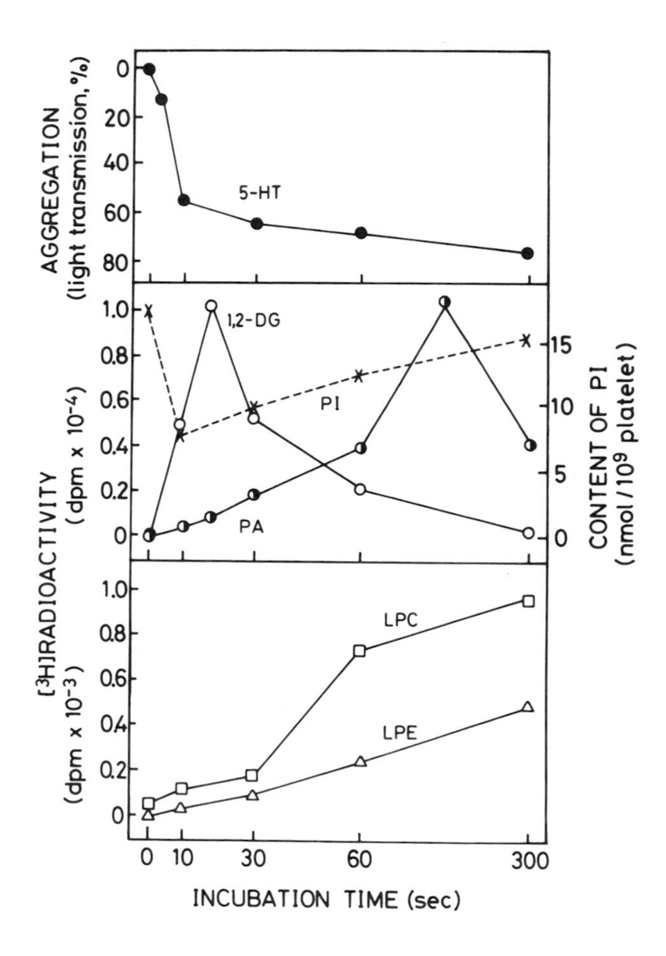

FIGURE 1. Secretion of serotonin and phospholipid metabolism in human platelets stimulated by thrombin. DG, diacylglycerol; PI, phosphatidylinositol; LPC(PE), lysophosphatidylcholine (phosphatidylethanolamine); PA, phosphatidic acid.

the intracellular stores (4). Thus, it appears reasonable to consider that these novel second messengers, DG and IP_3, generated by action of phospholipase C, play pivotal roles in the membrane signal transduction.

In this chapter I will focus attention mainly on the stimulus-induced enhancement of phospholipid turnover and Ca^{2+} mobilization in the human platelet, but from the comparative viewpoint I also will review briefly the initial events in some other cell types.

II. RECEPTOR-ACTIVATED PHOSPHOINOSITIDE TURNOVER IN THE HUMAN PLATELET

The implication of phospholipids in the aggregation and secretion of platelets has been extensively studied, and substantially rich information is present regarding the phosphoinositide metabolism and Ca^{2+} mobilization in platelets activated by various stimuli.

A. Phosphoinositide Breakdown in Thrombin-Activated Human Platelet

Activation of human platelets by thrombin caused the secretion of serotonin and aggregation, and the rapid hydrolysis of PI occurred in accordance with such cell responses (Fig. 1). When [^{3}H]arachidonate-labeled platelets were exposed to the secretagogue, the radioactivity in DG showed a transient increase. A marked enhancement of PA level followed a decrease of DG, indicating the conversion via DG kinase. As shown in Fig. 1, the mass content of PI was first decreased and then returned to the initial level (5). These findings appeared to be compatible with the PI cycle theory. The Ca^{2+} ionophore, A23187 induced an irreversible platelet aggregation, irrespective of the presence or absence of extracellular Ca^{2+}. However, the rate of increase of DG in A23187-treated platelets was much lower compared with that in thrombin-stimulated ones. Furthermore, a marked increase in PA formation observed during stimulation by thrombin did not occur in activation with A23187 (6). Ineffective production of DG and PA suggested the blockade of PI-turnover in A23187-activated platelets, which was reflected in inhibited incorporation of ^{32}P into PI. Evidence was thus presented that the PI-turnover was associated with the receptor-mediated stimulation but not with the bypassing stimulation of platelets (7).

On the other hand, it has been known for a long time that

acetylcholine stimulates the breakdown of polyphosphoinositides in brain, salt gland and iris smooth muscle. In most cases, the decrease of polyphosphoinositides was found to occur prior to the decrease of PI. Eventually, it was suggested that the primary event was breakdown of polyphosphoinositides in the activation sequence and that the decrease in the PI level was rather secondary to phosphorylation via PI kinase to replenish polyphosphoinositides. The agonist-induced rapid hydrolysis of polyphosphoinositide takes place in the plasma membrane where phospholipase C is located, and PIP_2 generates second messengers, DG and IP_3.

Following activation of $[^{14}C]$glycerol-labeled platelets by thrombin, the radioactivity in PI was progressively decreased, whereas a transient loss in PIP_2 was observed within 10 sec with a concurrent increase of DG (8,9,10), indicating evidence that some of the DG first arose from the phosphodiesteric cleavage of PIP_2. However, it would also be possible that the rapid loss of label from PIP and PIP_2 results from dephosphorylation to lower inositides. Recent studies have shown that the rapid decrease of PIP_2 coincides with production of water-soluble compound, IP_3. The addition of thrombin to human platelets labeled with $[^3H]$inositol induced production of inositol phosphates: inositol 1-phosphate (IP), inositol 1,4-bisphosphate (IP_2) and IP_3 (Fig. 2). The earlier formation of IP_2 and IP_3 in stimulated platelets implies that polyphosphoinositides are broken down before PI by phospholipase C. The phospholipase C involved in the phosphoinositide hydrolysis is considered to

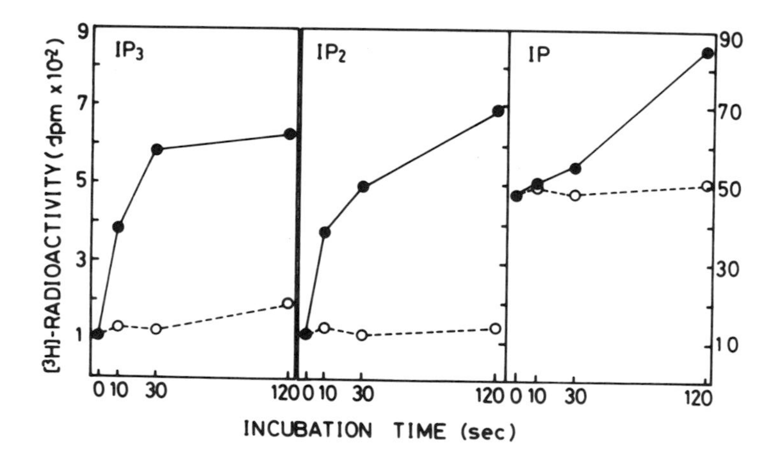

FIGURE 2. Formation of inositol phosphates in thrombin-activated human platelets. IP_3, inositol 1,4,5-trisphoshate; IP_2, inositol 1,4-bisphosphate; IP, inositol 1-phosphate.

be a single soluble phosphodiesterase capable of attacking all three phosphoinositides (11).

Although there is much information demonstrating the receptor-activated polyphosphoinositide breakdown in the human platelet, the recent results obtained by Wilson et al. (12) have shown that the conversion of PI to PIP was too slow to account for the loss of PI during stimulation, and that the direct PI breakdown is responsible for the majority of DG generated by stimulation.

B. Inhibitory Effects of Cyclic Nucleotides

It is well known that cyclic AMP prevents platelet aggregation and secretion, but the underlying mechanism of the inhibitory actions remains to be unexplored. When [^{14}C]serotonin-labeled platelets were incubated with various concentrations of dibutyryl cAMP and then activated by thrombin, the serotonin release was inhibited in a dose-dependent manner (13). Figure 3 demonstrates that preincubation of [^{3}H]arachidonate-labeled platelets with dbcAMP results in a marked decrease in thrombin-induced DG formation. In addition, phosphatidic acid (PA) also failed to increase by raised dbcAMP concentration. These findings are agreeable with the results of Billah et al. (14) who observed the inhibitory effect of dbcAMP on phospholipase C in deoxycholate-treated horse platelets. Takai et al. (15)

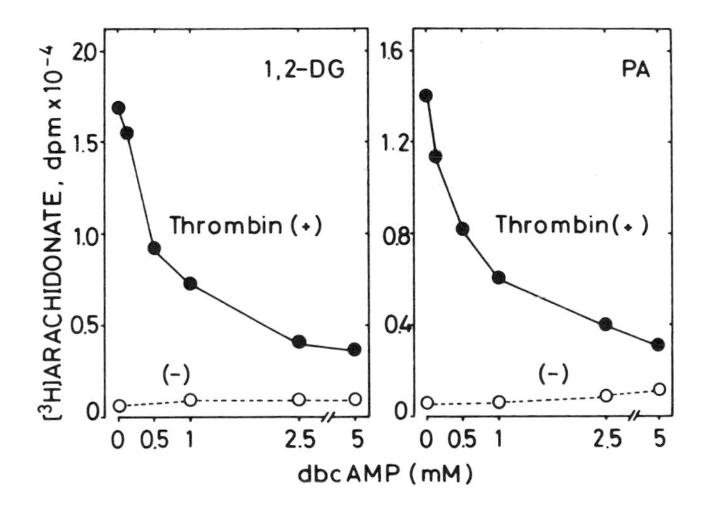

FIGURE 3. Effects of dibutyryl cAMP on thrombin-induced production of 1,2-diacylglycerol (1,2-DG) and phosphatidic acid (PA) (13).

showed that the impaired formation of DG caused by added dbcAMP was accompanied by 40 K protein phosphorylation as well as serotonin release, indicating decreased activity of protein kinase C due to insufficient DG. The rise of radioactivity in lysophosphatidylcholine (LPC) and lysophosphatidylethanolamine (LPE) of [^{3}H]glycerol-labeled platelets is a measure of phospholipase A_2 activity. The dbcAMP-pretreatment lead to inhibited formation of these two lysophospholipids in thrombin-activated platelets, and this suggests inhibitory effects of cAMP on hydrolysis to liberate arachidonic acid via phospholipase A_2 from the major membrane phospholipids.

Although less information is available regarding the effect of cGMP on platelet functions, two apposing theories are persent. In the permeabilized platelets, cAMP blocked and cGMP enhanced the thrombin-induced serotonin release (16). 8-Bromocyclic GMP was reported to act as a negative messenger in both 40 K protein phosphorylation and serotonin secretion (15). Our recent data showed the suppressed generation of all three inositol phosphates which was caused by inhibition of phospholipase C (unpublished data). However, the inhibitory action of cGMP was weaker compared with that of cAMP. These observations provide evidence that cyclic nucleotides, either cAMP or cGMP, subserve a negative feedback control over the agonist-induced hydrolysis of minor phosphoinositides to generate DG and IP_3, and of major phospholipids (PC, PE) to produce arachidonic acid.

However, it is of interest to note that cAMP can stimulate incorporation of [^{3}H]glycerol into almost all lipids including triacylglycerol (TG). This suggests us to consider an enhanced biosynthesis of the lipid components. Forskolin pretreatment caused dose-dependent increases in cAMP content and [^{3}H]glycerol incorporation into total lipids (17). There are several reports suggesting enhancement by cAMP of de novo synthesis of lipids in other cells, for example, hepatocytes, lung cells, and adrenal cells. Although the mechanism whereby cAMP increases lipid synthesis from glycerol is not clear, it does not seem unlikely that cAMP enhances the de novo synthesis of phosphatidic acid by activating glycerol kinase and/or glycerophosphate acyltransferase.

C. Dual Actions of Phorbol Myristate Acetate

Phorbol myristate acetate (PMA) is known as a potent tumor promotor and its target site is located on membranes. Nishizuka et al. (18) have found that this compound, which has DG-like structure, can activate protein kinase C both in vitro and in vivo. PMA increases the affinity of the enzyme for Ca^{2+} in the 10^{-7} M range, leading to its full activation without changing the intacellular Ca^{2+} concentration. It was also shown that protein kinase C activation and Ca^{2+} mobilization act in a synergistic manner to elicit the full secretion (19). Such synergistic effects by DG and Ca^{2+} ionophore (A23187) were also observed in other cell system such as mast cells, neutrophils, islet cells, hepatocytes, and lymphocytes.

PMA has been reported to alter lipid metabolism in intact cells and it stimulates breakdown as well as synthesis of PC. Recently, evidence was presented that PMA, an incomplete secretagogue, stimulated polyphosphoinositide formation by activating kinases for PI and PIP immediately after addition to human platelets (20). On the other hand, Watson and Lapetina (21) have demonstrated that pretreatment of human platelets with either PMA or DG suppresses thrombin-induced phosphoinositide hydrolysis via phospholipase C and also serotonin release. Furthermore, Ca^{2+} mobilization also was observed to be inhibited by pretreatment with PMA (22). Although the exact mechanism for such inhibitory action of PMA is unclear, a most likely interpretation is involvement of protein kinase C activation. To ensure this possibility, we have done experiments with the use of an inhibitory agent H-7 for the enzyme (23). The addition of thrombin to

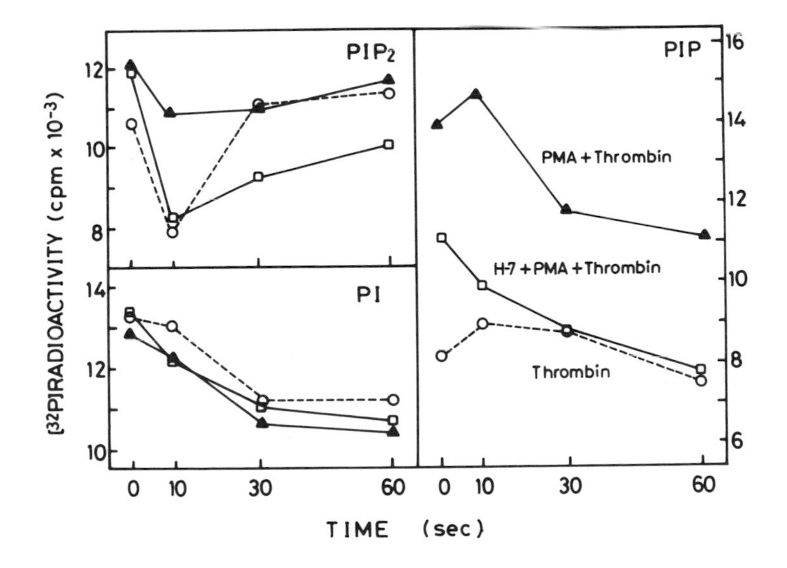

FIGURE 4. Effects of phorbol myristate acetate (PMA) and protein kinase C inhibitor H-7 plus PMA on phosphoinositide metabolism in human platelets activated by thrombin.

^{32}P-labeled platelets caused a transient decrease in radioactivity of PIP_2, but a 5-min pretreatment with PMA inhibited the PIP_2 breakdown (Fig. 4). However, this inhibition did not occur in platelets which had been incubated with the enzyme inhibitor before addition of PMA (unpublished data). Such distinct reversal of the inhibitory effect provides evidence supporting the theory that PMA-induced suppression of PIP_2 hydrolysis may be mediated by the activation of protein kinase C. These results suggest that protein kinase C, which initially fosters secretion and aggregation, may subsequently subserve a negative feedback role in the PIP_2 hydrolysis involving the generation of second messengers, DG and IP_3.

D. Phosphoinositide Metabolism in Other Secretory Cells

Among numerous secretory cells, the platelet is one of the most typical cells to display marked alterations of membrane phospholipid metabolism. However, many other secretory cells also undergo the rapid inositol phospholipid turnover upon stimulation.

Mast cells isolated from rat peritoneal cavity are known to secrete histamine by exposure to various stimuli, either receptor-mediated (antigen, ConA), membrane-perturbing (compound 48/80, mastoparan), or bypassing (A23187, PMA). Addition of antigen to the sensitized mast cells labeled with [^{3}H]arachidonate resulted in a rapid transient decrease in the levels of polyphosphoinositides (24). When ^{32}P-labeled cells were stimulated with mastoparan, the radioactivity of PA showed a progressive increase with incubation time and a transient decline of PIP_2 level followed by an abrupt and profound enhancement. The histamine release induced by compound 48/80 was accompanied with hydrolysis of polyphosphoinositides. The breakdown of the phospholipids caused by various stimuli was also confirmed by the formation of their corresponding inositol phosphates (25,26).

The release of lysosomal enzymes from granules in neutrophils constitutes a secretory response to various stimuli. Upon stimulation of [^{3}H]arachidonate-labeled rabbit neutrophils by a chemotactic peptide, formyl-methionyl-leucyl-phenylalanine (fMLP), the significant loss of PIP_2 and PIP occurred as the earliest measurable event (27). Resynthesis of polyphosphoinositides was ascertained by measuring ^{32}P incorporation into these lipids in response to fMLP. The stimulation by fMLP caused 2-3 fold increases of ^{32}P incorporation into PIP and PIP_2. On the other hand, enhancement of incorporation of ^{32}P into PI and PA appeared to be smaller than that into polyphosphoinositides. This can be explained by the fact that the metabolic turnover of the monophosphate groups attacked on the inositol rings of polyphosphoinositides is very rapid. In contrast, ^{32}P incorporation into other phospholipids was not significantly enhanced by fMLP. Both breakdown and synthesis of polyphosphoinositides were reported by Cockcroft et al. (28); the predominant reaction was breakdown of PI to PA by phospholipase D and only a small fraction of PI was converted to polyphosphoinositides. Two separate pathways for production of DG were proposed. An early phase of DG generation originates from PIP_2 breakdown by phospholipase C and a second late phase derives from PA formed via phospholipase D on PI.

Neuroblastoma x glioma hybrid NG 108-15 cells were stimulated by bradykinin and produced a sustained depolarization preceded by a transient hyperpolarization (29). Bradykinin also increased the frequency of miniature end-plate potentials recorded from cultured striated muscle cells which had been innervated by NG 108-15 cells. This indicates that bradykinin caused an enhanced synaptic transmission from NG 108-15 due to depolarization. The vasoactive peptide induced selective incorporation of ^{32}P into PA and PI without affecting [^{3}H]glycerol incorporation into these lipids. The addition of bradykinin to the hybrid cells prelabeled with ^{32}P caused a transient decrease in the radioactivity from PIP_2 followed by the accumulation of radioactivity in PA and PI (Fig. 5). The degree of PIP_2 breakdown was dependent on the concentration of bradykinin.

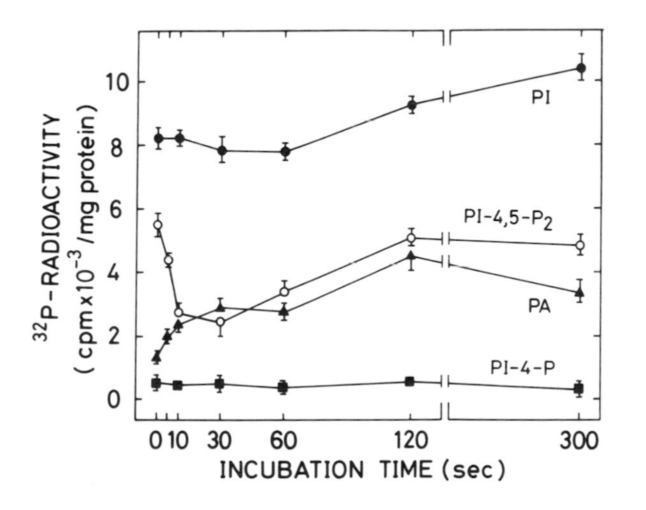

FIGURE 5. Bradykinin-induced phosphoinositide turnover in neuroblastoma x glioma hybrid NG 108-15 cells (29).

The PIP_2 hydrolysis was ascertained by detection of the degraded product IP_3 in [^{3}H]inositol-labeled cells (30). There was a subsequent accumulation of IP_3. The levels of these two inositol phosphates gradually decreased and returned to control levels, whereas the increase of IP was small at the initial phase but became greater thereafter. The progressive enhancement of IP may be the result of the sequential degradation of IP_2 and IP_3 by their specific phosphatases. Furthermore, evidence was presented that the bradykinin-stimulated PIP_2 hydrolysis in NG 108-15 cells was not dependent on elevation of intracellular Ca^{2+} concentration evoked by high K^+ depolarization which allowed Ca^{2+} influx through voltage-dependent Ca^{2+} channels (31).

III. MEMBRANE PERTURBATION ASSOCIATED WITH PLATELET ACTIVATION

While many studies suggested changes in the organization of platelet membrane lipids following thrombin activation, the nature of thrombin-induced microenvironmental alterations in the membrane is not clear. When human platelets labeled with stearate spin probe (5-SAL) were stimulated by thrombin, electron spin resonance spectra exhibited marked changes and the order parameters decreased (32). The prior addition of aspirin, a potent inhibitor of aggregation, did not cause any change in order parameter upon subsequent exposure to thrombin. These results indicate that the decrease of order parameter may be tightly correlated with aggregation of platelets. Data by ESR analysis were not compatible with those by fluorescence polarization study; the former showed a decrease and the latter indicated an increase in microviscosity (33). At any rate, the alteration of membrane fluidity may modulate the platelet activation by modifying activity of some plasma membrane-bound enzymes.

As described above, the rapid phospholipid metabolism is elicited in agonist-stimulated platelets. Based on the finding of the abrupt conversion of the acidic inositol phospholipids to the non-polar DG, one would expect that the physical property in the microdomain of the plasma membrane may be modified. Thus the artificial membranes (PC/PS) containing PI and DG were studied by ESR and freeze fracture electron microscopy (34). DG was prepared from PI by phospholipase C from _B. cereus._ The phase transition of phospholipids was characterized by freeze-fracture technique and jumbled (J), banded (B) and terrace (T) patterns on the fractured faces were identified to be L_α, $P_{\beta'}$ and $L_{\beta'}$ phase inferred by X-ray diffraction. The main transition from the

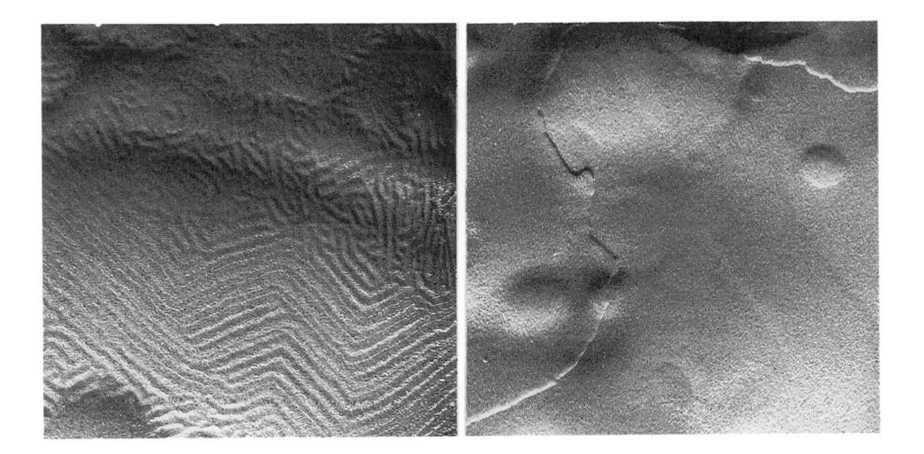

FIGURE 6. Freeze-fracture elecron micrographs of liposomes of (a) dipalmitoylphosphatidylcholine (DPPC) / phosphatidylserine (PS) / phosphatidylinositol (PI) (3:1:1) and (b) DPPC / PS / diacylglycerol (DG) (3:1:1).

liquid crystalline to the crystalline phase was depicted as a transformation from the jumbled to the banded pattern, and the pretransition as a transformation from the banded to the terrace pattern. Figure 6 exhibits both jumbled and banded patterns of the liposome (PC/PS/PI, 3:1:1) in the presence of 1 µM Ca^{2+}. The liquid crystalline and crystalline phases coexist (phase separation). However, this distinct phase separation was found to disappear upon replacement of PI by DG, indicating the transformation from the coexisting state of L_{α} and $P_{\beta'}$ to the crystalline state of $L_{\beta'}$. The effect of the conversion from PI to DG on phase separation was examined. The degree of phase separation was quantified by spin-spin exchange broadening in ESR spectra. In the PC/PI (1:1) liposome system, phase separation was not observed below 10 µM Ca^{2+} and even at 10 mM Ca^{2+} only about 30 % of total PI was segregated. In contrast, in the PC/PS (1:1) liposome, Ca^{2+}-induced phase separation of PS was remarkable below 1 µM Ca^{2+}. The liposomes containing these two acidic phospholipids (PC/PS/PI, 3:5:2) showed a slight enhancement of phase separation compared with the above binary liposomes (Fig. 7). However, the replacement of PI by DG induced a great increase in the degree of phase separation at lower Ca^{2+} concentration (Ca^{2+}-sparing effect) (35, 36). For example, at 1 µM Ca^{2+}, the liposome (PC/PS/DG, 3:5:2) underwent a much greater separation (73 %) than the liposome

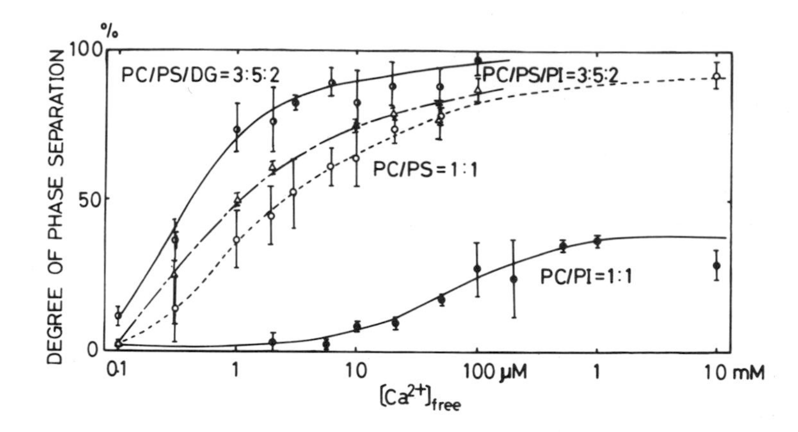

FIGURE 7. Degree of phase separation vs. concentration of free calcium ions for phosphatidylcholine (PC) / phosphatidylinositol (PI) (1:1) (●), PC / PS (1:1) (○), PC / PS / PI (3:5:2) (◑), and PC / PS / DG (3:5:2) (△) (36).

(PC/PS/PI, 3:5:2). These findings provide evidence suggesting that the transient phase separation might be induced by conversion from PI to DG by phospholipase C in the receptor-stimulated platelet plasma membrane.

Recently it was shown that incorporation of DG induced a lamellar to haxagonal transition in the PC liposome (37,38). In addition, suggestive evidence was provided that DG potentiated the ability of phospholipase A_2 and C to hydrolyze phospholipids in both solution and microsomes (38,39), thereby postulating physiological significance for formation of the non-bilayer lipid structure in the membrane. Protein kinase C is considered to interact with membrane lipids through its hydrophobic domain and alters the interaction upon activation by DG. It would be possible that the receptor-activated perturbation in the plasma membrane may act as a trigger for activation of protein kinase C, though its mechanism is still undefined. Boni and Rando (40) have demonstrated with the liposome (PC/PS, 4:1) that the specificity of protein kinase C activation is directed toward the glycerol backbone but not toward fatty acyl chains of DG.

IV. Ca^{2+} MOBILIZATION INDUCED BY RECEPTOR ACTIVATION

A. Regulation of Intracellular Ca^{2+} Level

The molecular mechanisms involved in cell responses to cAMP have been known for many years, whereas the messenger functions of Ca^{2+} have been appreciated largely in the past decade. The Ca^{2+} messenger system is more complicated than cAMP system and also the means for translating changes in Ca^{2+} into metabolic responses are more diverse than those by which cAMP acts. Since levels of calcium in the cytoplasm of cells are low in the resting state, its small increase is sufficient enough to induce such various cell functions as contraction, secretion and adhesion. Several machineries operate in regulation of the intracellular Ca^{2+} level (Fig. 8). The active Ca^{2+} transport dependent on ATP hydrolysis pumps Ca^{2+} out of the cytosolic compartment against a large electrochemical Ca^{2+} gradient. These Ca^{2+} pumps are present on the plasma membrane and also on sarcoplasmic reticulum of muscle cells. Analogous Ca^{2+} pumps are known to sequester Ca^{2+} within endoplasmic reticulum of many other cell types. Also the plasma membrane excludes Ca^{2+} by a Na^{+}-Ca^{2+} exchange mechanism. On the other hand, the influx of Ca^{2+} across the plasma membrane represents the most extensively studied and important source of activatable cytosolic Ca^{2+}. This Ca^{2+} mobilizing mechanism is the voltage-dependent Ca^{2+} channel which is activated by membrane depolarization and indeed acts in excitatory and secretory cells. The other Ca^{2+} entry

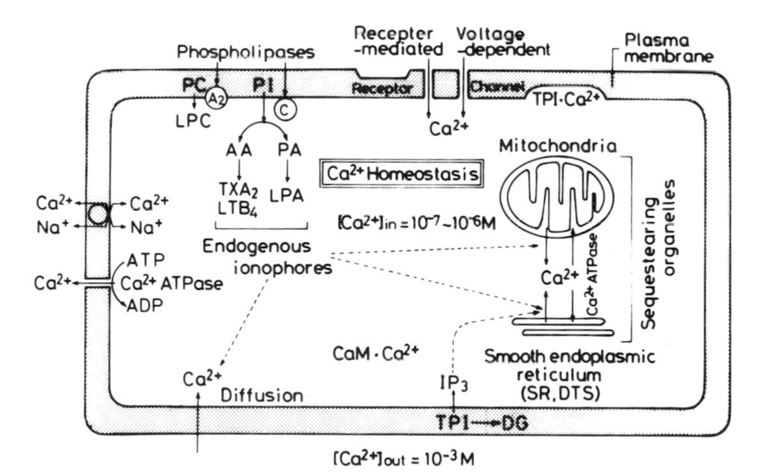

FIGURE 8. Proposed mechanisms of regulation of cytosolic calcium ions in cells. TXA_2, thromboxane A_2; LTB_4, leucotriene B_4; TPI, triphosphoinositide; C, phospholipase C; A_2, phospholipase A_2; AA, arachidonic acid; PA, phosphatidic acid; LPA, lysophosphatidic acid; PC, phosphatidylcholine; LPC, lysophosphatidylcholine.

system into the cell is the receptor-operated Ca^{2+} channel and, as its possible mechanisms, increased membrane permeability due to perturbation and formation of calcium ionophores (PA, TXA_2, LTB_4) have been proposed. In addition to these Ca^{2+} mobilization systems mainly through the plasma membrane, it has become clear that the release of Ca^{2+} from intracellular stores is one of the most important Ca^{2+} sources in cell stimulation. However, any messenger was not found between the receptor on the plasma membrane and the internal Ca^{2+} stores. Recently, by use of a fluorescent Ca^{2+} cnelator Quin 2, Berridge and Irvine (4) have proposed a hypothesis that IP_3 is the missing link.

B. Receptor-Coupled Ca^{2+} Mobilization

Since the dense tubular system of platelet is considered to be analogous to muscle sarcoplasmic reticulum, several lines of evidence have suggested that the Ca^{2+} release from this specific site may play a crucial role in regulation of the intracellular Ca^{2+} level, by either release or sequesteration. The agonists including thrombin, ADP, platelet-activating factor (PAF) and thromboxane A_2 can induce elevation of $[Ca^{2+}]i$ by increased influx across the plasma membrane as well as discharge from internal stores. At present, there is no convincing data lending support to the presence of voltage-dependent Ca^{2+} channel in platelet. Therefore the idea is favored that the Ca^{2+} entry may occur as a result of increased permeability induced by membrane lipid perturbations upon stimulation. When human platelets loaded with Quin 2 acetoxymethyl ester were exposed to thrombin in the presence of Ca^{2+}, the rapid rise of cytosolic Ca^{2+} content was observed (Fig. 9). The chelation of extracellular Ca^{2+} with EGTA reduced the level of increase in $[Ca^{2+}]i$ but the transient increase of $[Ca^{2+}]i$ still occurred which was sufficient to provoke almost full physiological responses, indicating that the mobilized Ca^{2+} arose from the release from the dense tubular system. Moreover, it was noted that activation of platelets with thrombin also induced a significant accumulation of cellular $[^{45}Ca^{2+}]$ reaching a plateau on further incubation. This finding demonstrated that an increase in cytosolic free Ca^{2+} may in part be due to an enhanced influx of extracellular Ca^{2+}. The temporal observations of Ca^{2+} influx, phospholipid turnover, and serotonin release suggested us to conclude that the formation of PA is closely coupled with the enhancement of Ca^{2+} uptake (5), which is in good agreement with the hypothesis that PA may act as an endogenous calcium ionophore (41, 42). Taken with the findings that receptor activation increased the PA

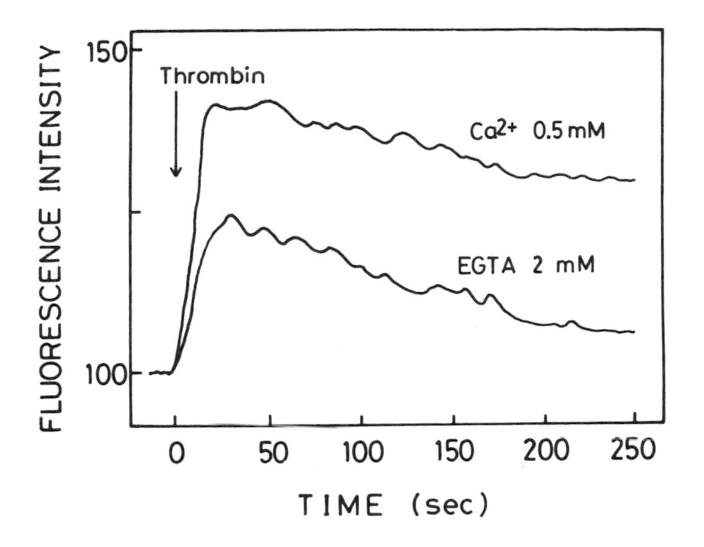

FIGURE 9. Thrombin-induced increase of intracellular calcium concentration in human platelets.

content in the parotid and that exogenous PA can stimulate Ca^{2+}-mediated responses, data showing the correlation between inhibiton of partitioning activity of PA and apparent affinity for the Ca^{2+} influx mechanism, suggested that PA may play an important role in calcium gating (42). This concept was supported by several experiments with liposomes. The rate of Ca^{2+} translocation across the PC liposome containing PA was almost 3 times larger compared with the liposome containing PI or DG. The ionophoretic activity of dipalmitoyl-, dioleoyl-, dilinoleoyl-, and dilinolenoyl-PA was found to depend on the degree of unsaturation of fatty acyl chains (unpublished data). The ability to transport Ca^{2+} was smallest in dipalmitoyl-PA with a lowest fluidity.

As for the release of Ca^{2+} from the non-mitochondrial stores, probably endoplasmic reticulum and also the plasma membrane, substantial experimental evidence has been provided by Berridge and Irvine (4), who proposed the IP_3-calcium-mobilizing hypothesis. The Ca^{2+}-releasing property of IP_3 was first demonstrated in rat pancreatic acinar cells permeabilized by saponin (43). This basic observation was verified in a variety of cell types including fibroblast, insulinoma, neutrophil, macrophage, artery smooth muscle, platelet, pituitary cell, hepatocyte, and mastocytoma. For example, mouse mastocytoma P815 cells were suspended in the medium supplemented with Quin 2, MgATP and low calcium, and

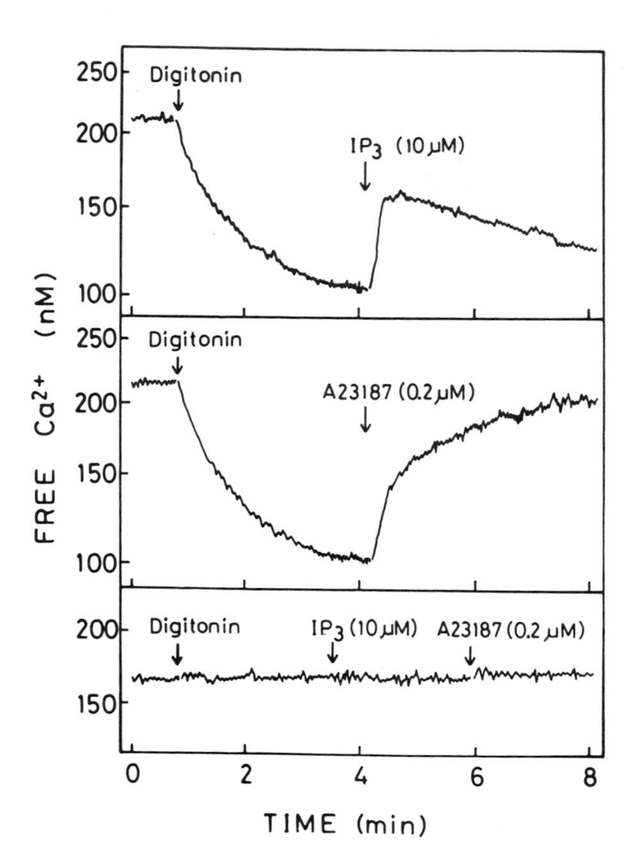

FIGURE 10. Mobilization of calcium from permeabilized mastocytoma P815 cells by inositol 1,4,5-trisphosphate.

then permeabilized by adding digitonin. Immediately after the onset of permeation, cells began to sequester calcium into the intracellular store, which was reflected in a rapid decline in the ambient calcium concentration to a steady value, where the fluxes into and from the internal stores did not occur (Fig. 10). Addition of IP_3 resulted in a rapid release of Ca^{2+} followed by a slower re-uptake via ATP-driven pump (unpublished data). It would also be possible that the apparent gradual disappearance of released Ca^{2+} may be due to degradation of IP_3 by a phosphatase.

It has also been known that cAMP is implicated in Ca^{2+} regulation of the platelet and that it exerts inhibitory effects on Ca^{2+} mobilization and secretion. Thrombin-induced DG formation by phopholipase C was repressed by addition of dbcAMP and forskolin capable of activating adenylate cyclase in a dose-dependent fashion (13, 17, 44). Subsequent failure to produce PA appeared to parallel the decrease in Ca^{2+} uptake activity. Alternatively, cAMP may act primarily through stimulation of Ca^{2+} translocation into the store by the enhanced activity of Ca^{2+} pump, thereby removing from the cytoplasm Ca^{2+} which previously had been released by thrombin. As described above, it was recently shown that tumor promotors such as PMA and phorbol dibutyrate, when added before the natural receptor-mediated stimuli, inhibited Ca^{2+} mobilization in human platelets. However, in the presence of protein kinase C inhibitor H-7, the increase in $[Ca^{2+}]i$ induced by thrombin was restored to nearly the normal level, irrespective of PMA-pretreatment (unpublished data). This may imply that protein kinase C would subserve a negative feedback on Ca^{2+} mobilization. Much extensive work will be required for complete understanding of PMA-mediated inhibition of Ca^{2+} mobilization.

REFERENCES

1. Hokin, M. R., and Hokin, L. E., J. Biol. Chem. 203:967 (1953).
2. Michell, R. H., Biochim. Biophys. Acta 415:81 (1975).
3. Nishizuka, Y., Trends Biochem. Sci. 9:163 (1984).
4. Berridge, M. J., and Irvine, R. F., Nature 312:315 (1984).
5. Imai, A., Ishizuka, Y., Kawai, K., and Nozawa, Y., Biochem. Biophys. Res. Commun. 108:752 (1982).
6. Imai, A., and Nozawa, Y., Biochem. Biophys. Res. Commun. 105:236 (1982).
7. Imai, A., Yano, K., Nakashima, S., Ishizuka, Y., Hattori, H., Takahashi, M., Okano, Y., and Nozawa, Y., in "Transmembrane Signaling and Sensation" (F. Oosawa, T. Yoshioka, and H. Hayashi, eds.), p. 3. Japan Scientific Societies Press, Tokyo, 1984.
8. Imai, A., Nakashima, S., and Nozawa, Y., Biochem. Biophys. Res. Commun. 110:108 (1983).
9. Siess, W., and Binder, H., FEBS Lett. 180:107 (1985).
10. Agranoff, B. W., Murthy, P., and Seguin, E. B., J. Biol. Chem. 258:2076 (1983).
11. Rittenhouse, S. E., Proc. Natl. Acad. Sci. USA, 80: 5417 (1983).
12. Wilson, D. B., Neufeld, E. J., and Majerus, P. W., J. Biol. Chem. 260:1046 (1985).
13. Imai, A., Hattori, H., Takahashi, M., and Nozawa, Y.,

Biochem. Biophys. Res. Commun. 112:693 (1983).
14. Billah, M. M., Lapetina, E. G., and Cuatrecasas, P., Biochem. Biophys. Res. Commun. 90:92 (1979).
15. Takai, Y., Kaibuchi, K., Sano, K., and Nishizuka, Y., J. Biochem. 91:403 (1982).
16. Knight, D. E., and Scrutton, M. C., Nature 309:66 (1984).
17. Imai, A., Hattori, H., Takahashi, M., Nakashima, S., S., Okano, Y., Hattori, T., and Nozawa, Y., Thromb. Res. 35:539 (1984).
18. Castagna, M., Takai, Y., Kaibuchi, K., Sano, K., Kikkawa, U., and Nishizuka, Y., J. Biol. Chem. 257: 7847 (1982).
19. Yamanishi, J., Takai, Y., Kaibuchi, K., Sano, K., Castagna, M., and Nishizuka, Y., Biochem. Biophys. Res. Commun. 112:778 (1983).
20. Halenda, S. P., and Feinstein, M. B., Biochem. Biophys. Res. Commun. 124:507 (1984).
21. Watson, S. P., and Lapetina, E. G., Proc. Natl. Acad. Sci. USA 82: 2623 (1985).
22. Zavoico, G. B., Halenda, S. P., Sha'afi, R. I., and Feinstein, M. B., Proc. Natl. Acad. Sci. USA 82:3859 (1985).
23. Kawamoto, S., and Hidaka, H., Biochem. Biophys. Res. Commun. 125:258 (1984).
24. Ishizuka, Y., Imai, A., and Nozawa, Y., Biochem. Biophys. Res. Commun. 123:875 (1984)
25. Okano, Y., Takagi, H., Tohmatsu, T., Nakashima, S., Kuroda, Y., Saito, K., and Nozawa, Y., FEBS Lett. 188: 363 (1985).
26. Okano, Y., Ishizuka, Y., Nakashima, S., Tohmatsu, T., Takagi, H., and Nozawa, Y., Biochem. Biophys. Res. Commun. 127:726 (1985).
27. Yano, K., Nakashima, S., and Nozawa, Y., FEBS Lett. 161:296 (1983).
28. Cockcroft, S., Barrowman, M. M., and Gomperts, B. D., FEBS Lett., 181:259 (1985).
29. Yano, K., Higashida, H., Inoue, R., and Nozawa, Y., J. Biol. Chem. 259:10201 (1984).
30. Yano, K., Higashida, H., Hattori, H., Nozawa, Y., FEBS Lett. 181:403 (1985).
31. Yano, K., Higashida, H., and Nozawa, Y., FEBS Lett. 183:235 (1985).
32. Ohki, K., Imai, A., and Nozawa, Y., Biochem. Biophys. Res. Commun. 94:1249 (1980).
33. Nathan, I., Fleischer, G., Livne, A., Dulansky, A., and Parola, A. H., J. Biol. Chem. 254:9822 (1979)
34. Ohki, K., Sekiya, T., Yamauchi, T., and Nozawa, Y., Biochim. Biophys. Acta 693:341 (1982).

35. Ohki, K., Yamauchi, T., Banno, Y., and Nozawa, Y., Biochem. Biophys. Res. Commun. 100:321 (1981)
36. Ohki, K., Sekiya, T., Yamauchi, T., and Nozawa, Y., Biochim. Biophys. Acta 644:165 (1981).
37. Das, S., and Rand, R. P., Biochem. Biophys. Res. Commun. 124:491 (1984).
38. Dawson, R. M. C., Irvine, R. F., Bray, J., and Quinn, P. J., Biochem. Biophys. Res. Commun. 125:836 (1984).
39. Hofmann, S. L., and Majerus, P. W., J. Biol. Chem. 257: 14359 (1982).
40. Boni, L. T., and Rando, R. R., J. Biol. Chem. 260:10185 (1985).
41. Serhan, C., Anderson, P., Goodman, E., and Weissman, G., G., J. Biol. Chem. 257:2736 (1982)
42. Putney, J. W., Weiss, S. J., Van De Walle, C. M., and Haddas, R. A., Nature 284:345 (1980).
43. Streb, H., Irvine, R. F., Berridge, M. J., and Schulz, I., Nature 306:67 (1983).
44. Feinstein, M. B., Egan, J. J., Sha'afi, R. I., and White, J., Biochem. Biophys. Res. Commun. 113:598 (1983).

MODELS FOR ION TRANSPORT IN GRAMICIDIN CHANNELS: HOW MANY SITES?

S. B. Hladky[1]

Department of Pharmacology
University of Cambridge
Cambridge, England

I. INTRODUCTION

We make kinetic measurements to obtain information about the physical properties of a system. Kinetic models serve as intermediaries in this process. For gramicidin the most common measurements are of the electrical current. It is now well known that the current can be resolved into increments each of which flows through a single pore, and there are several kinetic models which claim to describe the mechanism of the ion-transport. More recently similar single-channel records have become available for other more complicated pores, and they are also being analyzed in terms of models very similar to those used for gramicidin (see e.g., Ref. 1-4). It should therefore be useful to discuss both what the models can tell us about gramicidin and what gramicidin has taught us about the uses and limitations of the models.

In the initial studies on gramicidin it was inferred that the conducting unit is a pore which contains water (5-8) as has since been confirmed directly (see Dani and Levitt, (9) and Finkelstein and Andersen (10)). Thus gramicidin molecules form the wall of a hole. The inside surface of the wall is hydrophilic so that ions and water can be accommodated and the outside is hydrophobic so that the structure can remain in contact with the lipid chains in the core of the membrane. The exclusion of anions, polyvalent cations,

[1]Supported by SERC Grant GR/B18522

and ions larger than dimethylammonium suggests that the hole has a narrow portion probably in the range of 4 to 7 A. From the primary structure of gramicidin, one can also infer that the hole must be lined with the peptide backbone because there are no other hydrophilic groups. Similarly the side chains must be oriented towards the membrane material. These properties are consistent with both the single and double helical structures that have been proposed on the basis of spectroscopic measurements in various solvents. Subsequently kinetic and spectroscopic studies (see Bamberg *et al.* (11) and Weinstein *et al.* (12)) using modified gramicidin in membranes have strongly supported the end to end single-stranded dimer structure that was originally proposed by Urry (13).

II. KINETIC MODELS

The measurement of the single channel currents obviously provides information about the mechanism of ion transport through the pore. At least five different sorts of models have been used to try to extract this information.

With the most ambitious type of modelling, called molecular dynamics, an attempt is made to portray all significant features of the pore in terms of the positions of the atoms and the forces between them. The movements of the atoms are then calculated from Newton's laws of motion, i.e., by brute force. At least in principle this type of modelling is far superior to the others, because it allows the prediction of the kinetics from other forms of information. At present gramicidin is right at the limits of what can be done, and we have yet to discover whether the approximations required to make the calculations tractable allow prediction of the measured properties of the ion transport. It should be noted that this type of model requires specification of literally hundreds of parameters. Molecular dynamics modelling is discussed in much more detail by Pullman elsewhere in this volume.

Models of the second type, which we can call Eyring or absolute rate theory models, are much less ambitious, but they still attempt to refer to definite structural features of the pore. They do so by invoking the existence of binding sites that are separated by energy barriers (Fig. 1). The ions are assumed to be almost always bound to the sites, but occassionally they jump from one to another. The rates of these movements are calculated from the positions and energy levels using the formalism of absolute rate theory. In this formalism the rate of each step varies exponentially with the energy difference between the site of origin and the barrier

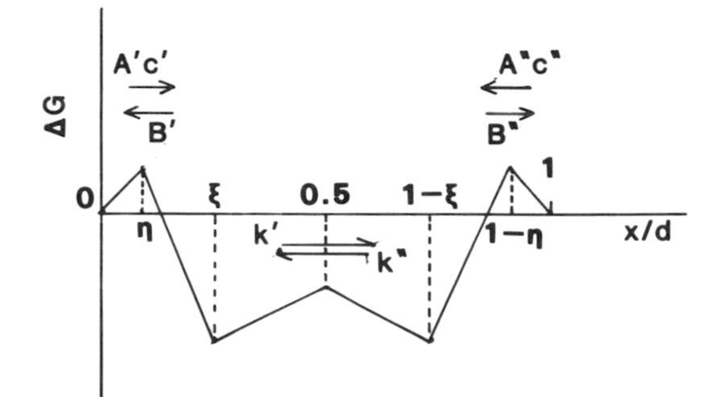

FIGURE 1. A schematic barrier profile for a 2-site Eyring model. The membrane surfaces are located at 0 and d.

which must be crossed. If there are enough sites in the model and we allow enough to be occupied, the resulting equations can fit almost any variation of conductance with ion concentration. Furthermore it was shown back in 1949 (14) that these theories predict an exponential current-voltage relation if transport occurs by a single jump, but that much flatter relations can be obtained if transport occurs by a sequence of jumps as in aqueous diffusion. Indeed by adjusting the fraction of the applied potential that affects each of the jumps these models can be made to fit almost any current-voltage data.

The use of specific binding sites, energy barriers and absolute rate theory in the interpretation of data has the advantage that it provides an explanation that is relatively simple, concrete and easy to understand. There are, however, at least two disadvantages. Firstly, no argument has yet been found to show that the particular choice of sites and barriers used to describe a set of kinetic data corresponds to actual sites and barriers in the pore. Thus the model may mislead. Secondly, unless there literally are just three rate limiting jumps, the number of sites must exceed two. Even if there are only two preferred binding sites in the pore, as argued for gramicidin in lysophosphatidylcholine micelles by Urry _et al_. (15,16), additional sites may be required if the kinetic model is to portray the potential dependence of the currents. In models with more than two sites the number of constants which must be determined from the data exceeds what we can actually calculate.

Levitt (17) has proposed a model that partially avoids these difficulties. It retains the Eyring description for the steps in which ions enter and leave the pore, but describes transfer through the pore as an electrodiffusion process. The ions move along the pore down the resultant of their concentration and potential energy gradients as described by the Nernst-Planck equation. Levitt estimates the potential energy gradient using electrostatic calculations that he had presented earlier (18). There is much to commend Levitt's approach, for one instance the ingenious way in which double-occupancy of the pore has been incorporated, and for another, the computational advantages of using differential rather than difference equations. However, I should remind you that any result which can be derived using the Nernst-Planck equation can, at least formally, also be derived using absolute rate theory with an appropriate choice of sites and barriers. Furthermore, the use of the Nernst-Planck equations completely ignores local variations in the potential energy. It is not clear to me that it's assumptions are any more justified than are those of absolute rate theory.

In a fundamentally different sort of model, ion binding induces changes in the conformation of the channel that occur on a time scale which is much slower than the transit time for an ion, but much faster than the resolution of the single-channel recording. Both Läuger, who introduced this type of model (19), and Eisenman consider it in some detail elsewhere in this volume. Undoubtedly if enough binding sites and conformations are used, models of this type will be capable of describing all of the kinetic data that can be obtained. However, there is at present little if any evidence for gramicidin that the conformation changes required in fact occur.

All of the preceding types of models attempt to interpret the data in terms of more or less definite structural features of the pore. The data can also be interpreted in another less concrete fashion (Fig. 2). This alternative description speaks not of specific sites and conformations but rather of distinguishable states of the pore. For gramicidin, the model which I have used allows the pore to contain 0, 1 or 2 ions. For a single species of permeant ion in the solutions it has four states. I have thought of these as the pore containing only water, the pore occupied by an ion on the left, the pore occupied by an ion on the right, and the pore occupied by two ions, one in each half. That view of the states implies certain rules for the transitions between them. For instance an empty pore becomes occupied at the left end by an ion entering from the left. The simplest assumption is to say that the rate of this process is proportional to the product of the probability the pore is empty, X_{00}, and the concentration of ions on the left, c'. The proportionality constant is denoted A'. To take another example, for the movement of an ion from the left to the right the rate is proportional to the probability there is an ion

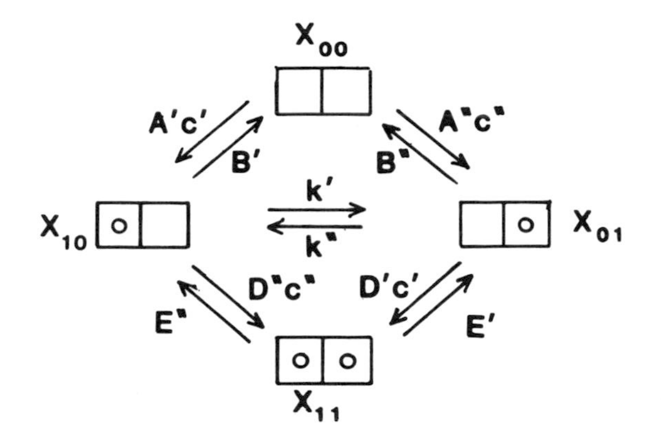

FIGURE 2. A four-state model for a pore. Only one ion can be found at each end at a time and the probability an ion leaves an end in either direction is independent of how it entered that end.

on the left and none on the right, X_{10}, and a constant, k'. There are also assumptions about the transitions which can not occur, e.g., ions are not allowed to enter an end of the pore which is already occupied. This prohibition could arise from either electrostatic repulsion, which may be important for an ion trying to enter the pore, or from the presence of water molecules between the ions which would prevent them approaching each other within the pore. The model is completely defined by the list of states and the rules for the transitions. It differs from an Eyring model in several important respects.

Firstly it does not specify the number of actual, physical sites. An end of the pore might have only one site, it might have additional sites that are only very transiently occupied, or it might contain several sites with an ion capable of moving rapidly from one to another. Thus so long as no more than double-occupancy can occur the four-state model includes as special cases the standard two-site Eyring model and the four-site model approximations considered by Sandblom *et al.* (20), Hägglund *et al.* (21), and Hladky (22). Levitt's calculation (17) for narrow pores is a little different.

Secondly the use of the state model makes no assumptions about the potential dependence of the rate constants, except to require that the theory be consistent with microscopic reversibility. The ability to adjust the potential dependence of the constants is both an advantage and an embarrassment. It is an advantage in that it allows data to be fitted. It is an embarrassment because that means for each ion we have at least two constants and three functions of potential to determine from the data.

Perhaps I can best illustrate the difference between a two-site model and the four-state model by giving you an example of how they can be used in the fitting of a current-voltage relation. I will illustrate the process by considering very low ion concentrations where the analysis is simplest. In such a model one draws a potential energy profile for the ion as in Fig. 1. The model allows one to adjust the fraction of the applied potential which affects each step and the sizes of the rate constants. The ratio k/B is particularly important. If transfer through the pore, k, is slow compared to the rate of exit, B, then the ion binding at each end stays at equilibrium with the aqueous phase. These binding constants will vary exponentially with the fraction of the potential between the site and the adjacent solution. Since in an Eyring model the rate constant for transfer also varies exponentially, we would then have a steeply increasing current-voltage relation. The precise prediction is a hyperbolic sine. By contrast if transfer through the pore is fast compared with exit, that is, k/B is large, the current becomes limited by the rate at which ions can be brought up to the pore. This access process depends only weakly on the applied potential, and this leads to a current-voltage rela-

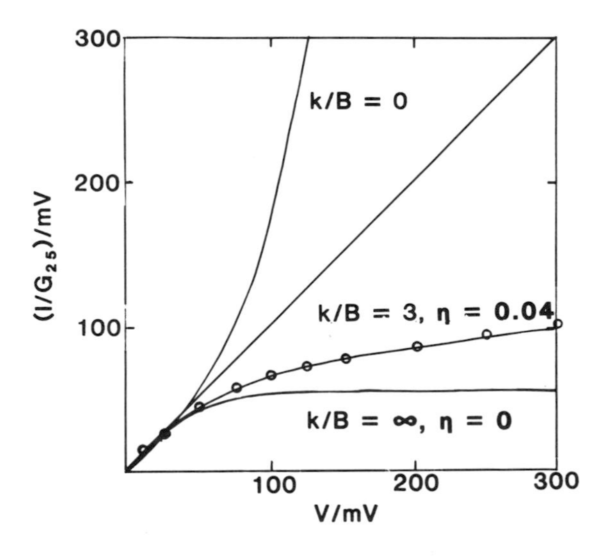

FIGURE 3. I-V relations for a two-site Eyring model and low ion concentrations. The data shown are for 0.1 mM CsCl with either 25 mM $MgSO_4$ or 99 mM CholineCl added to increase the ionic strength (Hainsworth, A. and Hladky, S. B., unpublished).

tion which increases very slowly at high potentials. These theoretical relations and an example of real data are shown in Fig. 3. For convenience, because that is the way we have measured it, the currents are plotted normalized to the conductance at 25 mV. The fitting parameters differ slightly from those found by Eisenman et al. (23). The difficulties begin as soon as one tries to consider data at higher concentrations. For caesium the conductance-activity relation requires values of k/B much larger than those which fit the current-voltage relation shown in Fig. 3.

Two ways have been used to avoid this contradiction between theory and experiment. One is to use an Eyring model that has more than two sites. Eisenman and his coworkers have pursued this approach the furthest and have found that the addition of just two more sites allows one to fit the available data (see e.g. Eisenman and Sandblom (24)). I have not used that approach for the simple, practical reason that a four-site model, even when it is strictly symmetrical and

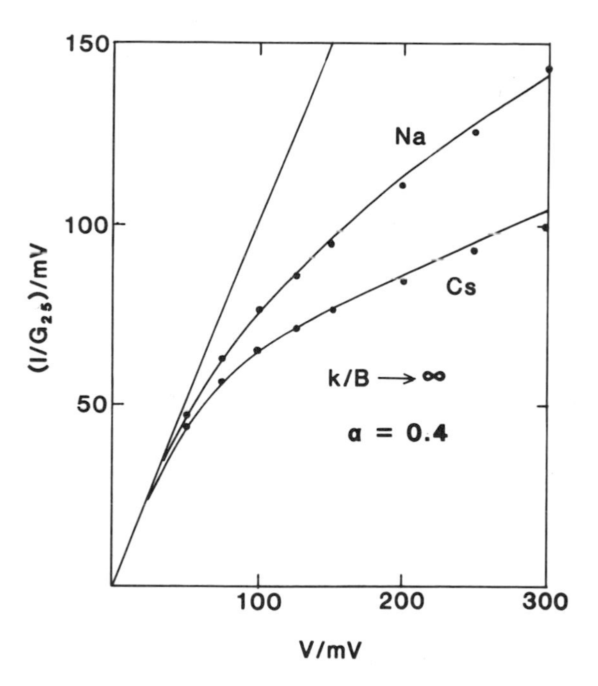

FIGURE 4. Current-voltage relations calculated from the four-state model. The access rate constant varies with potential as $A' = A\ (1 + \alpha)/(1 + \alpha\exp(-n\phi))$ where $\phi = FV/RT$ and n is 0.18 for sodium and 0.1 for caesium. The potential dependence of A/B and k is the same as in Fig. 3. (Hainsworth, A. and Hladky, S. B., unpublished).

restricted to no more than double occupancy, has 11 states and 18 arbitrary rate constants to which must still be added the additional constants, like n, which tell us how the others vary with the applied potential.

The approach I have used is to preserve the four states of the two-site model, but to drop the specific assumptions of Eyring rate theory. In the four-state model the formal expression relating the current to the rate constants is the same as in the two-site model, but we no longer presume to know how the constants vary with potential. As a result the theory no longer obviously contradicts the data, but for the negative reason that the data at any single concentration can no longer determine the ratio k/B. For instance, as shown in Fig. 4, for low concentrations of sodium and caesium the data can still be fitted using the values calculated with the Eyring assumptions, but they can also be fitted for very large k/B simply by adjusting the apparent potential dependence of the rate of entry, A. It thus appears that by abandoning the Eyring theory we lose the ability to calculate constants. I should emphasize that in fact we really could not calculate the constants even in the Eyring models. In those models, if the number of sites or the assumed division of the potential difference between them is changed, the values of the calculated constants change. We do not know how many jumps there are in the transport process, hence we do not know how many Eyring sites there should be in the model, and as a result we can not calculate unique values for the constants. For example, the particular potential dependence for A' and the other constants chosen for illustration in the slide can also be derived from a four-site model with two "real" sites and two transiently occupied outer sites at the mouths of the pore (22). Thus even within the class of four site models, k/B cannot be determined from the data at a single concentration.

Faced with these difficulties the obvious way forward for an experimentalist is to search for aspects of the data which can be interpreted without requiring specification of sites and the detailed potential dependence of the constants. I would now like to turn to these.

III. CONDUCTANCE CONCENTRATION RELATIONS

Estimates of the occupancy of the pore *in situ* in a membrane have been obtained by several groups using a variety of data including conductance *versus* concentration curves, permeability and conductance ratios, blocking effects, the variation of water permeability with ion concentration, channel duration *versus* ion concentration, equilibrium binding by dialysis, and the flux ratio *versus* ion concentration. I have reviewed these results elsewhere (25,26). I would like to consider here the information about the transport process which can be obtained from the conductance-concentration relation.

Even the earliest single-channel data contained clear indications that ions compete for occupancy of the pore. If there were no competition the conductances would increase linearly with the ion concentrations. They clearly do not (6). The curves for sodium and potassium each increase with concentration up to about 1 M but they then level off, and those for caesium and thallium increase up to similar concentrations but then take a bit of a nose-dive.

The evidence for interaction is more clearly displayed, if we replot the data using an Eadee-Hofstee plot, which is just a Scatchard plot by another name (see Fig. 5). If the rate of a process is proportional to the binding to a single site, then this plot should give a straight line. The y intercept, corresponding to high concentrations, is the maximum response, which here is the maximum conductance. The x intercept, corresponding to low concentrations, is the maximum response times the affinity constant. The slope is the reciprocal of the affinity constant. Shallow slopes mean large binding constants, that is, high affinity.

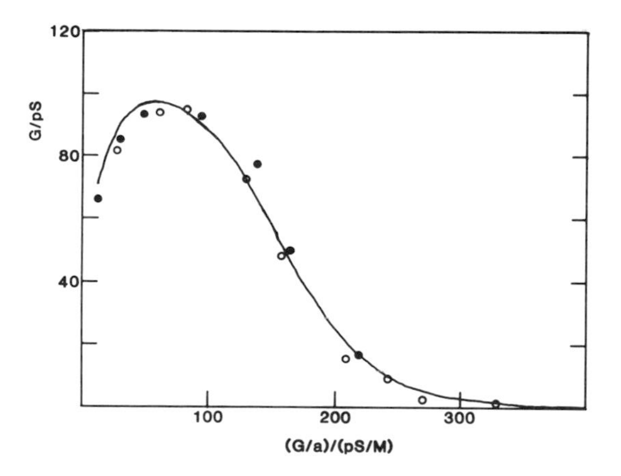

FIGURE 5. Scatchard plot of the conductance-concentration relation for caesium at 50 mV. The solid points are taken from Urban *et al.* (27); the open points from Neher *et al.* (28). Urban *et al.* report data primarily at 100 mV for the full range of concentrations. The curve is calculated from the four-state model with $A = D = 1.4 \times 10^8\ M^{-1}\ s^{-1}$, $B = 1.4 \times 10^6 s^{-1}$, $E = 1.2 \times 10^8 s^{-1}$, and $k = 1.2 \times 10^8$.

The data for caesium in Fig. 5 do not fit a straight line and display the need for a transport model which is more complicated than entry and exit steps from a single-site pore. If we look just at the data for low concentrations, the slope is small in magnitude which indicates binding with an apparent affinity constant of the order of 10 mM. Yet peculiarly when the concentration of caesium is increased above 10 mM the current does not reach a limiting value, it increases further, at least another 10 fold. Only at the highest concentrations do the conductances reach a limit and then fall.

This behaviour can be explained using models which allow ions to enter pores which can remember that another ion has been bound. That occurs most simply if the other ion is still there. At very low concentrations, 1 mM and lower, an ion comes up, goes through the pore and leaves from the other end without ever seeing another ion. If we increase the concentration, the current increases simply because ions arrive more often. If we increase the concentration yet further we reach a stage where the pore is usually occupied by an ion. If only one ion can be in the pore at a time, this will correspond to the maximum conductance. As we have seen, with that restriction the model does not fit the data. For instance for caesium it would predict that the maximum conductance would be reached for concentrations well below 100 mM. However, if leaving rather than transfer from end to end is limiting, entry of a second ion can increase the current in a simple manner. When the second ion enters, the two ions presumably repel each other either electrostatically or because water is compressed between them. This repulsion can greatly speed up the rate at which ions leave the pore. At least some of those which leave come out at the far end, which contributes to the current. Thus the rate at which ions leave goes up as the rate of second entries increases, which means that the current increases even though the pore is already usually occupied by one ion. Allowing double-occupancy also provides one possible explanation for the fall in conductance at high concentrations. For sufficiently high concentrations the pore will usually be doubly-occupied. Furthermore, if one ion comes out, the void is immediately refilled by entry of an ion from the same side. Because in the model two ions cannot be in the same end at once, this rapid refilling precludes transfer from end to end and thus blocks the current. I should note in passing that this model can explain concentration-dependent permeability ratios, selective block, anomalous mole fraction effects, and enhancement of fluxes by addition of impermeant ions which can enter the end of the pore (29). At least qualitatively it explains every type of kinetic data of which I am aware.

By contrast to the ease with which the four-state model

has provided qualitative explanations, I think it is fair to say that it had proved more difficult to determine the rate constants which describe the transport process than anyone in the field had expected. The difficulties have arisen because there are so many constants to determine. The conductance-concentration relation, such as that shown in Fig. 5, determines the values of all of these constants only in very favourable circumstances. Roughly we get an estimate of the rate constant for entry to empty pores, A, from the low concentration intercept, of the first ion affinity constant, A/B, from the low concentration slope, of the rate constant for entry to singly occupied pores, D, from the x intercept of the straight line at higher activities, of the composite rate constant for transfer and exit, kE/(k+E), from the slope of the straight segment, and the product kE from the down turn at the highest concentrations. The values, indicated in Fig. 5, are consistent, at least semi-quantitatively, with the other forms of data mentioned above. The values of D and k are least ambiguous, those of A/B and E the most uncertain (see Ref. 26).

I have discussed the analysis of the conductance-concentration relation in some detail because it is relatively insensitive to the limitations of site models. However, it does still make assumptions about the factors which can affect the transport process, and the type of analysis given here has been challenged on the grounds that it ignores two important physical effects that account for significant features of the data. These effects are double-layer polarization and diffusion limitations. I would now like to consider these in turn.

IV. DOUBLE-LAYER POLARIZATION

To apply a potential to the membrane a positive charge must be created close to one membrane surface and a negative charge close to the other, which requires accumulation and depletion of ions. Thus the ion concentrations near the surfaces must differ from their bulk values. The tendency for the ions to diffuse towards or away from the surfaces must therefore be balanced by a gradient of electrical potential as shown in Fig. 6. The region where the concentrations differ from their solution values and there is a gradient of potential is called the electrical double-layer. The changes which occur within it when a potential is applied are called polarization. As a consequence of this double-layer polarization, the potential difference across the membrane is less than the applied potential. These effects are very marked

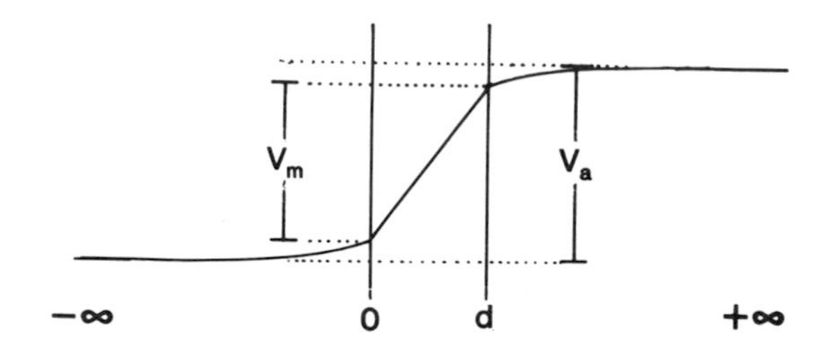

FIGURE 6. The profile of potential at low ionic strength. The membrane extends from 0 to d. V_a is the applied potential and V_m, the potential difference across the membrane.

when there are few ions in the solutions because the ions which must be accumulated to charge the membrane capacity then represent a large percentage change in the ion concentrations. Conversely the effects are minor when there are many ions present. Thus double-layer polarization can be greatly reduced by increasing the ionic-strength of the solution.

Double-layer polarization affects the rate of ion transport both because there are changes in the concentration of the permeant ions near the surfaces and as a result of the difference between the applied potential and the potential difference across the membrane. For low permeant ion concentrations, where the conductance is proportional to concentration, the changes in permeant ion concentration are more important and double-layer polarization increases the conductance, an effect which Walz *et al.* (30) called ion-injection. The size of the effect increases with the applied potential. Thus double-layer polarization will make current-voltage curves bend more towards the current-axis. Andersen (31) has proposed that double-layer polarization accounts for the apparent potential dependence of the access process for gramicidin at low concentrations. He has also noted that since it increases conductances at low concentrations but not at high concentrations, it will change the conductance-concentration curve in the manner which has been interpreted here as implying ion binding at low activities.

Fortunately the size of this effect can be estimated. When the permeant ion concentration is so low that the true conductance-concentration curve is linear, the Gouy-Chapman theory for the double-layer provides simple quantitative prediction for the changes in ion transport produced by double-layer polarization. These predictions are compared with experimental current-voltage relations for 1 mM NaCl in Fig. 7. The lower set of data were obtained with the ionic

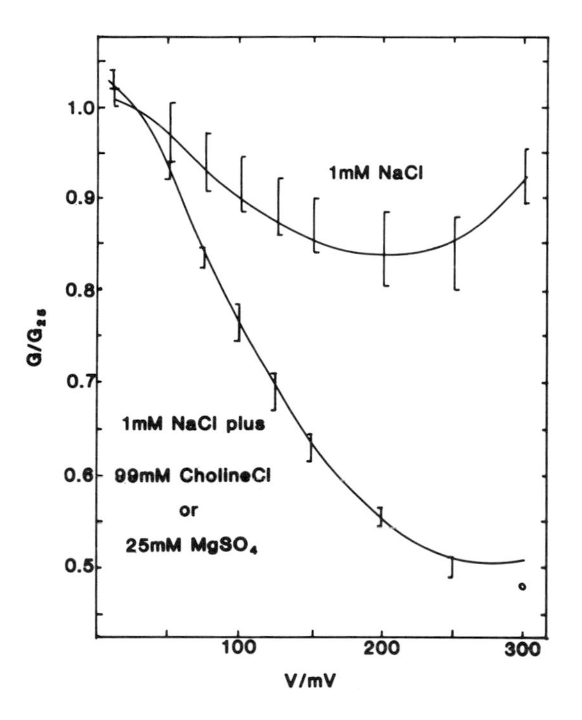

FIGURE 7. The conductance _versus_ potential measured with the solutions indicated in the figure. The bars represent mean ± s.d. The data for both choline chloride and magnesium sulphate are included within the lower bars. The lower curve was fitted to the points. The upper curve was calculated from the lower curve, the known membrane capacitance, 0.59 $\mu F/cm^2$, and the Gouy-Chapman theory applied to double-layer polarization. (Hainsworth, A. and Hladky, S. B., submitted.)

strength raised to 100 mM by adding either choline chloride or magnesium sulphate. No difference was observed between these two salts. At this high ionic strength double-layer polarization produces almost negligible effects. The upper set of data were obtained with 1 mM NaCl alone. The lower curve was arbitrarily fitted to the points. It is used to tell the theory which ion transport process is present. From that curve and the known capacitance of the membrane, it is then possible to calculate the shape of the current-voltage curve at any other ionic strength. The theoretical curve for 1 mM is shown and it goes through the bars. Similar data for intermediate ionic strengths were fitted equally well, and the theory also works with valinomycin. Thus in agreement with Andersen, we find that double-layer polarization substantially alters the shape of the current-voltage relation. If such relations are to be interpreted correctly, then either ionic strength must be raised with an inert salt, or double-layer polarization must be incorporated explicitly into the theory.

As Andersen has noted the effects of double-layer polarization on the form of the conductance-concentration curve could in principle be quite complex. But fortunately the effects are small at high concentrations and, once the Gouy-Chapman theory has been confirmed, we can calculate them for low concentrations. At 1 mM for either sodium or caesium, the error in conductance at 50 mV is less than 6%. An error of this size does not compromise the analysis given by either Neher _et al_. (28) or Hladky and Haydon (26). Largely for this reason, I did not immediately pursue the matter after C. L. Lawson and I obtained preliminary results in 1981.

V. DIFFUSION LIMITATIONS

Double-layer polarization produces effects which must be considered even if only to conclude that in particular circumstances they are small enough to ignore. The consequences of diffusion limitations for the analysis are quite different.

The conductances at low permeant ion concentrations are limited by the rate of entry of the ions into the pore and this process is only weakly potential dependent which suggests that it is limited near the pore mouth. Andersen and Eisenman have confirmed this result and Andersen (32) has gone on to show that the rate limiting step is affected by sucrose which cannot enter the lumen of the pore. A similar result was found for glycerol by Urban (33). Thus the rate limiting step probably occurs outside the lumen proper. It is thus quite proper to consider the possibility that the rate limiting step might be aqueous diffusion and to investigate how diffusion limitation and more generally external access steps can be incorporated into the kinetic models.

Diffusion of solutes up to the mouth of a pore has been considered in some detail in connection with the exchange of large solutes such as proteins between blood and interstitial fluid (34). It is now accepted that most polar substances cross the capillary walls through aqueous pores (35), and that for large solutes the transport is partly limited by diffusion up to one end of the pore and away from the other. These radial diffusion processes have been treated successfully by solving the equations for diffusion from infinity to a target disk located at the centre of the pore's mouth. The radius of the target is taken as the difference between the radius of the pore and the radius of the solute, so that when the solute reaches the target it is entirely clear of the pore's rim. The overall rate of transport is calculated by considering the radial diffusion and pore transport processes to be in series. For the capillary wall all the transport processes increase linearly with the solute concentration and the equations are easily solved. The application of this method to narrow, saturable pores like gramicidin is not so straightforward. The difference is easiest to explain by considering an example.

We consider a pore and the diffusion of ions up to the mouth on one side and away from the mouth on the other (see Fig. 8). The pore can hold only one ion at a time, and thus at high concentrations the flux saturates, that is, it approaches a limiting value. To keep this illustration simple the permeability of the pore and the permeability of each of the diffusion processes are taken as equal to each other. A potential is then applied so that for a given concentration at the mouth the flux from left to right is 10 times larger than for no applied potential, and the flux from right to left is 10 times smaller. If there were no diffusion barriers, then the concentrations at the mouths would both be C,

$$J_{nd} = \frac{10C - 0.1C}{1 + 2KC} = \frac{9.9C}{1 + 2KC}$$

the flux at low concentrations would be 9.9C, and the current would approach a limiting current, 4.95/K, at high concentrations. The current is half maximal when the concentration equals 1/2K.

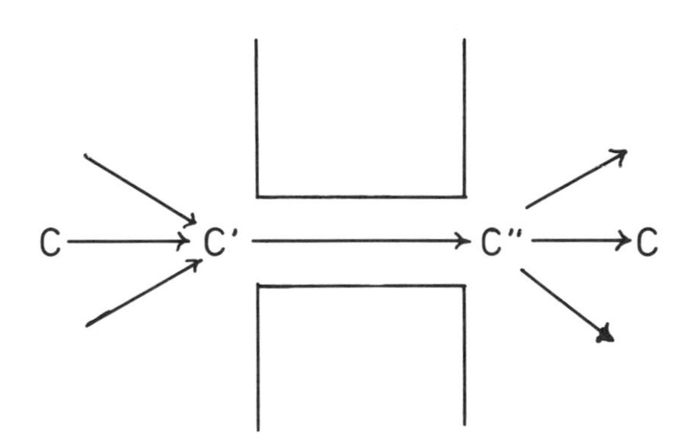

FIGURE 8. The diffusion steps in series with ion transport through a pore.

There is an obvious requirement that diffusion up to and away from the pore must be occurring at the same rate as transfer through the pore. It would thus appear, by analogy with the equations used for capillaries, that we can solve for the fluxes in the presence of diffusion limitations by setting the rates of transfer through the three steps equal to each other,

$$J = C - C' = \frac{10C' - 0.1C''}{1 + K(C'' + C'')} = C'' - C$$

$$= \frac{9.9C}{11.1\ (1 + \dfrac{2KC}{11.1})}$$

These equations predict that diffusion limitation reduces the flux 11.1 times at low concentrations as we know it must. At high concentrations, the equations reflect the view that the pore process saturates while the aqueous diffusion steps do not. Thus at high concentrations diffusion no longer appears important, and the flux must approach the original limiting value. In the equation this occurs since the 11.1's cancel. The 11.1 under the C means that the apparent affinity constant of the ions for the pore has been reduced 11.1 fold! If this result were correct for narrow pores like gramicidin, the failure to consider diffusion limitations explicitly in the flux equations would introduce very severe errors indeed. It is not correct.

The method of calculation used above makes an important unstated assumption. The rate of diffusion up to the pore depends upon the overall average concentration gradient between the mouth of the pore and the bulk of the aqueous phase. Thus in the diffusion equation C' is the average concentration at the mouth. By contrast the rate of entry into the pore depends upon the concentration at the mouth at those times when an ion can enter the pore. Thus for the present example C' in the pore transport equation is the concentration at the mouth averaged over only those times when the pore is empty. The equations used above have assumed that these two averages of the concentration near the mouth are equal. For a narrow pore they are not. To see why they are not we need to consider the size of the region at the mouth of the pore which can be depleted of ions when transport through the pore is rapid.

Enough is known about convergent aqueous diffusion that we can calculate with confidence that almost all the concentration difference which drives diffusion to the mouth of a gramicidin pore occurs within 0.1 to 0.2 nm of the mouth (see Ref. 22). What that means is that the volume of solution which can be depleted of ions by transport through a single pore is only a tiny volume right at the mouth. That volume will usually be empty--it will never contain more than one ion. Furthermore when an ion enters that volume we know that if it makes one more jump, it will have left it again. That means any one ion stays at the mouth for only the time interval between jumps, which is of the order of 10^{-10} second or less. That has two important consequences. Firstly it means that ions which arrive at the mouth of the pore when the pore is already occupied are not available for transport, they will not still be there when the pore becomes empty. Put the other way round, almost all ions which do enter the pore arrive at the mouth while the pore is empty. Secondly it means that when the pore becomes occupied, the concentration just outside the mouth comes to equilibrium with the rest of the aqueous phase in substantially less than 10^{-9} second.

If diffusion is rate limiting, we know, by definition that for a flux from left to right the overall average concentration at the left end of the pore is less than the bulk value. However, by the argument just given, the average concentration calculated only for those times when the pore is occupied is equal to the value in the bulk of the aqueous phase. In order for the overall average to be less than this bulk value, while that for part of the time equals it, the average over the remaining time, when the pore is empty, must be less than the overall average. To a good approximation (22) we can calculate how much less by setting the net rate at which ions diffuse up to the mouth of the pore while the pore is empty equal to the rate of transport through the pore. Thus

$$J = (C - C')X_o = \frac{10C' - 0.1C''}{1 + K(C' + C'')} = (C'' - C)X_o$$

where the probability that the pore is empty is given by

$$X_o = 1/(1 + K(C' + C'')) = 1/(1 + 2KC).$$

Ions approach empty pores at a rate $C'X_o$, but of these $C'X_o$ leave it again where C' is now the conditional average concentration. The flux through the pore is described in the same way as before. These equations now lead to

$$J = \frac{9.9C}{11.1\ (1 + 2KC)}\ .$$

Thus for a narrow pore a diffusion limit appears precisely the same as a reduction in the rates of both entry and exit. It does not change the apparent affinity constant of ions for the pore. In hindsight, this result should have been obvious. The flux equations cannot possibly differ in kind if a rate limiting step occurs one jump inside the mouth of the pore or one jump outside it. Therefore so long as the diffusion limitation occurs close to the mouth of the pore, it must be expressed in the equations in the same manner as a slow internal access step. The analysis of the conductance-concentration relation is therefore unaffected. External access steps do of course figure in the interpretation of the rate constants.

VI. CONCLUSIONS

From the flux data for all of the alkali cations and for thallium, we can reach several conclusions. Firstly the conductances are largely limited by the rates of ion entry into the pore, the rate of entry is not much different when there is an ion at the far end, and the rate of the entry process varies only a little with the applied potential (6,26). The rate of this process is about one-tenth of the rate at which ions can diffuse up to first contact with the pore and roughly half the rate at which their centres could reach the "diffusion" target at the mouth (22,32). Almost certainly the rate limiting step is in fact the dehydration of the ion necessary for it to fit into the pore. Rate limitation by partial dehydration is also fundamental to aqueous diffusion. Thus the limitation of the conductances by the access process explains why the conductances fall into a series which is remarkably like aqueous diffusion even though the ions are in fact binding quite strongly to the pore.

Secondly for all ions investigated the fits yield almost the same value for the rate of transfer of ions from one end to the other. That strongly suggests that this process is limited by something which does not depend on the identity of the particular ion. Both Andersen and Procopio (36) and Dani and Levitt (9) have suggested that the ion movements are limited by the necessity of moving the water out of the way so that the ion can pass through. They also note that the rates of transfer are so fast that it appears almost certain that when an ion moves within the pore, it can force water to move with it (see Ref. 26 for discussion).

The data do not specify the number of binding sites for ions, nor the heights of particular energy barriers. A number of descriptions in terms of site and barrier positions and heights appear possible, but at present there is little evidence to allow the conclusion that any of these reflects the actual state of the pore. My own prejudice is that no Eyring model with fewer sites than there are water molecules can be expected to show any close correspondence to the physical structure of the pore. If indeed such a large number of sites is required, it will be largely a matter of taste whether the pore is described using an Eyring of Nernst-Planck formalism. I have preferred to use neither and to adopt instead a formal description in terms of the distinguishable states of the pore.

REFERENCES

1. Hille, B., and Schwarz, W., J. Gen. Physiol. 72:409 (1978).
2. Hess, P., and Tsien, R. W., Nature 309:453 (1984).
3. Almers, W., and McCleskey, E. W., J. Physiol. (Lond.) 353:585 (1984).
4. Benham, C. D., Bolton, T. B., Lang, R. J., and Takewaki, T., J. Physiol. (Lond.) in press (1985).
5. Hladky, S. B., and Haydon, D. A., Nature 225:451 (1970).
6. Hladky, S. B., and Haydon, D. A., Biochim. Biophys. Acta 274:294 (1972).
7. Hladky, S. B., in "Ion Conduction by Gramicidin A". Ph.D. Thesis, University of Cambridge, 1972.
8. Haydon, D. A., and Hladky, S. B., Q. Rev. Biophys. 5:187 (1972).
9. Dani, J. A., and Levitt, D. G., Biophys. J. 35:485 (1981).
10. Finkelstein, A., and Andersen, O. S., J. Memb. Biol. 59: 155 (1981).
11. Bamburg, E., Alpes, H., Apell, H. J., Bradley, R., Härter, B., Quell, M. J., and Urry, D. W., J. Membrane Biol. 50:257 (1979).
12. Weinstein, S., Wallace, B. A., Morrow, J. S., and Veatch, W. R., J. Mol. Biol. 143:1 (1980).
13. Urry, D. W., Proc. Nat. Acad. Sci. U.S.A. 68:672 (1971).
14. Zwolinski, B. J., Eyring, H., and Reese, C. E., J. Physiol. Colloid Chem. 53:1426 (1949).
15. Urry, D. W., Venkatachalam, C. M., Spisni, A., Läuger, P., and Khaled, M. A., Proc. Nat. Acad. Sci. U.S.A. 77: 2028 (1980).
16. Urry, D. W., Prasad, K. U., and Trapane, T. L., Proc. Nat. Acad. Sci. U.S.A. 79:390 (1982).
17. Levitt, D. G., Biophys. J. 37:575 (1982).
18. Levitt, D. G., Biophys. J. 22:209 (1978).

19. Läuger, P., Stephan, W., and Frehland, E., Biochim. Biophys. Acta 602:167 (1980).
20. Sandblom, J., Eisenman, G., and Hägglund, J., J. Memb. Biol. 71:61 (1983).
21. Hägglund, J., Eisenman, G., and Sandblom, J., Bull. Math. Biophys. 46:41 (1984).
22. Hladky, S. B., Biophys. J. 46:293 (1984).
23. Eisenman, G., Sandblom, J., and Hägglund, J., Biophys. J. 37:253a (1982).
24. Eisenman, G., and Sandblom, J. P., Biophys. J. 45:88 (1984).
25. Hladky, S. B., in "Membranes and Transport." (A. N. Martonosi, ed.), Vol. 2, p. 295. Plenum Press, New York, 1983.
26. Hladky, S. B., and Haydon, D. A., Current Topics in Membranes and Transport 21:327 (1984).
27. Urban, B. W., Hladky, S. B., and Haydon, D. A., Biochim. Biophys. Acta 602:331 (1980).
28. Neher, E., Sandblom, J., and Eisenman, G., J. Memb. Biol. 40:97 (1978).
29. Urban, B. W., and Hladky, S. B., Biochim. Biophys. Acta 554:410 (1979).
30. Walz, D., Bamberg, E., and Läuger, P., Biophys. J. 9:1150 (1969).
31. Andersen, O. S., Biophys. J. 41:135 (1983).
32. Andersen, O. S., Biophys. J. 41:147 (1983).
33. Urban, B. W., in "The Kinetics of Ion Movements in the Gramicidin Channel". Ph.D. Thesis, University of Cambridge, 1978.
34. Ferry, J. D., J. Gen. Physiol. 20:95 (1936).
35. Landis, E. M., and Pappenheimer, J. R., in "Handbook of Physiology Circulation", Vol. 2, p. 961. American Physiological Society, Washington D. C., 1963.
36. Andersen, O. S., and Procopio, J., Acta Physiol. Scand. Suppl. 481:27 (1980).

ON THE MOLECULAR MECHANISM OF ION TRANSPORT THROUGH THE GRAMICIDIN A TRANSMEMBRANE CHANNEL

Dan W. Urry[1]

Laboratory of Molecular Biophysics
The University of Alabama at Birmingham
Birmingham, Alabama

I. INTRODUCTION

Our approach to the mechanism of ion transport through the Gramicidin A Transmembrane channel is predicated upon several axioms: 1) that a detailed knowledge of structure is required and that such can be derived in a membrane without the direct aid of diffraction methods; 2) that under the usual circumstances rate limiting processes for channel facilitated ion movement across the membrane are due to channel rather than extrachannel factors; 3) that a kinetic mechanism requires characterization of elemental steps in the process of ion progression through the channel by physical methods independent of the measurement of current; and finally 4) that a refined understanding is going to be limited by the state of the art in theoretical descriptions of the process and that the theoretical descriptions themselves can achieve validation by their capacity to calculate the fundamental experimentally derived quantities such as the location and energetics of ion binding and the energetics of barriers. The precise location of barriers, for example, would seem to be a refinement only achievable by theoretical calculations.

The initial development of channel structure was by conformational analyses. These conformational analyses, which are described elsewhere (1-5), were three: development from an L-D, D-L β-turn perspective to give the helix of highest

[1]Supported by NIH grant GM-26898

pitch with 4.4 residues per turn (2), development using the concept of cyclic conformations with linear conformational correlates with the enniatins being the cyclic structure and the channel being the linear correlate (3,4), and development from a β-chain structure with the necessary inversion of optical isomers (L→D) at alternating residues (5). The efforts to verify structure have involved both chemical and physical methods. Derivatives such as the N-acetyl desformyl- and the malonyl-bis-desformyl gramicidins of this Laboratory (2,3) followed by the dramatic O- and N-pyromellityl derivatives of the Bamberg and Läuger group (6,7) clearly demonstrated a single-stranded head to head (formyl end to formyl end) dimeric channel. Physical characterization of ion induced carbonyl carbon chemical shifts made possible by selected carbonyl carbon-13 labelling confirmed the single-stranded, head to head nature of the channel (8) and provided convincing evidence for the left-handed helical sense in the bilayer membrane (9-11). Most recently analogs wherein an aliphatic side chain is replaced by a larger (12) or smaller (13) aliphatic side chain to result in channels exhibiting substantially larger currents demonstrate that in the parent molecule, Gramicidin A, even at low ion concentrations (14),

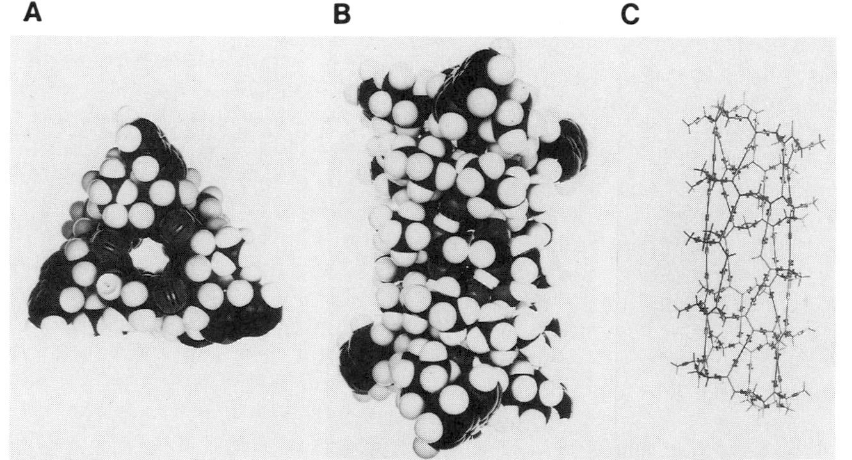

FIGURE 1. Molecular structure of the Gramicidin A transmembrane channel comprised of two molecules associated head to head (formyl end to formyl end) by means of six intermolecular hydrogen bonds. The two formyl hydrogens are seen at the center of the side views in (B) and (C). (A) View showing approximately 4Å diameter channel. Reproduced with permission from reference 5.

channel factors rather than extrachannel factors are rate limiting. The molecular structure of the Gramicidin A (15) transmembrane channel (HCO-L•Val^{1}-Gly^{2}-L•Ala^{3}-D•Leu^{4}-L•Ala^{5}-D•Val^{6}-L•Val^{7}-D•Val^{8}-L•Trp^{9}-D•Leu^{10}-L•Trp^{11}-D•Leu^{12}-L•Trp^{13}-D•Leu^{14}-L•Trp^{15}-$NHCH_2CH_2OH)_2$ is shown in space-filling models in Fig. 1A (channel view) and B (side view) and in wire model Fig. 1C (side view).

Efforts using independent physical means to locate the binding site (8,9) and to estimate binding constants and elemental rate constants have for the most part utilized nuclear magnetic resonance methods assisted in the case of the rate constant for the central barrier by dielectric relaxation methods. This has been done most thoroughly for the sodium ion (16-18) but data has now been achieved for the other alkali metal ions (19-23) and sufficiently so for K^+ and Rb^+ to calculate currents and to calculate conductance ratios with respect to Na^+. Such detailed comparison can provide a stringent test of the NMR approaches. The data for the alkali metal ions will be reviewed in the present report and the resulting information on energy profiles for ion movement through the channel will provide an appropriate test for the theoretical methods currently being applied to the Gramicidin A channel. With regard to theoretical descriptions it should also be mentioned that additional analogs such as those in which the length is varied demonstrate diffusional length not to be the limiting process (24) but rather the barrier heights derivable from the physical studies related here provide the appropriate description.

II. LOCATION OF THE ION BINDING SITES WITHIN THE CHANNEL

By means of Gramicidin A syntheses in which one carbonyl carbon per synthesis was carbon-13 enriched, it has been possible to locate the ion binding sites by observing the ion induced carbonyl carbon chemical shifts. The data are plotted in Figure 2 for the sodium ion and thallium ion (9,25). The data for the odd numbered residues are plotted on the left-hand ordinate and that for D•Val^{8} and D•Leu^{14} are plotted on the right-hand ordinate. The latter data along with the divalent ion induced carbonyl carbon chemical shifts (10,11) are the basis for indicating a left-handed helix. The low chemical shift of the Trp^{15} residue places the binding sites within the channel with the ion being localized between the carbonyl oxygens of residues 11 and 13. Because of the slow reorientation of the channel-membrane complex in the magnetic field, it was necessary to carry out the study at 60-70°C in order to

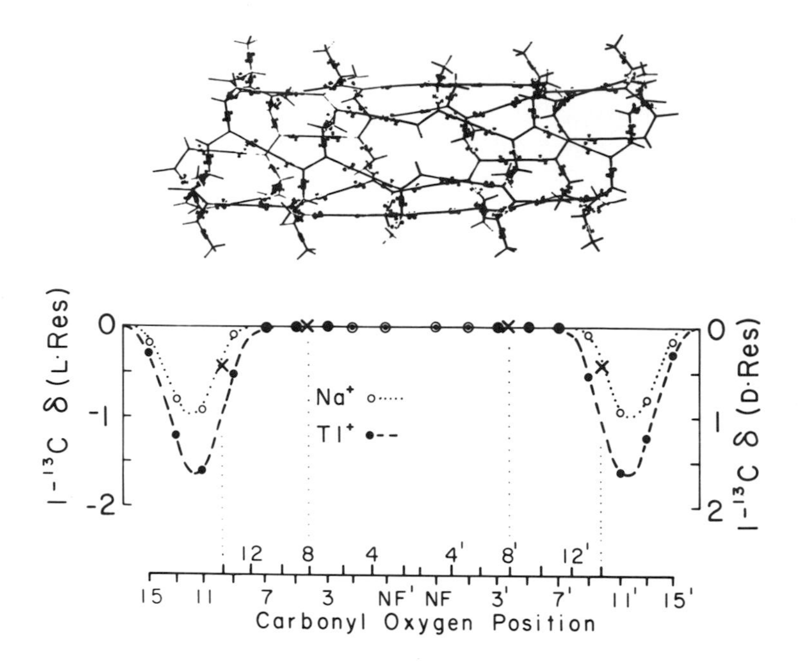

FIGURE 2. Ion induced carbonyl carbon chemical shift plotted for the single-stranded, left-handed β^6-helical conformation with two molecules hydrogen bonded head to head. The chemical shifts indicate two binding sites separated by about 22 Å with the ions localized between the carbonyls of residues 11 and 13. The x's, which are plotted with respect to the right-hand ordinate, are for the $D{\bullet}Val^8$ and $D{\bullet}Leu^{14}$ carbonyls. The absence of a shift for $D{\bullet}Val^8$ but the observation of a shift for $D{\bullet}Leu^{14}$ are consistent with the left-handed helical sense. Adapted with permission from reference 9.

obtain sufficiently narrow carbonyl carbon resonance line widths. This is actually the condition used for incorporating the channels into a lipid bilayer system by heat incubation of Gramicidin A with lysophosphatidyl choline.

III. TWO BINDING PROCESSES SEEN FROM WITHIN THE CHANNEL

It is possible to use ion induced carbonyl carbon chemical shifts to characterize ion binding from within the channel. This is shown in Figure 3A for Tl^+ (20) and in Figure 3B for

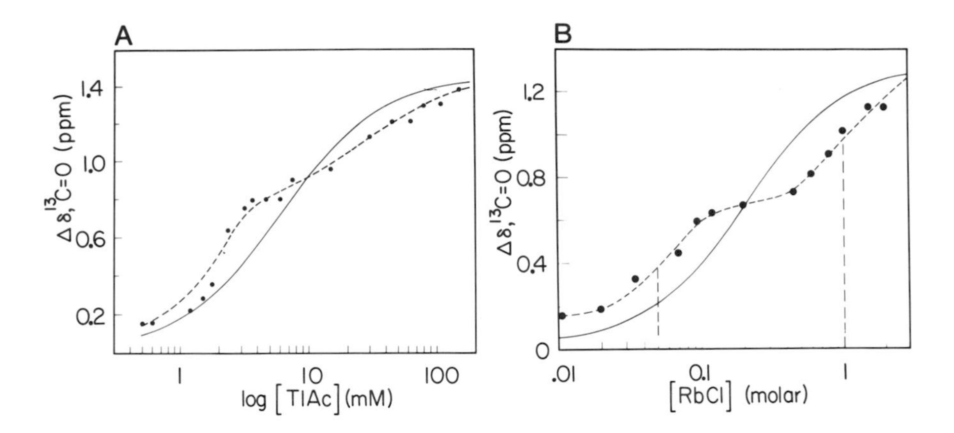

FIGURE 3. Plots of Tl^+ (A) and Rb^+ (B) ion induced carbonyl carbon chemical shifts using carbonyls of either Trp^{11} or Trp^{13}. The data points have a standard deviation of ± 0.06 ppm. In both cases superimposed on the data is the sigmoid binding curve for a single binding process. This data clearly argue for two binding processes within the channel. The first binding process has a tight binding constant, K^t_b, and the second a weak binding constant, K^w_b. A. Adapted with permission from reference 20. B. Submitted for publication (23).

Rb^+ (23). In both cases two binding processes are clearly observed; at lower ion concentration is the binding process for entry of the first ion in the channel and at high concentration is the binding process for entry of the second ion. The magnitudes of the chemical shifts for each process are also indicative of half ion occupancy (one ion in the two site channel) and full occupancy (two ions in the two site channel). Temperature studies of the binding constants and the off-rate constant following the ion rather than the channel show the processes to be well-behaved from 15°C to 75°C (18).

IV. TWO BINDING PROCESSES SEEN FROM ION NMR DATA

The tight binding process has been followed by means of ion NMR both using the concentration dependence of the ion resonance chemical shift and of the longitudinal relaxation times of the quadrupolar alkali metal ions (18-23). The estimates for the tight binding constants at 30°C are: Li^+(14/M);

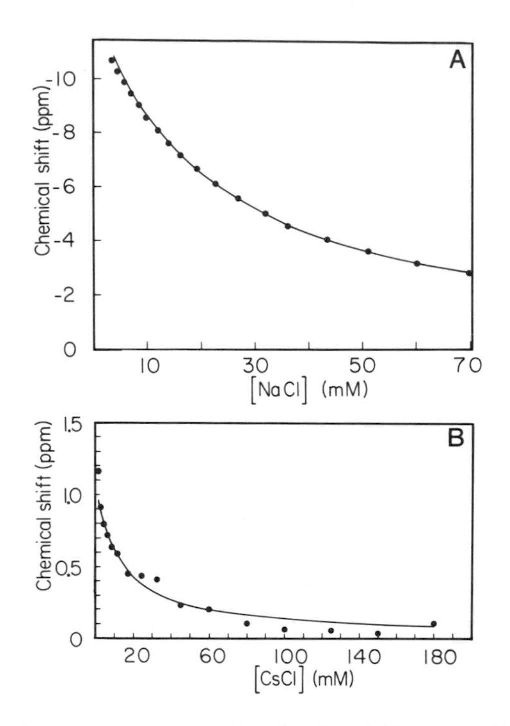

FIGURE 4. Ion resonance chemical shift for Na^{+}(A) and Cs^{+}(B) at low ion concentrations where the binding process for entry of the first in the channel is apparent. The values are $K^{t}_{b}(Na^{+})$ = 30/M and $K^{t}_{b}(Cs^{+})$ = 60/M. The same values (more accurately so for Cs^{+}) are obtained at low ion concentrations from the excess longitudinal relaxation rate studies in Figures 5A and D.

Na^{+}(30/M); K^{+}(> 30/ M); Rb^{+}(>30/M) and Cs^{+}(60/M). Because of the sensitivity problems for potassium-39 and rubidium-87 at low ion concentrations only greater than values of these tight binding constants have been established. Ion resonance chemical shift data in the low ion concentration range are shown in Figure 4A for sodium ion and Figure 4B for cesium ion. The chemical shift data of these figures give a tight binding constant of 30/M for Na^{+} and 60/M for Cs^{+}. As will be seen below in Figures 5A and 5D, the same values are obtained from the low concentration range of the longitudinal relaxation data.

What becomes most important with respect to calculating currents and relative conductances for these ions, however, is an estimate of the magnitudes of the weak binding constant, K^{w}_{b}, and of the off-rate constants from the doubly occupied sites. Both of these quantities are now reasonably well

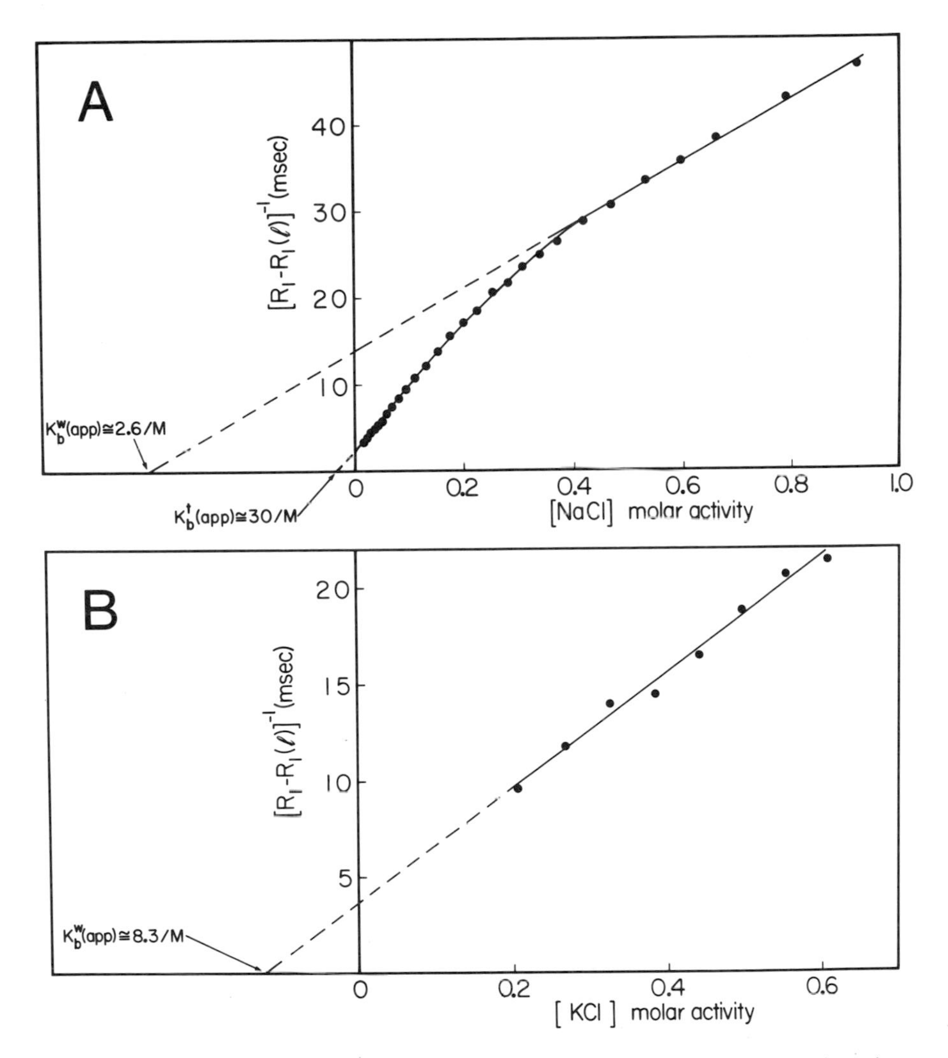

FIGURE 5. A. Excess longitudinal relaxation rate (ELR) plot for Na^+ where the negative x-axis intercept gives the reciprocal of a binding constant. The marked curvature clearly shows a complex process. The intercept from the initial low concentration slope gives an estimate of the tight binding constant ($K^t_b \simeq 30/M$ as also obtained from the low concentration chemical shift data in Figure 4A) and the limiting slope at high ion concentration gives the weak binding constant ($K^w_b = 2.6/M$).

B. An ELR plot for K^+ data has been obtained in the high concentration range. The negative x-axis intercept gives an estimate of 8.3/M for the weak binding constant. Adapted with permission from reference 22.

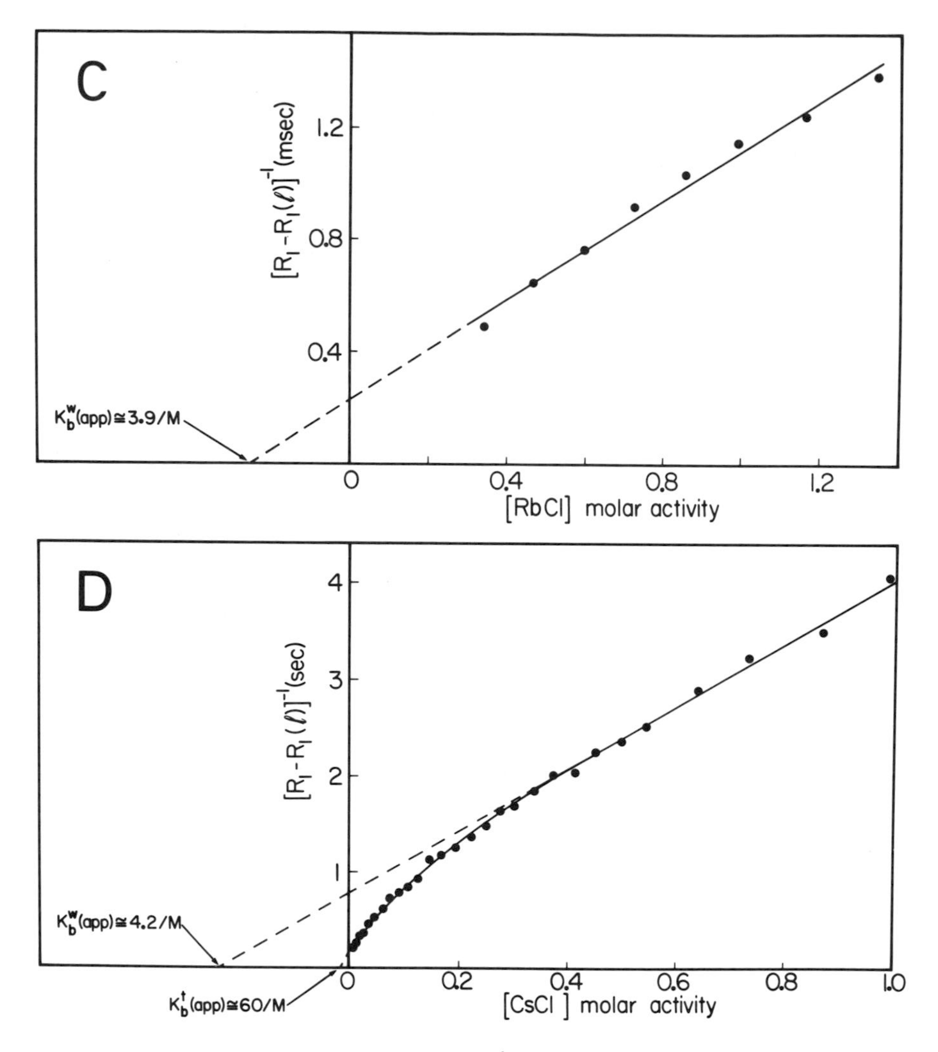

FIGURE 5. C. ELR plot for Rb^+ using the null method to determine T_1' (for the narrow component) of data such as that shown in Figure 9. The narrow component is the one for which the longitudinal relaxation time can be most accurately determined. The negative x-axis intercept gives a weak binding constant of 3.9/M.

D. ELR plot for Cs^+. The dramatic non-linearity requires the consideration of a complex binding process. The initial slope at low concentrations gives an intercept indicating a tight binding constant of 60/M as also estimated from the chemical shift data at low ion concentrations in Figure 4B. The limiting slope at high CsCl concentration gives an intercept indicating a weak binding constant of 4.2/M. Adapted with permission from reference 21.

determined for Na^+, K^+ and Rb^+. Data, from which weak binding constants are estimated, are presented in Figure 5 to similar approximations, that is, high concentration data is used and apparent weak binding constants are obtained from the reciprocal of the negative x-axis intercept. (It should be appreciated that in a complete analysis using also the low concentration data the values could be different because a thorough treatment of multiple binding would necessitate introduction of different relaxation rates for ions under conditions of single and double ion occupancy). The element here to be emphasized is that the data be treated similarly for the set of ions. When this is done, the values for binding of the second ion in the channel, i.e. K^w_b, are Li^+(0.9/M); Na^+(2.6/M); K^+(8.3/M); Rb^+(3.9/M) and Cs^+(4.2/M).

V. OFF-RATE CONSTANTS FOR THE DOUBLY OCCUPIED CHANNEL

Due to the very important contributions of Bull, Forsén and colleagues (26-28), it is possible to utilize transverse and longitudinal relaxation time data on spin 3/2 nuclei to determine ion correlation times. On consideration of the membranous peptide binding site and the temperature dependence of the correlation time, it is possible, for the membranous Gramicidin A channel system under study here, to interpret the ion correlation time as the occupancy time in the channel, the reciprocal of which is the off-rate constant from the doubly occupied channel. In what follows will be given the relevant data for sodium-23, potassium-39 and rubidium-87.

A. The Sodium Ion

In Figure 6 is the resolved resonance for sodium ion at high concentration (1M NaCl) and 30°C in the presence of channels (18). From the line widths at half intensity for the broad, $\nu_{1/2}'$, and the narrow, $\nu_{1/2}''$, components, the transverse relaxation times are obtained; $T_2' = (\pi\nu_{1/2}')^{-1} = 0.98$ msec and $T_2'' = (\pi\nu_{1/2}'')^{-1} = 19.9$ msec. When the occupancy time in the channel binding site, τ_b, is much shorter than the relaxation times at the site and $\tau_b^{-2} >> \Delta\omega^2$ where $\Delta\omega = 2\pi(\nu_W-\nu_f)$ with ν_W being the chemical shift for the ion in the doubly occupied channel and ν_f the chemical shift in the absence of channel, then the following equation due to Bull (26) can be used to estimate the correlation time, τ_c, i.e.

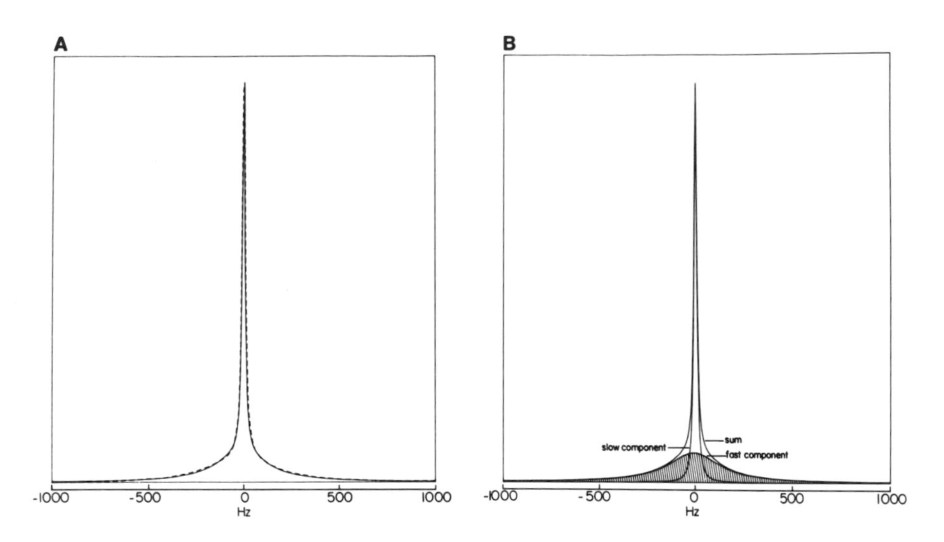

FIGURE 6. A. Sodium-23 NMR resonance for 1 M NaCl at 30°C in the presence of 3 mM Gramicidin A channels; broken line is the observed signal and the solid line is the calculated curve using the sum of two Lorentzians as shown in B. The quantities $\nu_{1/2}'$ and $\nu_{1/2}''$ i.e. the line widths at half intensity for the broad and narrow components, are 325 Hz and 16 Hz respectively to give $T_2' = 0.98$ msec and $T_2'' = 19.9$ msec. Reproduced with permission from reference 18.

$$\frac{1/T_2' - 1/T_{2f}}{1/T_2'' - 1/T_{2f}} \approx \frac{1 + (1 + \omega^2\tau_c^{\,2})^{-1}}{(1 + 4\,\omega^2\tau_c^{\,2})^{-1} + (1 + \omega^2\tau_c^{\,2})^{-1}} \qquad [1]$$

where for sodium-23, $\omega = 2\pi \times 26.3 \times 10^6$. For the above values of T_2' and T_2'' and for a T_{2f} of 31.8 msec obtained in the absence of channels, τ_c calculates to be 4.8×10^{-8} sec.

It is now necessary to analyze the origin of the correlation time for this channel system, that is, to identify the particular source of a fluctuating electric field gradient (EFG) that is experienced by the nucleus which accesses the doubly occupied channel. The expression of interest is the following (28,29)

$$\frac{1}{\tau_c} = \frac{1}{\tau_r} + \frac{1}{\tau_{int}} + \frac{1}{\tau_b} \qquad [2]$$

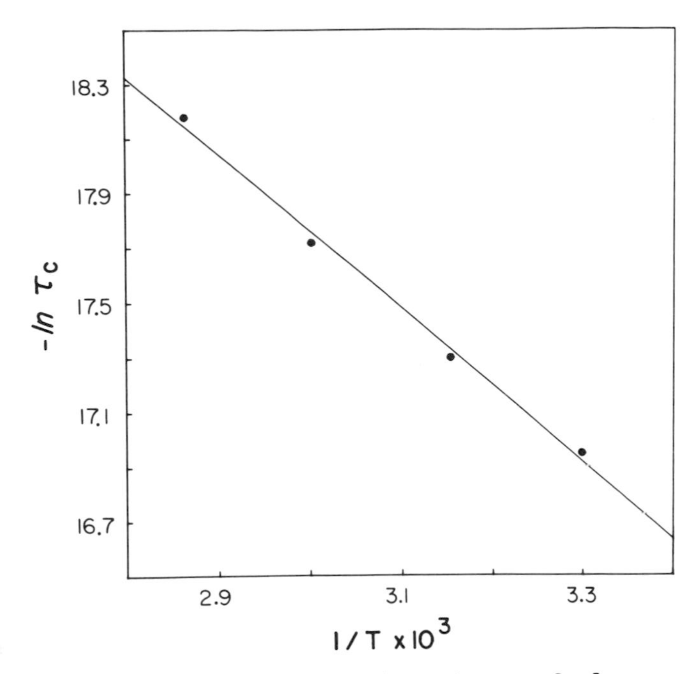

FIGURE 7. The temperature dependence of -log τ_c plotted versus T^{-1}(°K). The reasonably straight line suggests that the relaxation process is relatively simple and is characterized by an enthalpy of activation of 5.9 kcal/mole and an entropy of activation of -5.4 cal/mole-deg. These values are consistent with interpreting τ_c as an ion occupancy time. Adapted with permission from reference 18.

where τ_r, τ_{int} and τ_b are the reorientation correlation time for the site, the correlation time for oscillations of the ligand field, and the occupancy time, respectively. The reorientation of the channel binding site in the magnetic field, called the reorientation correlation time τ_r, could be responsible except that for such a large membranous system this quantity would be long compared to 48 nsec, more like microseconds. This slow reorientation correlation time made it necessary to raise the temperature to 70°C in order to narrow the carbonyl carbon resonance line width sufficiently to be observable. The temperature dependence of -log τ_c, plotted vs T^{-1}(°K), is given in Figure 7 from which the enthalpies and entropies are obtained i.e. $\Delta H^{\ddagger}$ = 5.9 kcal/mole and $\Delta S^{\ddagger}$ = -5.4 cal/mole-deg (18). If this entropy were due to a vibrational process, e.g. τ_{int}, as considered in terms of the entropy of a harmonic oscillator, it would be relevant to a vibrational process with a correlation time of tenths of a psec rather than 48 nsec, such that vibrating peptide carbonyl ligands do not appear responsible for the observed correlation time. Accordingly the magnitude of τ_c would seem to be due to an occupancy time, τ_b. Is this a reasonable result and what occupancy time is relevant? The occupancy time could relate to the time it takes to go from one site to the other within the channel because the electric field gradient experienced by the nucleus in this intrachannel ion translocation would fluctuate, that is, the ligand field is quite different for an ion at the head to head junction than at an ion binding site. This possibility is discounted for two reasons. The observation occurs at high ion concentration when there are two ions in the channel and the jump between sites is not possible, rather than at low concentration where maximal single occupancy would make optimal such a process. Also the rate over the central barrier for Gramicidin A and the malonyl dimer differ by more than an order of magnitude (30,31), yet the τ_c obtained for the two channels are essentially the same, 4.8×10^{-8} sec for Gramicidin A (18) and 4.3×10^{-8} sec for the malonyl dimer (17). Thus the occupancy time is due to the ion residence time in the doubly occupied channel before the ion returns to solution. This is the reciprocal of the off-rate constant, i.e. $k^{w}_{off} \simeq \tau_c^{-1} = 2.1 \times 10^7$/sec. Of particular interest, of course, is that this is just the magnitude for an off-rate constant that is required to give the experimental single channel currents (see below).

B. The Potassium Ion

The data for potassium-39 (22) is given in Figure 8 for 1 M KCl in the presence of 0.3 mM channels at 30°C. In this case with the relatively low observation frequency (4.65×10^6 Hz), two components are not immediately apparent. If the line widths are compared at one-eight height, $\nu_{1/8}$, and one-half height, $\nu_{1/2}$, the ratio $\nu_{1/8}/\nu_{1/2}$ is greater than three whereas for a single Lorentzian the value would be $\sqrt{7}$ or 2.65. Also with the correlation time in the range of ten nsec $\omega^2\tau_c^2$ would approach one. Under such circumstances a ratio involving the transverse and longitudinal relaxation times (26,28,32) can be used to calculate τ_c, i.e.

$$\frac{1/T_2 - 1/T_{2f}}{1/T_1 - 1/T_{1f}} \simeq \frac{0.6 + 0.4\,(1 + 4\omega^2\tau_c^2)^{-1} + (1 + \omega^2\tau_c^2)^{-1}}{1.6\,(1 + 4\omega^2\tau_c^2)^{-1} + 0.4\,(1 + \omega^2\tau_c^2)^{-1}} \quad [3]$$

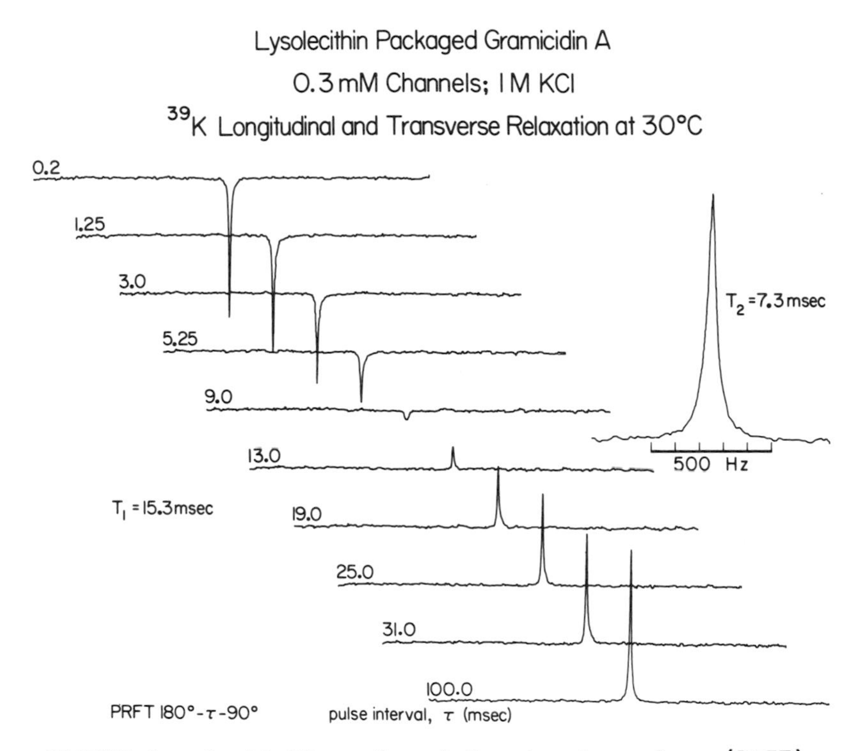

FIGURE 8. Partially relaxed Fourier transform (PRFT) spectra for 1 M KCl at 30°C in the presence of 0.3 mM channels. Given in the inset is the τ = 100 msec spectrum from which the line width at half intensity is used to calculate T_2. The resonance line is seen to change from negative to positive between the pulse intervals of 9 and 13 msec. The exact τ of the null point divided by ln 2 is one way to estimate T_1. The T_1 value given was obtained from the slope of a plot of ln (peak height) versus τ. Reproduced with permission from reference 22.

From Figure 8 T_1 is found to be 15.3 msec, T_2 is 7.3 msec and the values for the absence of channel are $T_{1f} \simeq 55$ msec and $T_{2f} \simeq 50$ msec. When these values are used in Eq 3, the correlation time is found to be 2.45×10^{-8} sec. As seen for the sodium ion, this quantity may be taken as the inverse of the off-rate constant for the doubly occupied channel, i.e. $k^{W}_{off} \simeq 4.1 \times 10^{7}$/sec.

C. The Rubidium Ion

The relaxation data for rubidium-87 is given in Figure 9

for 1.5 M RbCl, 30°C and 3.16 mM channels (23). For this ion a process different from curve resolution is illustrated to estimate the line widths of the narrow and broad components. By selecting the pulse intervals between the 180° and 90° pulses and collecting data for a longer period of time, it is possible to determine the spectrum for which the narrow component is largely nulled out, i.e., near a pulse interval of 0.50 msec. This leaves the broad component from which its line width at half intensity can be estimated. This gives for T_2' ($= 1/\pi\nu_{1/2}'$), a value of 0.10 msec. At a longer pulse

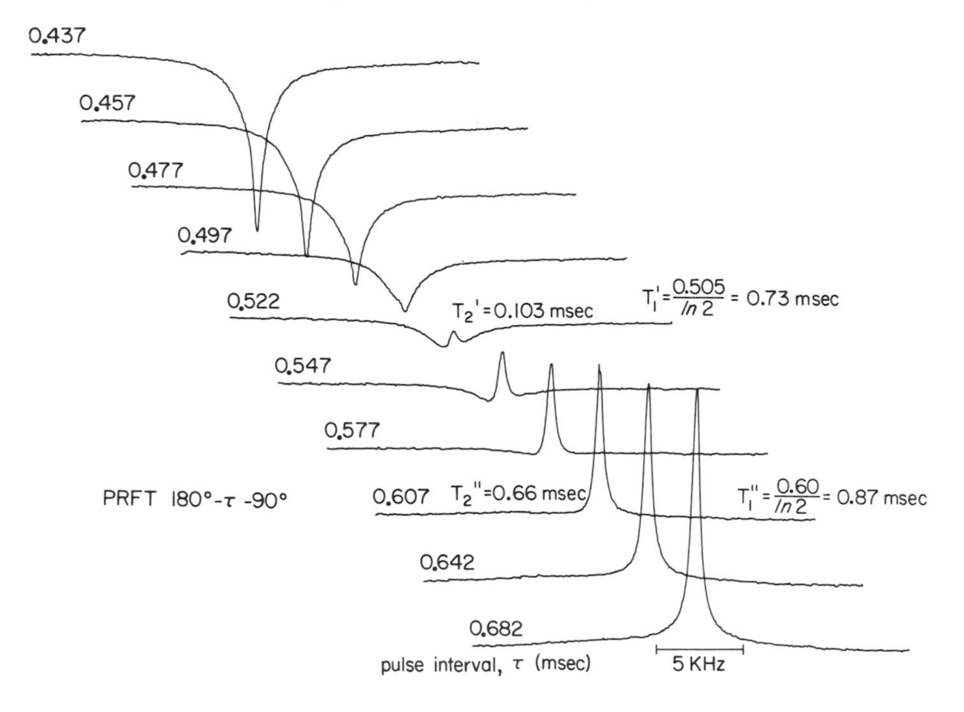

FIGURE 9. Longitudinal relaxation study of 1.5 M RbCl in the presence of channels at 30°C. Two components are apparent as the null region is scanned. When the narrow component is nulled, this leaves the broad component from which T_2' is obtained. Similarly when the broad component is nulled this leaves the narrow component from which T_2'' is obtained. Urry, Trapane, Venkatachalam and Prasad, unpublished data.

interval, i.e. 0.60 msec, the broad component is nulled out and the line width at half height of the remaining narrow component gives the value of $T_2'' = 0.66$ msec. With $\omega = 2\pi \times 32.6 \times 10^6$ and with $T_{2f} \simeq 1.9$ msec from the line width at half intensity for the resonance line obtained on the system without channels, Eq 1 is again used to estimate τ_c. The value is 1.53×10^{-8} sec. As with the sodium and potassium ions, this is interpreted as the inverse of the off-rate constant for the doubly occupied channel, i.e. $k^w_{off} \simeq \tau_c^{-1} = 6.6 \times 10^7$/sec.

VI. APPROXIMATE CALCULATIONS OF SINGLE CHANNEL CURRENTS AND CONDUCTANCE RATIOS

Having obtained estimates for binding constants for the process of going from the singly to the doubly occupied channel, K^w_b, and for the rate at which an ion leaves the doubly occupied channel, k^w_{off}, for the ions--Na^+, K^+ and Rb^+ -- it now becomes of interest to determine if these quantities are relatable to the single channel currents measured in the electrical studies on planar lipid bilayers doped with Gramicidin A channels.

In the initial approximate calculation advantage will be taken of the fact that the singly occupied state is largely "electrically silent" because the off-rate constant for this state is relatively small. For the sodium ion the off-rate constant from the singly occupied state of both Gramicidin A and the malonyl Gramicidin A was approximated from NMR data to be 3×10^5/sec (16,17). In this perspective it is on addition of the second ion to the channel that the off-rate constant from the channel becomes appreciable, almost two orders of magnitude greater, and that a current is measurable. Accordingly, to a simple instructive approximation, it is sufficient to consider explicitly only the probabilities of the singly and doubly occupied states of the channel and not of the empty channel. Subsequently the complete calculation will be used following the approach that has already been carried out for the sodium ion with the Gramicidin A (16) and malonyl Gramicidin A (17) channels. Continuing the approximate calculation, the entrance-exit barrier is taken to be rate limiting on the positive side of the membrane with an applied 100 mV potential; the probabilities of the singly and doubly occupied states are x_s and x_d, respectively, with $x_s + x_d = 1$; the rate constant for forming the doubly occupied state is $k^w_{on} = K^w_b k^w_{off}$; and Eyring rate theory is used to introduce voltage dependence. This gives for the single channel current, i, in ions per second,

$$i = \text{forward rate} - \text{backward rate} \quad [4]$$

$$I = [Me^+]k^W_{on}\chi_s e^{\ell_f zFE/2dRT} - k^W_{off}\,\chi_d e^{-\ell_b zFE/2dRT} \quad [5]$$

where ℓ_f and ℓ_b are the forward and backward distances from site minimum to barrier; z is the charge on the ion; F is the Faraday (23 kcal/mole-volt); E is the applied potential (0.1 volt); 2d is the total distance across the channel from solution site on one side of the membrane to solution site on the other; R is the gas constant (1.987 cal/mole-deg), and T is the absolute temperature. Rewriting Eq. 5 with $\chi_s = (1 + [Me^+]K^W{}_b)^{-1}$ and $\chi_d = K^W{}_b(1 + [Me^+]K^W{}_b)^{-1}$ gives

$$I = \frac{K^W b\ k^W off}{(1 + [Me+]K^W_b)}\left([Me+]\ e^{\ell_f zFE/2dRT} - e^{-\ell_b zFE/2dRT}\right) \quad [6]$$

Of course at 1 molar ion activity, $[Me^+]$ disappears from Eq 6 and with the same sites and voltage dependences, the coefficient $K^W{}_b k^W{}_{off}(1 + K^W{}_b)^{-1}$ gives the relative magnitudes of single channel conductances. These quantities are listed in Table I. As contained in Table I, using the values of $K^W{}_b$ and $k^W{}_{off}$ obtained from the data in Figures 5, 6, 8 and 9 for Na^+, K^+, and Rb^+, the calculated conductances, γ(=zei/V where $e = 1.6 \times 10^{-19}$ coulombs/ion), at V = 0.1 volt are $\gamma(Na^+) = 19$ picosiemens (pS); $\gamma(K^+) = 45$ pS and $\gamma(Rb^+) = 65$ pS. These values are high for phosphatidyl choline membranes but only slightly low for glycerylmonooleate membranes, that is, differences between calculated and experimental values are within the variation observed between membranes. The conductance ratios are $\gamma(K^+)/\gamma(Na^+) = 2.4$ and $\gamma(Rb^+)/\gamma(Na^+) = 3.4$; these values are satisfyingly close to the experimental values of Bamberg et al.(33) of 2 and 3.6, respectively. It might also be noted that the ratios of the weak off-rate constants themselves perhaps do even better than the calculated conductances in giving the values of the experimental conductance ratios and that the ratios also vary somewhat with membrane system (33-36).

TABLE I. Approximation of Conductances and Conductance Ratios Using NMR-Derived Binding and Rate Constants (for 1 molar ion activity)

Ion	K^w_b	k^w_{off}	$\frac{K^w_b k^w_{off}}{(1 + K^w_b)}$	$\gamma(Me+)$	$\gamma(Me+)/\gamma(Na+)$ calc'd	$\gamma(Me+)/\gamma(Na+)$ expt'l
Na^+	2.6/M	2.1×10^7/sec	1.52×10^7/sec	19 pS	1	1
K^+	8.3/M	4.1×10^7/sec	3.66×10^7/sec	45 pS	2.4	2
Rb^+	3.9/M	6.6×10^7/sec	5.25×10^7/sec	65 pS	3.4	3.6

VII. COMPLETE TWO-SITE CALCULATION OF SINGLE CHANNEL CONDUCTANCE USING CONSTANTS ESTIMATED FROM INDEPENDENT PHYSICAL METHODS

At this stage the complete two site channel calculation can be carried out using experimentally derived binding and rate constants as had been done for the sodium ion (17). The values required for each ion are K^t_b, k^t_{off}, K^w_b, k^w_{off} and k_{cb} (the rate constant for the central barrier, i.e. for the ion translocation within the channel). The additional values used are given in Table II. The dependence of the current on K^t_b is minimal; for Na^+ the experimental value is 30/M (18); for K^+ it is taken as 50/M based on the binding constant of 28/M at 70°C (20) and the temperature dependence for this binding constant for the sodium ion (18); for Rb^+ it is taken to be the same as for Cs^+ (21), that is, 60/M. The value of k^t_{off} has been experimentally estimated for the sodium ion (16,17); for the other ions it is taken to differ from that of Na^+ in the same ratio as the k^w_{off} values, i.e. $k^t_{off}(K^+) = (4.1/2.1)\ 3 \times 10^5$/sec and $k^t_{off}(Rb^+) = (6.6/2.1)\ 3 \times 10^5$/sec. The values for K^w_b and k^w_{off} are as experimentally obtained for each ion and are as listed in Table I. The rate over the central barrier has been estimated for the Gramicidin channel to be greater than 5×10^7/sec for the thallium ion (Henze and Urry, unpublished data). A value for Tl^+ of 4×10^6/sec was found from dielectric relaxation studies on the malonyl Gramicidin A channel (30). This experimental value was close to the value of 3.2×10^6/sec which was the "best fit" value when calculating the sodium ion single channel currents with k_{cb} as the only variable (17). Therefore, the values of k_{cb} appear to be similar for Na^+ and Tl^+. Consistent with this,

TABLE II. Complete Calculation of Single Channel Conductances (with additional required experimentally estimated constants)

Ion	K^t_b	k^t_{off}	k_{cb}	$\gamma(Me+)$ calc'd
Na^+	30/M	$3x10^5$/sec	~$5x10^7$/sec	21 pS
K^+	~50/M	~$5.86x10^5$/sec	~$5x10^7$/sec	30 pS
Rb^+	~60/M	~$9.43x10^5$/sec	~$5x10^7$/sec	46 pS

it is reasoned that the process of binding in the channel is a normalizing process such that each ion is expected to see a similar central barrier. Accordingly a value of 5 x 10^7/sec is used for each of the ions. With these values, the calculated conductances for one molar ion activity are 21 pS for Na^+, 30 pS for K^+ and 46 pS for Rb^+. Experimentally for diphytanoyl phosphatidyl choline membranes at 0.6 molar ion activity potassium ion and 100 mV, the single channel conductance is 25 pS for the mean of the most probable single channel conductance (37). Also for diphytanoyl phosphatidyl choline membranes with 1 M CsCl the single channel conductance is reported to be 48 pS (21) and since the conductance ratio $\gamma(Rb^+)/\gamma(C^+_s)$ is reported to be near one (33-36), a similar value is expected for Rb^+. Thus the values calculated from NMR data for Gramicidin A channels in lipid bilayers formed with lysophosphatidyl choline are remarkably close to the experimental values. This gives substantial credence for the use of NMR to obtain binding and rate constants relevant to channel transport.

VIII. SUMMARIZING COMMENTS

The work reviewed here shows the Gramicidin A transmembrane channel to contain within it two binding sites separated by about 22 Å. When monitoring the binding sites within the channel, two binding processes are observed, a single ion entering the channel at low concentration with rapid exchange of the ion between the two sites and a second ion entering the channel at higher concentration and with the

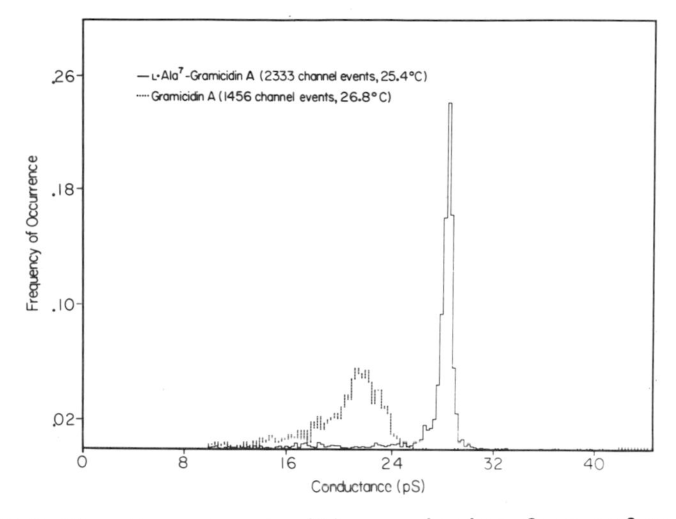

FIGURE 10. Reproduced with permission from reference 13.

two binding constants differing in magnitude by more than a factor of ten. The two binding processes have also been observed by means of the ion resonance chemical shift to characterize the initial ion entry, and by means of longitudinal relaxation time studies which can characterize both the first and second ion binding. The two binding processes have now been observed for Li^+, Na^+, Rb^+, Cs^+ and Tl^+. Transverse relaxation time studies on the quadrupolar ions have allowed relatively accurate measurement of the off-rate constants for the doubly occupied channel for Na^+, K^+ and Rb^+.

The above studies carried out on suspensions of channels in lipid bilayer membranes formed by the heat incubation of Gramicidin A with lysophosphatidyl choline provided binding and rate constants capable of calculating the single channel conductances and conductance ratios exhibited by Na^+, K^+ and Rb^+ in electrical measurements made on planar lipid bilayer membranes. These studies argue that the dominant conducting state is the doubly occupied channel and that the rate limiting process for permeation is the entrance-exit barrier at the mouth of the channel and not extrachannel factors.

These conclusions are supported by conductance studies on analogs of Gramicidin A. In Figure 10 is a conductance histogram comparing, under identical conditions, the distribution of conductance states of Gramicidin A and the L•Ala^7-GA analog, where it is seen that the most probable conductance state is at a substantially higher conductance value. This means that under these conditions accessing the channel cannot be rate limiting for Gramicidin A. A property of the channel itself is rate limiting. This is also the case for lower concentrations, e.g. 200 mM CsCl (Sigworth and Urry, unpublished data) and at 50 mM CsCl (14). Also consistent with the rate limiting process being described by an Eyring-type barrier rather than the view of a diffusional length limitation is demonstrated by des $L \cdot Val^7$-$D \cdot Val^8$-Gramicidin A in which the channel is shortened by 3 Å. If diffusional length were the process the conductance of this shortened channel at 100 mV would be expected to increase by 30 to 40%. Instead, the conductance of the most probable states decrease by some 40% (24). Accordingly the issue of concern is not the length that an ion must transverse but rather the height of the energy barrier. As shown by the above data, that barrier is the entrance-exit barrier at the mouth of the channel and, as is consistent with the decreased conductance of des $L \cdot Val^7$-$D \cdot Val^8$-GA, the doubly occupied channel is the state that contributes most significantly to the conductance.

With the above demonstration of the relevance of the NMR data to the conductance process, this data now provides the basis against which theoretical calculations may be considered, calculations which are essential for the refinement of our understanding of the details of channel transport. It is important to note in this regard that the calculations of A. Pullman and colleagues (38-40) have identified an energy minimum at the location of the experimentally defined binding site.

REFERENCES

1. Urry, D. W., Biochim. Biophys. Acta Reviews on Biomembranes, in preparation.
2. Urry, D. W., Proc. Natl. Acad. Sci. USA 68:672 (1971).
3. Urry, D. W., Goodall, M. C., Glickson, J. D. and Mayers, D. F., Proc. Natl. Acad. Sci. USA, 68:1907 (1971).
4. Urry, D. W., Proc. Natl. Acad. Sci. USA 69:1610 (1972).
5. Urry, D. W., Long, M. M., Jacobs, M., and Harris, R. D., Ann. NY Acad. Sci., 264:203 (1975).
6. Apell, H. J., Bamberg, E., Alpes, H., and Läuger, P., J. Membr. Biol. 31:171 (1977).
7. Bamberg, E., and Janko, K., Biochim. Biophys. Acta 465:486 (1977).
8. Urry, D. W., Trapane, T. L., and Prasad, K. U., Science 221:1064 (1983).
9. Urry, D. W., Walker, J. T., and Trapane, T. L., J. Membr. Biol. 69:225 (1982).
10. Urry, D. W., Trapane, T. L., Walker, J. T., and Prasad, K. U., J. Biol. Chem. 257:6659 (1982).

11. Urry, D. W., Trapane, T. L.,and Prasad, K. U., Int. J. Quantum Chem.:Quantum Biology Symp. 9:31 (1982).
12. Urry, D. W., Alonso-Romanowski, S., Venkatachalam, C. M., Trapane, T. L., and Prasad, K. U., Biophys. J. 46:259 (1984).
13. Prasad, K. U., Alonso-Romanowski, S., Venkatachalam, C. M., Trapane, T. L., and Urry, D. W., Biochemistry (in press).
14. Sigworth, F., unpublished results.
15. Sarges, R.,and Witkop, B., J. Am. Chem. Soc. 86:1862 (1964).
16. Urry, D. W., Venkatachalam, C. M., Spisni, A. Läuger, P., and Khaled, M. A., Proc. Natl. Acad. Sci. USA 77:2028 (1980).
17. Urry, D. W., Venkatachalam, C. M., Spisni, A., Bradley, R. J., Trapane, T. L., and Prasad, K. U., J. Membr. Biol. 55:29 (1980).
18. Urry, D. W., Trapane, T. L., Venkatachalam, C. M., and Prasad, K. U., Int. J. Quantum Chem.:Quantum Biology Symp. (in press).
19. Urry, D. W., Trapane, T. L., Venkatachalam, C. M., and Prasad, K. U., J. Phys. Chem. 87:2918 (1983).
20. Urry, D. W., Trapane, T. L., Venkatachalam, C. M., and Prasad, K. U., Canadian J. Chem. 63:1976 (1985).
21. Urry, D. W., Trapane, T. L., Brown, R. A., Venkatachalam, C. M., and Prasad, K. U., J. Magn. Reson. (in press).
22. Urry, D. W., Trapane, T. L.,and Venkatachalam, C. M., J. Memb. Biol. (in press).
23. Urry, D. W., Trapane, T. L., Venkatachalam, C. M., and Prasad, K. U., J. Am. Chem. Soc. (submitted).
24. Urry, D. W., Alonso-Romanowski, S., Venkatachalam, C. M., Trapane, T. L., Harris, R. D., and Prasad, K. U., Biochim. Biophys. Acta 775:115 (1984).
25. Urry, D. W., Prasad, K. U. and Trapane, T. L., Proc. Natl. Acad. Sci. USA 79:390 (1982).
26. Bull, T. E., J. Magn. Reson. 8:344 (1972).
27. Bull, T. E., Forsén, S.,and Turner, D. L., J. Chem. Phys. 70:3106 (1979).
28. Forsén, S., and Lindman, B., in "Methods of Biochemical Analysis" (D. Glick, ed.), p. 289. John Wiley & Sons, New York, 1981.
29. Marshall, A. G., J. Chem. Phys. 52:2527 (1970).
30. Henze, R., Neher, E., Trapane, T. L., and Urry, D. W., J. Membr. Biol. 64:233 (1982).
31. Henze, R.,and Urry, D. W., unpublished data.
32. Rose, K.,and Bryant, R. G., J. Magn. Reson. 31:41 (1978).
33. Bamberg, E., Noda, K., Gross, E., and Läuger, P., Biochim. Biophys. Acta 419:223 (1976).
34. Eisenman, G., Sandblom, J.,and Neher, E., Biophys. J.

22:307 (1978).
35. Hladky, S. B.,and Haydon, D. A., Biochim. Biophys. Acta 274:294 (1972).
36. Hladky, S. B.,and Haydon, D. A., Current Topics in Membranes and Transport 21:327 (1984).
37. Urry, D. W., Alonso-Romanowski, S., Venkatachalam, C. M., Bradley, R. J., and Harris, R. D., J. Memb. Biol. 81:205 (1984).
38. Etchebest, C.,and Pullman, A., FEBS Letters 170:191 (1984).
39. Etchebest, C., Pullman, A.,and Ranganathan, S., Biochim. Biophys. Acta, 818:23 (1985).
40. Pullman, A., this Symposium

THE STRUCTURE OF GRAMICIDIN, A TRANSMEMBRANE ION CHANNEL

B. A. Wallace

Department of Biochemistry & Molecular Biophysics
Columbia University
New York, New York, USA

The biological activities of membrane channels depend on the conformations these molecules adopt in the hydrophobic environments of lipid bilayers. Just as detailed structural studies of soluble proteins have provided insight into their modes of action, high-resolution information concerning the structure of proteins which are embedded in membranes may elucidate their function as channels. Until recently, there has been little molecular information available on membrane proteins because their hydrophobic nature has made it difficult to obtain sequence information, to purify them in large quantities, and to reconstitute them into forms which are both active and suitable for physical analyses. With the advent of molecular cloning techniques, a number of channel sequences have now been determined, but little or no three-dimensional structural information exists for these molecules. Hydropathy algorithms have been used to model the dispositions of membrane proteins relative to the bilayer (1), and their secondary structures have been predicted using empirical methods derived for soluble proteins (2). However, there is as yet no evidence that the former will actually predict the correct disposition, while there is a strong indication that the latter procedure does not give appropriate results for hydrophobic membrane proteins (3). This is not an unexpected result, given the functional asymmetries of membrane proteins, and the different propensities for polypeptide backbones to form intramolecular hydrogen bonds in the hydrophobic environment of a lipid bilayer as compared to aqueous solution (4). As a consequence, empirical predictions based on soluble protein data bases are unlikely to produce appropriate results for membrane channels. For such calculations to be meaningful, data bases derived from membrane proteins are needed.

Until now no membrane channel structure has been determined at the atomic level. Therefore, high-resolution information on any membrane channel would vastly increase our knowledge of this class of protein; such information will be especially valuable if the biochemical and physiological properties of the molecule are known and this information can be related to the structure. For this reason, we have chosen to study the physiologically well-characterized (5) ion channel formed by gramicidin A.

Gramicidin A is a hydrophobic linear polypeptide antibiotic which has the sequence (6):

HCO-L-val-gly-L-ala-D-leu-L-ala-D-val-L-val-L-trp-
D-leu-L-trp-D-leu-L-trp-D-leu-L-trp-$NHCH_2CH_2OH$.

It is extremely hydrophobic since it contains no charged or polar residues and both its N- and C-termini are blocked. As a result, gramicidin is insoluble in water, but very soluble in lipid bilayers and in a wide range of organic solvents. Its high partition coefficient in membranes over water means that virtually all of the polypeptide is associated with bilayers, even in very dilute solutions.

In membranes, gramicidin forms channels that are specific for monovalent cations (7). Fluorescence and conductance measurements have demonstrated that the active conducting form of the molecule is a dimer (8,9). The structure of that

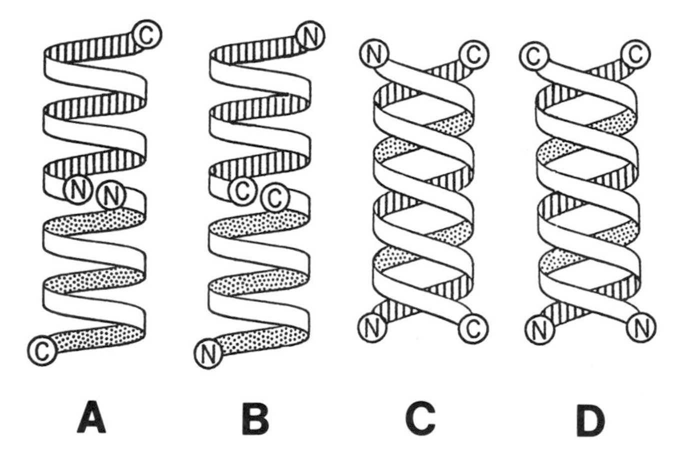

FIGURE 1. Schematic diagrams of models for the gramicidin A dimer. A, N-to-N helical dimer; B, C-to-C helical dimer; C, antiparallel double helix; D, parallel double helix.

dimer has been the subject of investigation in a number of laboratories.

Urry first proposed (10) a π(L,D) helical conformation for the channel in which two helical monomers were associated in the membrane via their N-termini (Fig. 1A). This model was based, in part, on circular dichroism (CD) and nuclear magnetic resonance (NMR) studies of the molecule in organic solvents. Conductance studies using chemically modified gramicidins in black lipid films formed from monoolein and decane (11,12) supported this model, and, to some extent, a similar helical dimer with, instead, the C-termini associated (13) (Fig. 1B). Based primarily on IR and CD measurements of gramicidin in alcohol and dioxane solutions, Veatch et al. (14) proposed an alternate family of models (Fig. 1C, D), the parallel and anti-parallel intertwined double helices. However, in none of the studies which gave rise to these models was the gramicidin in phospholipid bilayers, so it was not clear which structure was relevant for membranes.

While the molecular folds of the helical dimer and double helical models are completely different, they have many similar physical characteristics which make distinguishing between them difficult (Fig. 2). The lengths (∿30 Å) and outer (∿15 Å) and inner channel diameters (∿4 Å) of the two folding motifs are nearly indistinguishable. Although the pitch of the double helix is twice that of the helical dimer, the chain separation per turn along the helix axis and the hydrogen bonding patterns are similar. However, in the double helix, adjacent hydrogen bonds are intermolecular, while most are intramolecular in the helical dimer. One striking difference between the two models, however, is the location of their N- and C-termini relative to the bilayer surface. Model A has the C-termini exposed at the surface and the N-termini buried deep in the membrane. In contrast, model B has the C-termini buried and the N-termini exposed. Models C and D have both the N- and C-termini at the membrane surface. Hence, a method which differentially probes locations normal to the bilayer surface could be used to identify the gramicidin structure present in membranes. ^{13}C-and ^{19}F-NMR studies of gramicidin in phospholipid vesicles were able to make those distinctions (15,16). The C-termini were shown to be accessible to membrane-impermeable paramagnetic ions (Tm^{+3} and Mn^{+2}) which were located in the aqueous space surrounding the bilayer and not transported by the channel, while the N-termini were inaccessible to these probes, a result consistent with model A. However, the absence of effect for the N-terminus was inconclusive since it could have also been the result of steric hindrance to accessibility, rather localization at a distance from the membrane surface. To distinguish between these possibilities,

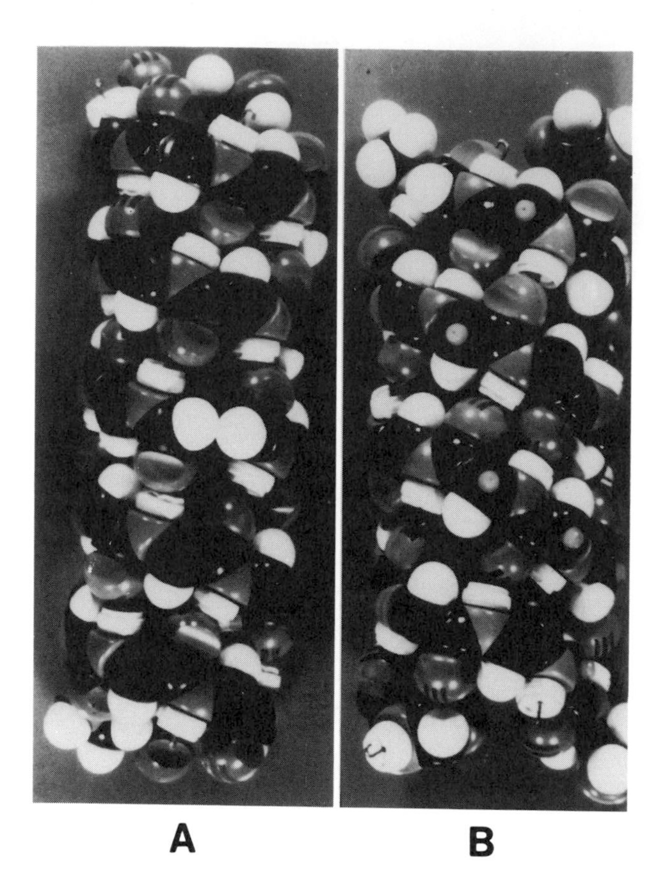

FIGURE 2. Space-filling models of the N-to-N helical dimer (A) and antiparallel double helix (B). The side chains have not been included so differences in the backbone folds can be more easily seen.

another paramagnetic probe, located on the distal end of a phospholipid molecule, was used to determine the location of sites at the bilayer center. This probe had the opposite effect of the surface probes: the N-termini, but not the C-termini of gramicidin, were affected, demonstrating that the amino-terminal ends of the polypeptide are indeed buried deep in the lipid bilayer, while the carboxyl termini are surface exposed, consistent only with model A.

The question then arises as to what type of structure exists in organic solvents, the environment for which the double helical models had been proposed. That is, are the structures adopted in isotropic, hydrophilic organic solvents and in anisotropic, hydrophobic lipid bilayer environments different? Previous studies in organic solvents and membranes were difficult to compare not only because the solvent environments were different, but also different methods of detection were used. A possible reason for the different models could have been in the interpretation of the data, rather than in actual conformational differences of the molecule. Therefore, it was important to do parallel experiments with gramicidin in the two environments using a single physical technique. CD spectroscopy was the method that permitted direct comparison between molecules in small unilamellar phospholipid vesicles and in organic solution (17). Veatch <u>et al</u>. (14) had shown that several different interconvertible species of gramicidin are present in alcohol solutions and that each has a unique spectrum. The net spectrum of an equilibrium mixture of these species in methanol is presented in Fig. 3. Neither the net spectrum nor the individual spectra correspond to spectra typical for α-helical or β-sheet type secondary structures. This is because the alternating L- and D-amino acids in the gramicidin sequence produce different ϕ, ψ angles for the polypeptide backbone, resulting in altered peptide transitions. Also, the high tryptophan content (4 out of 15 amino acids) means that there are large signals from the $\pi \rightarrow \pi^*$ transitions of the aromatic rings at wavelengths normally dominated by the $n \rightarrow \pi^*$ and $\pi \rightarrow \pi^*$ transitions of the peptide backbone. As a result, unique structural assignments have not been made based on these data. Comparisons between spectra of the molecules in two environ-

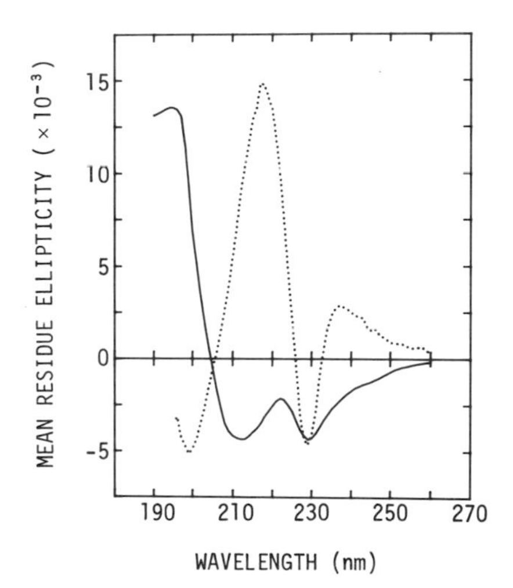

FIGURE 3. Net circular dichroism spectra of gramicidin A in methanol (——) and in phospholipid vesicles (•••).

ments were, however, possible and permitted distinction between folding motifs. The spectrum of gramicidin in lipid vesicles (Fig. 3) is very different from the spectra in organic solvents and cannot be represented as a linear combination of any of the individual solution spectra (17). The far UV region of the spectrum of gramicidin in vesicles has a rather unusual shape which suggests exciton coupling between tryptophans which are close in space and in different locations relative to each other other than in the organic solvent structure (for which no such exciton interactions are seen). The structures gramicidin adopts in membranes and organic solvents were, thus, shown to be very different. Such studies suggest that both of the general folding motifs (helical dimer and double helical) originally proposed by Urry (10) and Veatch _et al_. (14) might be found, depending on the solvent environment. To distinguish between the structures found in organic solvents and in membranes, they will hereinafter be referred to as "pore" and "channel" forms, respectively.

Conductance measurements have shown that gramicidin exhibits single file conductance and has two ion binding sites per channel. The relative locations of those sites have been suggested by channel profile and chemical modification studies (18) and by NMR studies in detergent solutions (19). The latter provide the more direct evidence, although the altered binding properties of ions in the detergent-bound form of the molecule (20) and the altered gramicidin peptide backbone conformation in the lysolecithin detergent relative to that in membranes (21) make the relevance of this data to the native channel structure unclear. The pore structure also contains two cation sites, the precise locations of which are seen in the crystal structure of this form (22). These are at different locations relative to each other and to the ends of the structure than the sites proposed for the channel. Solvent and ion sites are also organized in a single file in the center of the pore.

CD spectroscopy has been used (17,23,24) to examine the effect of binding monovalent cations to the pore and channel structures. A large change occurs when the pore binds cations. Not only is the magnitude of the spectrum (Fig. 4) increased two-fold when saturating amounts of cesium are added (a consequence of a change in pitch of the helix), but also the sign of the curve (and hence the hand of the helix) is reversed (23). The spectrum of gramicidin with bound cesium is the same as that obtained for dissolved crystals of a gramicidin/cesium complex. The calculated decrease in helical pitch upon binding cesium which is derived from these CD spectra corresponds well with the differences in helical repeat and subcell dimensions of the molecule in crystals

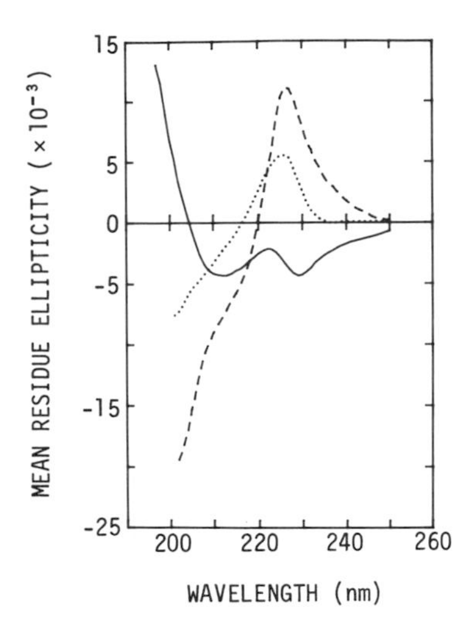

FIGURE 4. Circular dichroism spectra of gramicidin A in methanol in the absence of ions (——) and in the presence of saturating concentrations of Cs^+ (---) and Li^+ (•••).

formed with and without cesium ions (25-27). The spectrum obtained using saturating amounts of the smaller ion lithium is also opposite in sign when compared with the spectrum for the polypeptide without ions (Fig. 4), but is smaller in magnitude than that for the cesium form, indicating a larger pitch and suggesting that the pore containing the lithium ions is not as large. These results suggest an interesting possible mechanism for ion binding in which, upon binding ions, the molecules foreshortens and widens and, upon releasing the ions, it extends to the longer, narrower pore. One could envision this as a gating mechanism. However, no such changes are observed in the spectrum of the channel form upon addition of saturating concentrations (7) of ions (Fig. 5), suggesting that no change in pitch or in overall secondary structure occurs when the membrane channels bind ions. This is reasonable because the native channel should be sufficiently large to accommodate cesium ions. The estimated size of the uncomplexed pore form, as derived from crystallographic parameters (27), might have suggested that in this form, too, the central core would be large enough to accommodate the cesium ions without changing pitch. The reason it widens upon complexation, however, may be because the pore also contains solvent (methanol) and counterions (chloride), as seen from the crystal structure of a gramicidin/cesium complex (22,28). The differential effects of ion binding in the pore and channel forms are another demonstration that the conformations differ in membranes and in organic solvents (24), and provide information on a functionally-related aspect of the structure. Finally, a significant change has been observed in the CD spectrum of detergent-bound gramicidin (29). Although that change may be primarily a consequence of light scattering due to particle aggregation in high salt, this difference in behavior between gramicidin in lysolecithin and in membranes may be another indication that the detergent and channel forms of the molecule also are not equivalent.

To examine physiologically-relevant structural features of the channel, peptide/lipid interactions have been studied. First, the bilayer thickness was altered. Two possible results were envisioned for insertion of short (∿30 Å) gramicidin molecule into thick (ie., ∿45 Å) membranes: 1) puckering in of the lipid molecules to accommodate the invariant gramicidin structure (Fig. 6A) (30) or 2) elongation of the gramicidin by refolding to accommodate the invariant lipid molecules (Fig. 6B). These possibilities could be distinguished by CD measurements. The first case was expected to result in no difference between the spectra in thick and thin membranes, while the second would result in a spectrum with the same shape but a decreased magnitude, corresponding to an

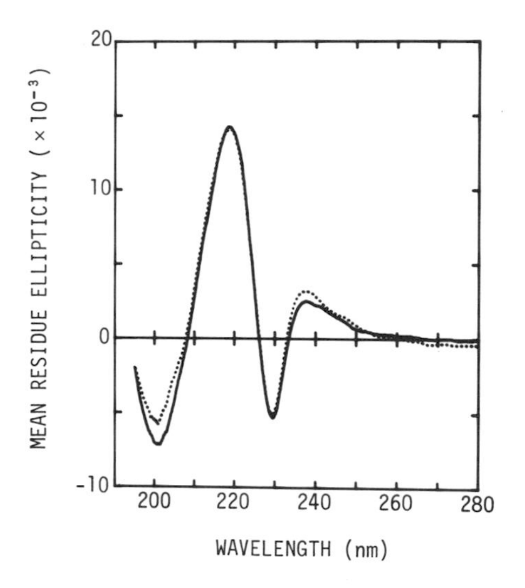

FIGURE 5. Circular dichroism spectra of gramicidin A in phospholipid vesicles in the absence of ions (——) and in the presence of 2 M CsCl (•••).

increased helical pitch in the thicker membranes. When the spectra of gramicidin in thin (dilauryl phosphatidylcholine) and thick (distearoyl phosphatidylcholine) membranes were compared, neither of these results were observed. Instead, a completely different spectrum was obtained (17) which corresponded to the spectrum of an inactive monomer, as confirmed by comparison with chemically modified molecules which could no longer form dimers (17). This result suggested that the consequence of thicker membranes is dissociation of dimers so they no longer form conducting channels (Fig. 6C). This conclusion is in accord with conductance measurements (31) which indicated that the gramicidin mean channel lifetime is significantly decreased in thicker membranes. Hence, an observed functional property could be directly correlated with a structural feature of the channel.

The effects of varying peptide/lipid rations were also examined. Very similar CD spectra were obtained for a wide range of protein-to-lipid ratios (1:15 to 1:363) (Fig. 6), if samples were examined under conditions which minimized differential light scattering. This result is in contrast to that reported by Ovchinnikov and Ivanov (32), who suggested that the very different spectra they obtained at low and high peptide ratios corresponded to completely different structures. However, the spectrum they reported for the low peptide ratio sample appears to be dominated by the transitions due to the lipid component (Fig. 7) rather than being due to a different gramicidin conformation. It is similar to the spectrum we obtained for the low ratio sample prior to subtraction of the lipid baseline. When corrected for this, the spectrum is very nearly the same as that of the higher grami-

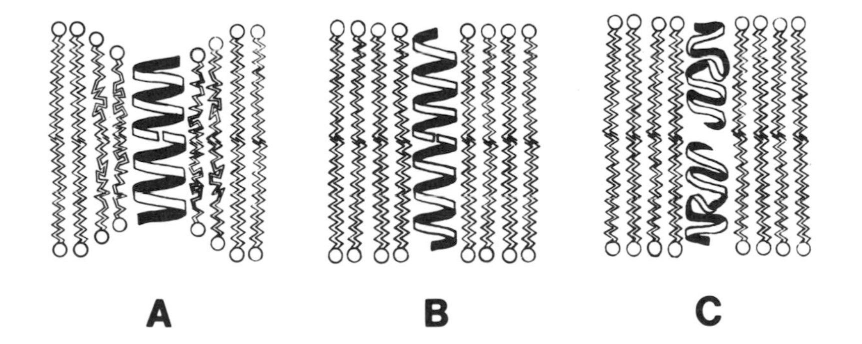

FIGURE 6. Schematic diagram showing possible consequences of insertion of gramicidin into thick bilayers. A, lipid puckering; B, gramicidin elongation; C, gramicidin dissociation. The experimental results indicate dissociation is the likely mechanism.

cidin ratio samples. Hence, there does not seem to be a dependence of the structure in membranes on the gramicidin-to-lipid ratio, and the structures inferred from NMR studies (15) using the high ratio samples must also reflect the membrane conformation of gramicidin at the lower ratios more near to those used in conductance measurements.

These studies demonstrate that spectroscopy can provide useful information on structural and functional features of the gramicidin molecule. Ultimately, however, one would like detailed molecular information on the polypeptide backbone, side chains, and binding sites, which the spectroscopic studies have not provided. The most suitable methods for determining these features at high resolution is X-ray diffraction of well-ordered single crystals of the molecule. Crystals of gramicidin were first prepared 37 years ago (33). Despite work by a number of groups (14,25,26,34-37), they remained intractable to solution. This may be because gramicidin falls in the difficult intermediate size range for crystallographic studies: too large for direct methods traditionally used on small molecules and rather small to produce heavy atom derivatives suitable for the multiple isomorphous replacement methods of macromolecular crystallography. Forming isomorphous heavy atom derivatives has been especially difficult since this molecule has only hydrophobic side

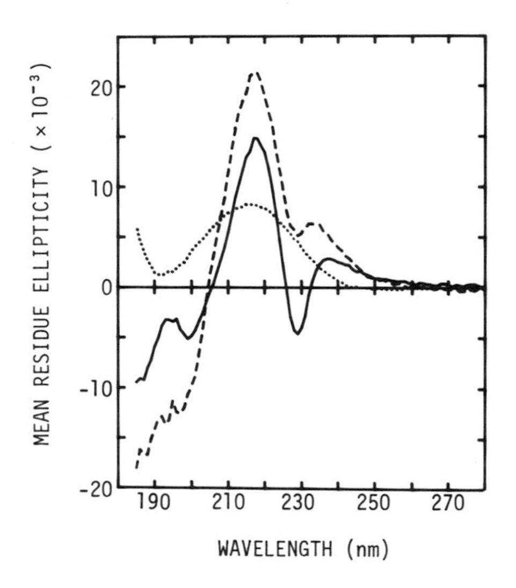

FIGURE 7. Circular dichroism spectra of gramicidin A in phospholipid vesicles at protein-to-lipid ratios of 1:30 (——) and 1:363 (---), and of DMPC vesicles without gramicidin (•••) at the same lipid concentration as in the 1:363 sample.

chains with limited capacities for binding heavy atoms and the crystals have relatively low solvent contents, which limits both the ability to diffuse in heavy atoms and the number of sites available for binding without stereochemical interference. In addition, the strong dependence of the gramicidin conformation on environment, even to the extent of changing helical pitch, results in a large change in unit cell dimensions upon binding heavy atoms (23,27). Koeppe et al. (26) did succeed in forming isomorphous gramicidin/CsSCN and gramicidin/KSCN crystals that diffracted to moderate resolution (∿3 Å). Those crystals, however, provided only single isomorphous replacement (SIR) phase information. The phase ambiguity from SIR in any region other than the central sections prevented calculation of an accurate three-dimensional Fourier map. Patterson analyses, however, which did not rely on phase information, indicated that the general shape of the molecule was a cylinder. More recently, neutron diffraction studies by Koeppe & Schoenborn (37) used hydrogen-deuterium exchange to provide SIR phases on an uncomplexed form of the molecule. These studies also indicated, at low resolution, that the molecule must be cylindrically shaped, but provided no information on the backbone fold or on side chain positions. What was needed in order to solve the structure was an alternate way to phase the data. We employed the method of single wavelength anomalous scattering (38) from cesium atoms in a gramicidin/cesium crystalline complex. Because in this method the phases are determined from a combination of the Bijvoet differences and partial structure of the cesium in a single crystal, there was no need to make isomorphous derivatives. The crystals used for these studies have 8 cesium sites (4 partially occupied) and 2 gramicidin dimers per asymmetric unit, so the cesium anomalous signal was calculated to be 17%, with a partial structure of ∿80%. Actual measured values were 8% and 43%, respectively, which was sufficient for phasing. In order to use this method, however, one needs very accurate and high-resolution data. A new crystal form of a gramicidin/cesium complex which diffracts to 1.5 Å resolution was prepared from cesium chloride (27). The solvent volume in these crystals was calculated to be 24%.

The gramicidin/cesium crystals contain four independent monomers (ie., two dimers) per asymmetric unit. All monomers appear to adopt similar backbone conformations, since the correlation coefficient between dimers is ∿0.8 to a radius of 8 Å, and within a dimer is ∿0.5. The dominant feature of the dimers is a tubular structure with a central cavity containing the cesium ions, much as was inferred from the Patterson analysis of Koeppe et al. (25). Since the anomalous data permitted phasing, a 1.8 Å resolution Fourier map was calculated and a preliminary structure fit to the electron density; it shows the polypeptide backbone structure and the positions of the amino acid sidechains and the ions in the pore (22,28; Wallace and Hendrickson, unpublished results), and resolves the ambiguity as to molecular fold. Crystallographic refinement of the structure is in progress.

The dimer appears to be a left-handed double helix formed from two anti-parallel β-strands with 6.3 residues per turn

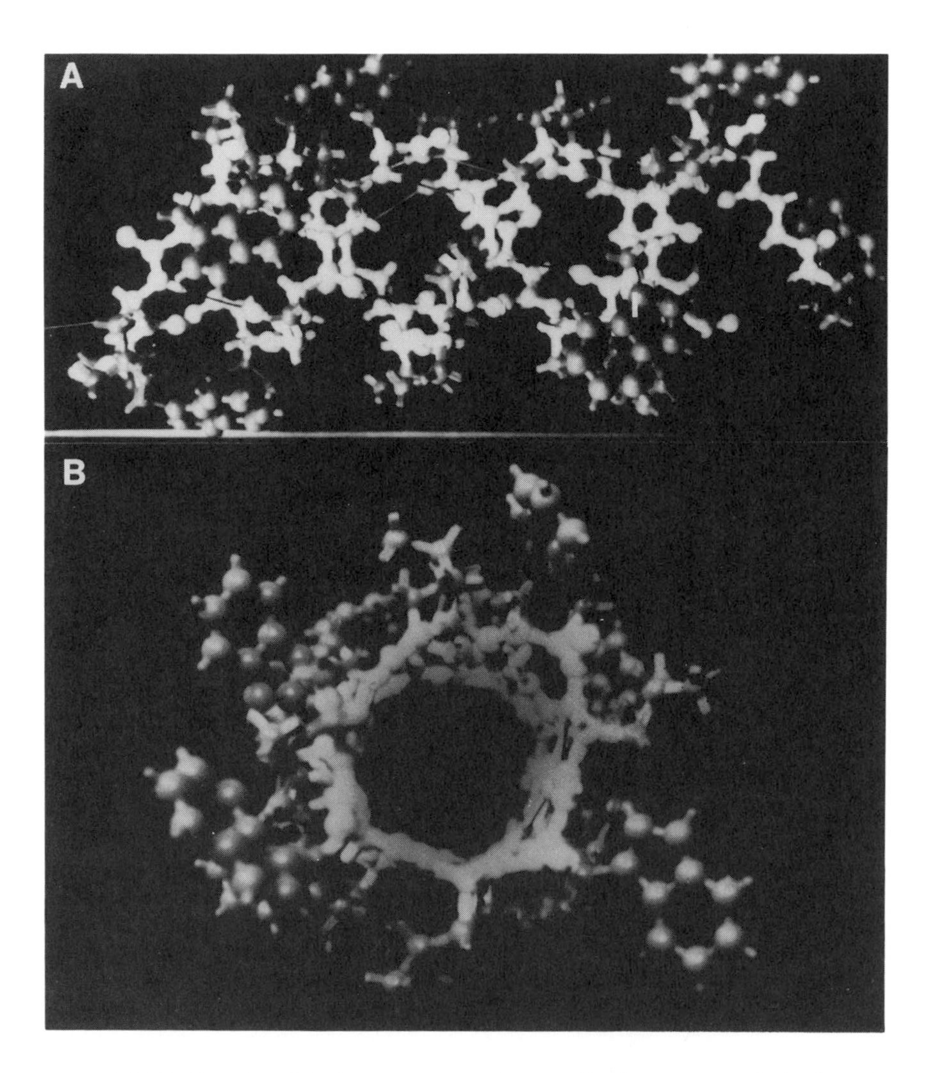

FIGURE 8. Preliminary structure of the gramicidin/cesium complex, based on the 1.8 Å electron density map. A, View along helix axis; B, view down helix axis.

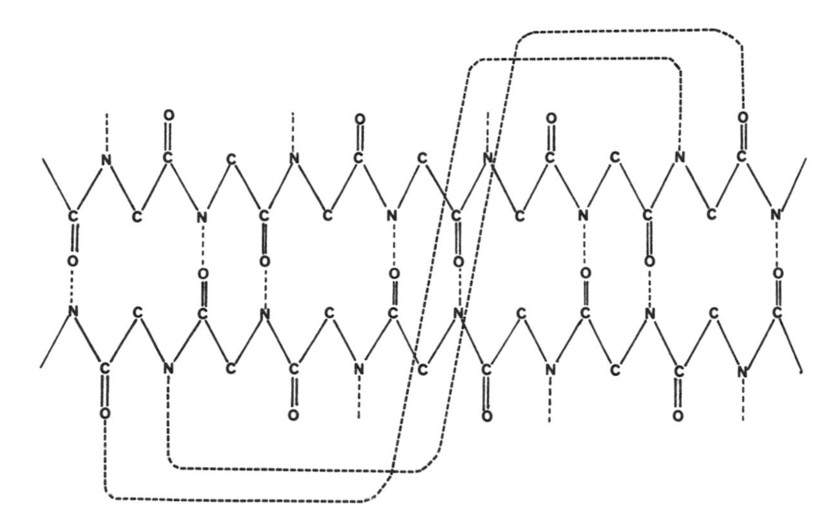

FIGURE 9. Diagram showing hydrogen-bonding patterns in the gramicidin/cesium complex.

and a helical repeat of 10.4 Å (Fig. 8). The number of residues per turn is different than reported for model calculations by Lotz et al. (39). Although the hydrogen bonding pattern is the same, the twist of the helix and the ϕ, ψ angles differ somewhat. The types of intermolecular hydrogen bonds involved between monomers are illustrated in Fig. 9. The relative offsets of the polypeptide chains are shown schematically in Fig. 10.

The dimer is 26 Å long with a central pore of ∿4.4 Å diameter. Side chains are located axially on the perimeter of the pore (Figs. 8B & 11B), as a consequence of the alternating pattern of L- and D- amino acids in the sequence. In a typical β-sheet type structure containing all L-amino acids, side chains protrude alternately from the top and bottom sides of the sheet. With an alternating L- and D-sequence, however, all side chains protrude from the same side of the sheet. Hence, when the sheet is rolled up to form a helix, all side chains are on the periphery and the center of the helix forms a pore that can accommodate the ions. This results in very efficient use of polypeptide chain, since forming an equivalent size pore from a bundle of α-helices (which have no such hole down their center) would use 5 to 6 times as much polypeptide chain. The tryptophan side chain rings do not appear to be stacked, consistent with the observed absence of substantial exciton interactions in spectra obtained in this environment. The structure resembles one of the double helical structures (Fig. 1C) originally

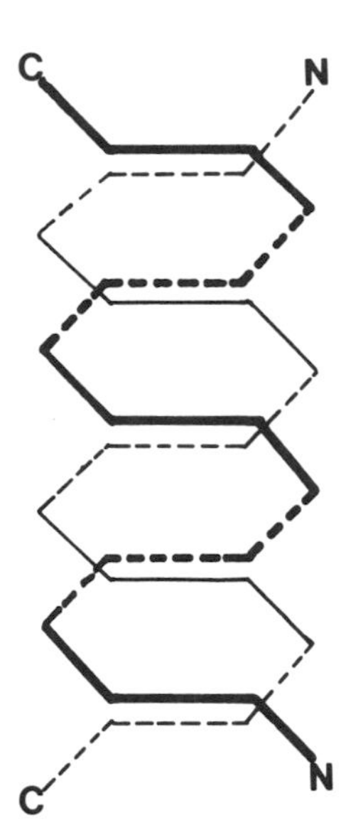

FIGURE 10. Schematic diagram showing the relative offsets of the two polypeptide chains in the double helical dimers found in the gramicidin/cesium crystals. The solid lines indicate the portions of the molecule at the front of the molecule in the orientation of the electron density maps given in Fig. 10, and the dashed lines indicate the portions of the molecule at the back in that the map.

proposed by Veatch et al. (14), although the pitch and alignment of the chains are somewhat different. The pitch and handedness are similar to those predicted by CD studies on the ion-bound form (23,40). The folding motif suggested by the recent vibrational spectroscopy studies of Naik and Krimm (41) is also very similar to the crystal structure. However, the model proposed by Arseniev et al. (42) based on 2-dimensional NMR studies of the cesium complex does not have the same handedness, alignment of polypeptide chains, nor helical twist.

The ion binding sites are clearly visible in the crystal structure. Two cesium sites are found in each pore located 7.1 Å from each end and separated by 11.8 Å (Fig. 11). There are also two partially occupied cesium sites between dimers in the crystal. Three chloride counterions are also seen in each dimer. The unexpected presence of chloride ions in the pore may be accounted for by competition between chlorides and solvent for the same sites at the high local ion concentrations present in the crystal (P. Jordon, personal communication). The cesium sites in this crystal form are different than those reported by Koeppe et al. (26) for a CsSCN complex which was prepared differently and diffracts only to 3 Å (26). However, the zero layer diffraction patterns from the CsCl complex (21) are identical to those from the CsSCN crystals, except that the chloride-containing crystals diffract to considerably higher resolution. This suggests that although the backbone conformation of the other crystal form is as yet unknown, the molecular folds in the two crystal struc-

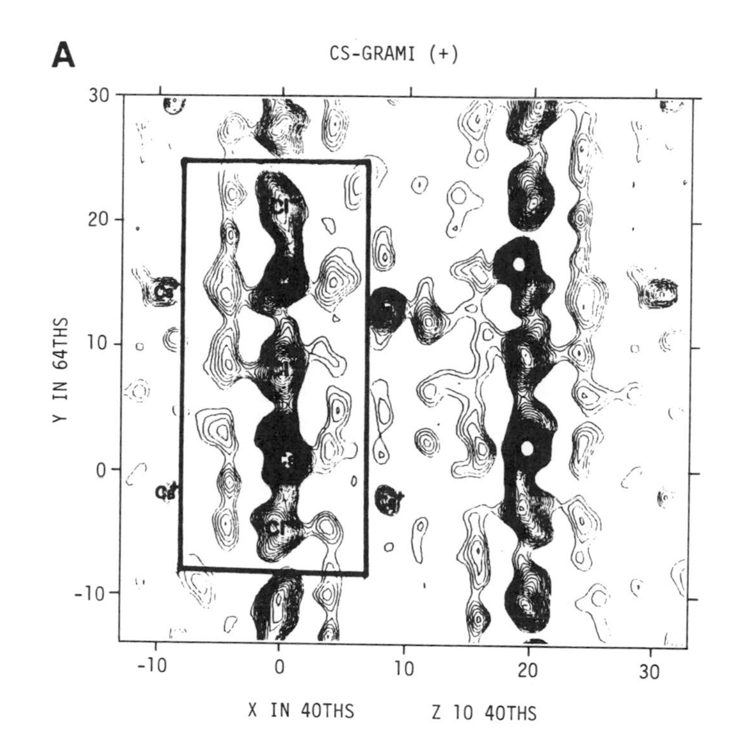

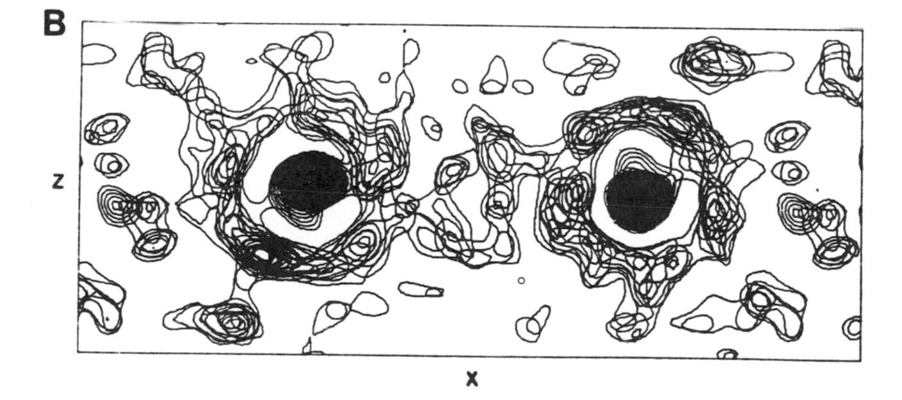

FIGURE 11. Section of the 1.8 Å resolution electron density map of the gramicidin/cesium complex. A, 2 dimers per asymmetric unit viewed along the helix axis (y), showing the center of the pore and the locations of the cesium and presumed chloride ions. The approximate boundaries of the dimer are indicated. B, View down the helix axis showing cesium ions in the channel and the axial arrangement of tryptophan side chains on the periphery of the channel.

tures must be very similar. A different space group has also been assigned for the chloride crystals, based on Patterson maps (22; Wallace and Hendrickson, unpublished results). The differences in space group and cesium sites of the two crystal forms, if real, could be a consequence of the different counterions present.

The gramicidin/cesium structure has a number of features which would be attractive for a membrane ion channel: the pore size and molecular length are compatible with the dimensions of the bilayer and those estimated for this structure based on conductance studies (7,43); two cation binding sites are found per dimer as suggested by conductance measurements (44); the outside of the dimer consists of hydrophobic amino acid side chains which would interact well with the hydrophobic lipid fatty acid chains in the membrane, and the pore is lined with the hydrophilic peptide backbone, permitting complex formation with the ions. However, these crystals were formed from an organic solution of gramicidin, and a wide variety of spectroscopic evidence (24), some of which has

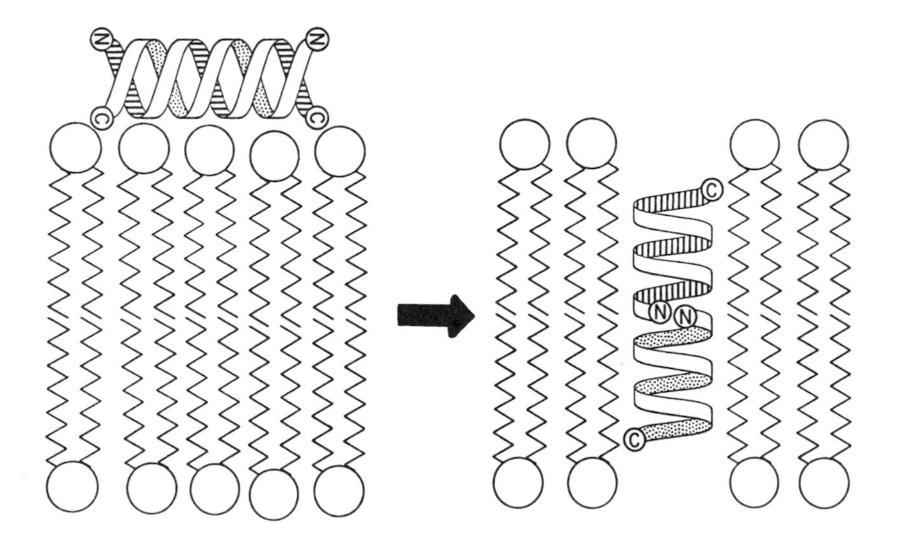

FIGURE 12. Schematic diagram indicating the process of insertion of the gramicidin molecule into a phospholipid bilayer. The molecular folds at two stages in the insertion process are shown: bound as the double helix and incorporated as the helical dimer. The orientation of the bound double helix form relative to the membrane surface has not yet been determined, and is shown here only for illustrative purposes. The process of insertion requires the breakage and formation of 28 hydrogen bonds.

been detailed above, indicates that this structure is not the same as the molecular fold in membranes.

What, then, is the role of the double helical form? While not the predominant conducting form in membranes, the pore form may, under some circumstances, be a conducting channel. Previous NMR studies which established the helical dimer as the major channel form could not rule out a minor component (up to ∿ 10%) of double helix. Recently Ovchinnikov and Ivanov (32) have shown that an N-terminal-to-C-terminal crosslinked dimer is capable of forming active channels, consistent with a conducting double helix form. One would expect double helical channels to have very long mean channel lifetimes. Another role for the double helix could be as an RNA polymerase inhibitor. Sarker et al. (45) have shown that gramicidin binds to that enzyme and exhibits activity in the absence of lipid molecules. Thus, the double helix may bear on the structure of the antibiotic, and this structure could be useful for drug modeling. Finally, CD studies (40) examining the process of gramicidin binding and insertion into membranes indicate that the pore form exists as the form initially bound to vesicles and that during the process of integration, the molecule is converted to the other (presumably helical dimer) form (Fig. 12). Hence, the gramicidin/cesium crystals provide the first detailed information on an intermediate in the process of insertion of a polypeptide into membranes. In the future it will be important to understand the dynamics and energetics of how the twenty-eight intermolecular hydrogen bonds in the membrane-bound double helix form of the molecule break and reform a new set of hydrogen bonds in the membrane-inserted helical dimer form of the channel. Hence, the double helix form, while not the predominant channel in membranes, may be important for understanding conduction, membrane insertion, and antibiotic activities.

Finally, it is of interest to understand why the pore form is not the predominant active channel in membranes, and to determine structural details of the channel form. Some progress has been made in both these areas: The types of interactions of gramicidin with lipid molecules needed to induce channel formation have been investigated. The mere presence of lipid in an organic or aqueous/organic solution is not sufficient to induce the channel formation (17), suggesting that the ordered juxtaposition of head group and hydrophobic tails is important. In addition, studies using bilayers formed from lipid molecules in which the ester linking the head group and fatty acid has been replaced by an ether linkage (Wallace, unpublished results) suggest that a specific interaction between the lipid carbonyl group and the polypeptide may be an important feature in stabilizing the membrane conformation. To address the question of the detailed conformation of the channel form of the molecule, crystals of a complex formed from gramicidin and DPPC have also been prepared (24). Preliminary characterization indicates that the unit cell, space group, and molecular transform in these crystals are different from any of the crystal forms without lipid. This finding suggests that at least the packing, and likely the molecular fold as well, are different. Raman spectroscopy (24,46) comparing these crystals and gramicidin in vesicles indicate that the backbone conformation of the gramicidin in the lipid co-crystals may be different than that in crystals without lipid. Thus, there is thus reason to believe that solution of this crystal form will lead to an understanding of the structure of the molecule as found in membranes.

In summary, using a combination of spectroscopic and diffraction techniques, it has been possible to examine the conformation and functionally-related structural features of the gramicidin molecule. The conformation of gramicidin has been the subject of study in a number of laboratories for many years. A variety of different structural models have been proposed to account for its physiological properties. CD spectroscopy has demonstrated that the molecule can adopt different conformations in different environments, but the actual structures found in these environments were not known. X-ray crystallography has now permitted examination of the structure of a gramicidin/cesium complex at high resolution. This is the first crystal structure of any form of gramicidin to be solved. The double helix in this crystal form corresponds to the pore form present in organic solvents and is a long-lived intermediate in the membrane-insertion process and possibly a minor conducting form. These studies have also demonstrated the importance of using a variety of complementary techniques in the determination of functionally-relevant membrane channel structures, especially for relatively small and flexible polypeptide molecules. Without the background of spectroscopic and conductance data from this and many other laboratories, we might have believed that the structure of the molecule in the gramicidin/cesium crystals represented the predominant channel form of the molecule, because it has a number of features which would be attractive for an ion channel in membranes. Instead, the gramicidin/cesium crystals have given us a view of one of the intermediates in the insertion process and should, when compared with the channel structure, provide a more complete understanding of the structure, function, and folding of this ion channel.

ACKNOWLEDGMENTS

I thank my collegues Drs. W. A. Hendrickson, E. R. Blout, and the late Dr. W. R. Veatch, for helpful discussions, and M. Cascio for drawing Fig. 6. This work was supported by NSF grants PCM 80-20063, PCM 82-15190, and DMB 85-17866 and NIH grant GM27292. The author was the recipient of a Hirschl Career Scientist Award during the course of these studies.

REFERENCES

1. Kyte, J., and Doolittle, R. F., J. Mol. Biol. 157:105 (1982).
2. Chou, P. Y., and Fasman, G. D., J. Mol. Biol. 155:135 (1977).
3. Mielke. D. L., Cascio, M., and Wallace, B. A., in "Macromolecular Interactions at High Resolution" (H. Vogel, and R. M. Burnett, ed.), in press. 1986.
4. Wallace, B. A., and Blout, E. R., Proc. Natl. Acad. Sci. USA 76:1775 (1979).
5. Finkelstein, A., and Andersen, O. S., J. Membr. Biol. 59:155 (1981).
6. Sarges, R., and Witkop, B., J. Amer. Chem. Soc. 87:2011 (1965).
7. Hladky, S. B., and Haydon, D. A., Biochim. Biophys. Acta 274:294 (1972).
8. Veatch, W. R., Mathies, R., Eisenberg, M., and Stryer, L., J. Mol. Biol. 99:75 (1975).
9. Veatch, W. R., and Stryer, L., J. Mol. Biol. 113:89 (1977).
10. Urry, D. W., Proc. Natl. Acad. Sci. USA 68:672 (1971).
11. Bamberg, E., Apell, H. J., and Alpes, H., Proc. Natl. Acad. Sci. USA 74:2402 (1977).
12. Szabo, G., and Urry, D. W., Science, 203:55 (1979).
13. Bradley, R. J., Urry, D. W., Okamoto, K., and Rapaka, R., Science, 200:435 (1978).
14. Veatch, W. R., Fossel, E. T., and Blout, E. R., Biochemistry, 13:5249 (1974).
15. Weinstein, S., Wallace, B. A., Blout, E. R., Morrow, J. S., and Veatch, W. R., Proc. Natl. Acad. Sci. USA 76:4230 (1979).
16. Weinstein, S., Wallace, B. A., Morrow, J. S., and Veatch, W. R., J. Mol. Biol. 143:1 (1980).
17. Wallace, B. A., Veatch, W. R., and Blout, E. R., Biochemistry 20:5754 (1981).
18. Andersen, O. S., Barrett, E. W., and Weiss, L. B., Biophys. J. 33:63a (1981).

19. Urry, D. W., Prasad, K. U., and Trapane, T. L., Proc. Natl. Acad. Sci. USA 79:390 (1982).
20. Whaley, W. L., Shungu, D., Hinton, J. F., Koeppe, R. E., and Millett, F. S., 8th Int. Biophys. Congress Proc. p. 274 (1984).
21. Wallace, B. A., Biophys. J. 49:295 (1986).
22. Wallace, B. A., and Hendrickson, W. A., Biophys. J. 47: 173a (1985).
23. Kimball, M. R., and Wallace, B. A., Biophys. J. 37:318a (1982).
24. Wallace, B. A., Biopolymers 22:397 (1983).
25. Koeppe, R. E., Hodgson, K. O., and Stryer, L., J. Mol. Biol. 121:41-54 (1978).
26. Koeppe, R. E., Berg, J. M., Hodgson, K. O., and Stryer, L., Nature 279:723 (1979).
27. Kimball, M. R., and Wallace, B. A., Ann. N.Y. Acad. Sci. 435:551 (1984).
28. Wallace, B. A., and Hendrickson, W. A., Acta Cryst. A40: c49 (1984).
29. Urry, D. W., Trapane, T. L., and Prasad, K. U., Science 221:1064 (1983).
30. Urry, D. W., Long, M. M., Jacobs, M., and Harris, R. D., Ann. N.Y. Acad. Sci. 264:203 (1975).
31. Kolb, H. A., and Bamberg, E., Biochim. Biophys. Acta 464:127 (1977).
32. Ovchinnikov, Yu. A., and Ivanov, V. T., in "Conformation in Biology" (R. Srinivasan, and R. H. Sarma, ed.), p. 155. 1983.
33. Hodgkin, D. C., Cold Spring Harbor Symp. Quant. Biol. 14:65 (1949).
34. Cowan, P. M., and Hodgkin, D. C., Proc. Royal Soc. Ser. B. 141:89 (1953).
35. Olesen, P. E., and Szabo, L., Nature 183:749 (1959).
36. Veatch, W. R., Ph.D. Thesis, Harvard University, Boston, Mass. 1973.
37. Koeppe, R. E., and Schoenborn, B. P., Biophys. J. 45:503 (1984).
38. Hendrickson, W. A., and Teeter, M. M., Nature 290:107 (1981).
39. Lotz, B., Colonna-Cesari, F., Hertz, F., and Spach, G., J. Mol. Biol. 106:915 (1976).
40. Wallace, B. A., Biophys. J. 45:114 (1984).
41. Naik, V. M., and Krimm, S., Biochem. Biophys. Res. Commun. 125:919 (1984).
42. Arseniev, A., Barukov, I., and Bystrov, V., FEBS Lett. 180:33 (1985).
43. Myers, V. B., and Haydon, D. A., Biochim. Biophys. Acta 27:313 (1972).
44. Hladky, S. B., Urban, B.W., and Haydon, D. A., in "Ion

Permeation through Membrane Channels" (C. Stevens and R. W. Tsien, ed.), p. 89. Raven Press, New York, 1979.
45. Sarker, N., Langley, D., and Paulus, H., Biochemistry 18:4536 (1979).
46. Short, K. W., Wallace, B. A., Myers, R., Fodor, S. P. A., and Dunker, A. K., Biochemistry, in press.

THE EFFECT OF MOLECULAR STRUCTURE AND OF WATER ON THE ENERGY PROFILES IN THE GRAMICIDIN A CHANNEL[1]

Alberte Pullman
Catherine Etchebest

Institut de Biologie Physico-Chimique
Laboratoire de Biochimie Théorique associé au CNRS
13, rue Pierre et Marie Curie
Paris, France

I. INTRODUCTION

The pentadecapeptide gramicidin A (Fig. 1), which carries a formyl group at its N terminal ("head") and an ethanolamine group at its C terminal ("tail"), forms ion-transmembrane channels made of dimers (1) which, most likely (2,3) adopt, in the conducting state, a head-to-head left-handed β-helical dimeric structure (4-6).

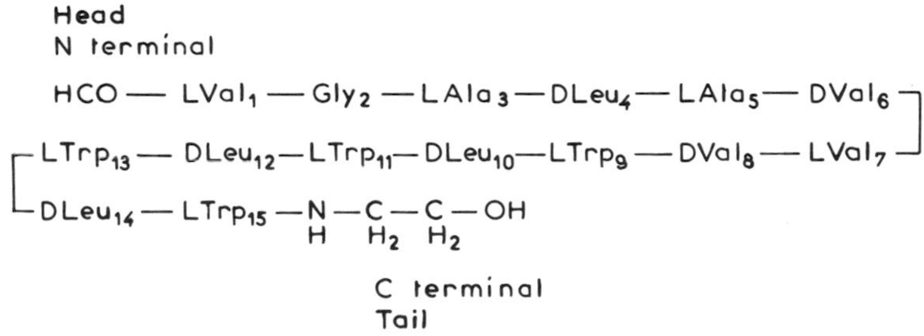

Head
N terminal

HCO — $LVal_1$ — Gly_2 — $LAla_3$ — $DLeu_4$ — $LAla_5$ — $DVal_6$ —
— $LTrp_{13}$ — $DLeu_{12}$ — $LTrp_{11}$ — $DLeu_{10}$ — $LTrp_9$ — $DVal_8$ — $LVal_7$ —
— $DLeu_{14}$ — $LTrp_{15}$ — $\underset{H}{N}$ — $\underset{H_2}{C}$ — $\underset{H_2}{C}$ — OH

C terminal
Tail

FIGURE 1. Molecular structure of gramicidin A.

[1]Work supported by the National Foundation of Cancer Research (Bethesda, U.S.A.).

Owing to its relative simplicity, this well-characterized channel provides an ideal model for understanding, at the microscopic level, the mechanisms involved in ion transport and, more particularly, the role of the structure-dependent features of this transport. A fundamental problem in this field is the determination of the energy profile felt by an ion inside the channel. Such a determination was undertaken in our laboratory some time ago (7) using a precise geometry of the channel and a refined methodology for calculating intra- and intermolecular interactions, including all the components of the energy (electrostatic, polarization, repulsion, charge transfer, and dispersion) [see Pullman (8), for a detailed description].

In order to delineate the respective roles of the different structural characteristics of the channel in determining the energy profile, our study has proceeded in successive stages which are described below. In order, furthermore, to discriminate between the "intrinsic", structural determinants of the profile and the effect of the molecular environment, we have first calculated the profiles "in vacuo", then considered the effect of water. We follow this order in the present summary.

For convenience, we shall call GA the dimer of gramicidin A.

II. ENERGY PROFILES COMPUTED IN VACUO

A. Na^+ in Urry's Head-to-Head Dimer Backbone

In our early study (7) the constituents of the channel were limited to all the atoms of the polypeptide backbone of GA, adopting Urry's proposed geometry with the published ϕ and ψ angles (6), standard bond length and angles and orienting the two monomers in the dimer so as to obtain the best geometrical fit for their head-to-head junction. The amino-acid side chains (which point towards the exterior of the channel) were omitted and the ethanolamine tails replaced by $-NHCH_3$ (Fig. 2).

The energy profile for single occupancy of this structure by a Na^+ ion computed in different approximations showed i) the importance of letting the ion adopt its optimal position at each step of the progression, instead of imposing on it to stay on the channel axis, ii) the importance of taking into account all the energy terms in the calculation (as opposed to the pure ion-dipole interactions only) (Fig. 3). The optimal profile, A, computed with all the terms, shows a succession of local minima associated with the binding to

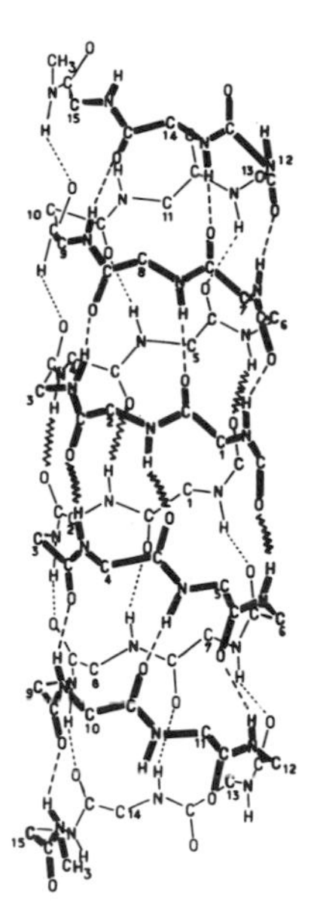

FIGURE 2. Urry's head-to-head β-helical dimer backbone used in the calculation of curve A of Fig. 3.

successive carbonyls. The deepest minimum is at about 4.5 Å from the center of the channel. A central energy barrier exists, of about 8 kcal/mole and two other barriers can be seen external to the "mouth" of the channel.

The location of the global minimum so obtained does not

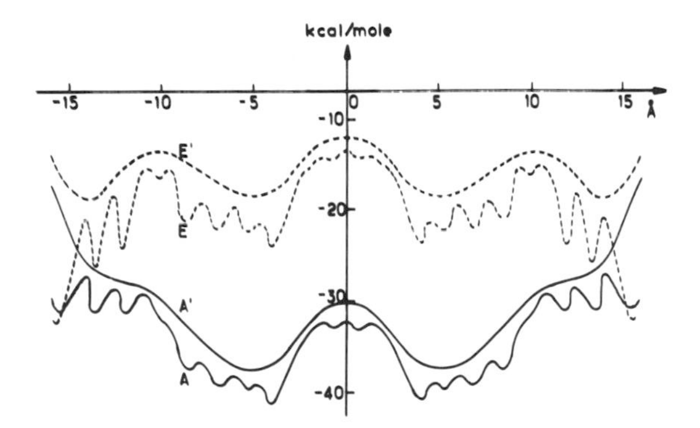

FIGURE 3. Energy profile computed for Na^+ in the GA backbone. A, total energy; E, electostatic component only. Corresponding profiles A' and E' for Na^+ constrained to remain on the channel axis.

appear satisfactory: measurements of ^{13}C NMR chemical shifts for individually labelled carbonyl groups (9) indicated the presence of a "binding site" in the neighborhood of the Trp 11 carbonyl group, that is, about 11 A from the center.

B. The effect of Including the Ethanolamine End in the Calculation

No information being available on the precise configuration adopted by the ethanolamine tail of the molecule, we determined by calculation its most stable conformation with respect to the rest of the backbone, optimizing the values of the three dihedral angles defined in Fig. 4A. This yielded the doubly hydrogen bonded structure of Fig. 4B (10).

Then the profile for Na^+ was recomputed in two fashions: a) fixing the tail in its most stable conformation, b) letting the tail be free to adopt its preferred conformation at each step of the progression of the ion, itself optimizing its position. The corresponding profiles (Fig. 5) show a dramatic difference with respect to curve A of Fig. 3, namely the appearance of a new energy minimum at 10.5 Å from the channel center. This minimum, nearly as deep in curve B (tail fixed) as the global minimum of curve A, is strongly deepened, so as to become the minimum minimorum when the tail is allowed to move. The appearance of this new minimum and its deepening are due to the fact that the Na^+ ion adopts constantly a position where it interacts favorably, not only with the surrounding carbonyls of GA, but also with the

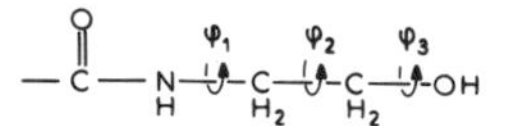

(b)

FIGURE 4. Energy optimization of the conformation of the ethanolamine end. A, the three dihedral angles optimized; B, the most stable conformation obtained (at the end of GA).

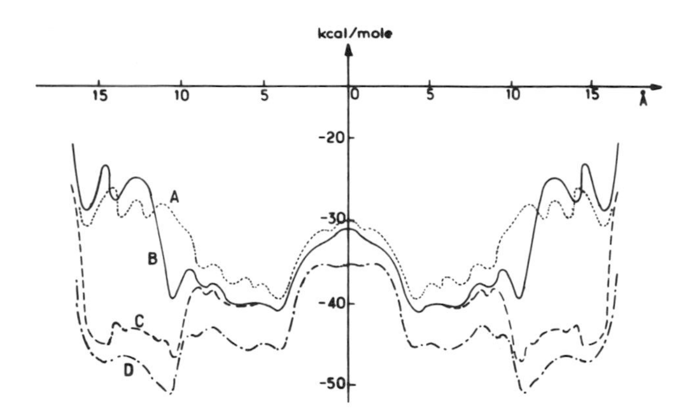

FIGURE 5. Energy profiles computed for Na^+ in the GA backbone including the ethanolamine tail. A, same as in Fig. 3; B, tail fixed; C, tail mobile; D, same including the side chains.

hydroxyl oxygen of the tail. This oxygen, in the initial conformation of Fig. 4B, is turned so as to provide an extra binding energy for the cation, thus creating a new minimum at 10.5 Å. When the tail is free to rotate, it uses the considerable lability conferred by its three single bonds to permit its oxygen to follow at best the ion until the limit set by the attachment of the tail, limit attained at the rejoining of curves C and B, a situation where the initial conformation of the tail is recovered (see ref. 11 for details).

The observed evolution brings the theoretical results in agreement with the experimental suggestion of a "site" in the region 10.5-11 Å and points to an important role of the ethanolamine end of the molecule in the location of the site.

The role of the position of the ethanolamine end in the location of the site has been put into evidence further recently (12) by a comparison of our tail-optimized geometry and of the geometry of the tail corresponding to the (unpublished) Urry's coordinates (Urry, private communication). Those were utilized in recent theoretical calculations (13) which failed to find the site in the appropriate region. The two structures shown side-by-side in Fig. 6 indicate very clearly the reason of this failure: in Urry's structure (based on model building) the hydroxyl oxygen is on the edge of the GA cylinder, turned towards the exterior of the channel at about the same height as the Trp 15 carbonyl oxygen. In this conformation, when the cation approaches from outside, it feels the strong attraction of the oxygen of the tail and

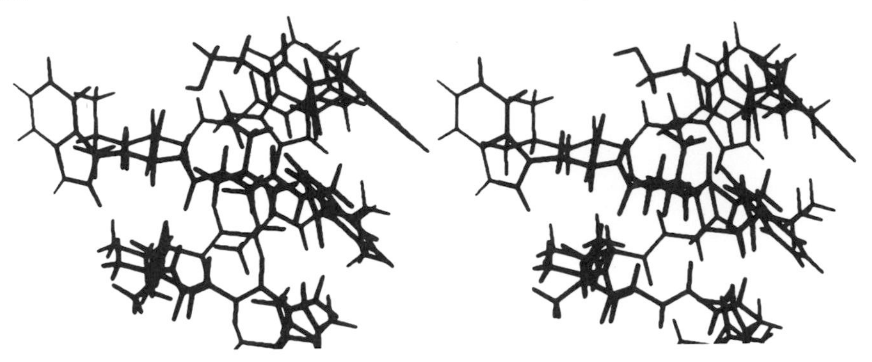

FIGURE 6. The disposition of the ethanolamine end. Left, in our optimized conformation; right, in Urry's conformation.

of the Trp 15 carbonyl oxygen and places itself in a deep minimum of 60 kcal/mole nearly equidistant of the two atoms, well outside the channel entrance. In our geometry the oxygen of the tail manifests its attraction farther inside, producing, in conjunction with the rest of the backbone, the minimum at 10.5 Å.

C. The Effect of the Side-Chains in the Profile for Na^+

The next step in the completion of the structure of the GA channel was the inclusion of the side-chains (14). This was done by adding on the α carbons of the backbone, the appropriate side-chains, fixing them in the conformations calculated recently as the most stable ones (15). The calculation of the profile for Na^+ was done in the same way as before, using i) the frozen conformation of the tail (after verifying by energy optimization that it was the same as that without the side chains); ii) letting the tail free to change its conformation at each step of the sodium progression.

In both cases it was found that the presence of the side chains produces a global lowering of the profile, without change in its overall shape, thus without modifying the general previous conclusions (Fig. 5, curve D). An analysis of the factors responsible for this energy lowering has indicated that it is due essentially to the electrostatic and polarization components of the interaction, which interplay differently, however, in different parts of the channel (Fig. 7): the polarization (ion-induced dipole) term is everywhere favorable, increasing slowly upon progression of the ion, due to the increasing number of polarizable groups surrounding the ion; the pure electrostatic contribution of the side chain is negative (favorable) in the early part of the pro-

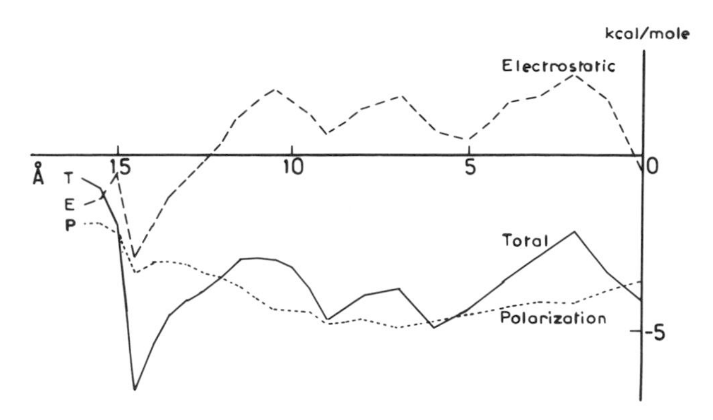

FIGURE 7. The total contribution, T, of the side chains to the energy profile for Na^+ and its electrostatic, E, and polarization, P, components.

file and positive afterwards, a behavior which can be related to the nature of the side chains along the channel. Among these residues, tryptophans 15, 13, 11 and 9 are the sole polar ones, thus the only ones expected to possibly give rise to an increase in electrostatic attraction for the ion. The role of each tryptophan was individualized by a separate calculation of their contribution in the early part of the profile. The results have shown that they fall into two groups, tryptophans 13 and 11 contributing an attractive effect, the two others (15 and 9) being repulsive. The inverse effects of the two groups of tryptophans have been related to the inverse orientation of their dipole moments in the conformation adopted (see ref. 16 for details). This study opens the way to an interpretation of the effects observed on the conductance properties (17-19) of gramicidin A in which different side chains have been substituted by others.

D. The Case of Other Cations

1. Potassium and Cesium. The energy profiles for K^+ and Cs^+ have been calculated in the same successive states of complication of the structure (14). Figure 8 gives the results for these two ions in the case corresponding to that of curve C of Fig. 5 for Na^+ (reproduced for convenience). Comparison of the three curves indicates that, in the zone extending from the entrance to about 9.5 Å, the deepest profile is that of Na^+, the less deep one that of Cs^+, whereas this order is reversed from 9.5 Å to the channel center, K^+

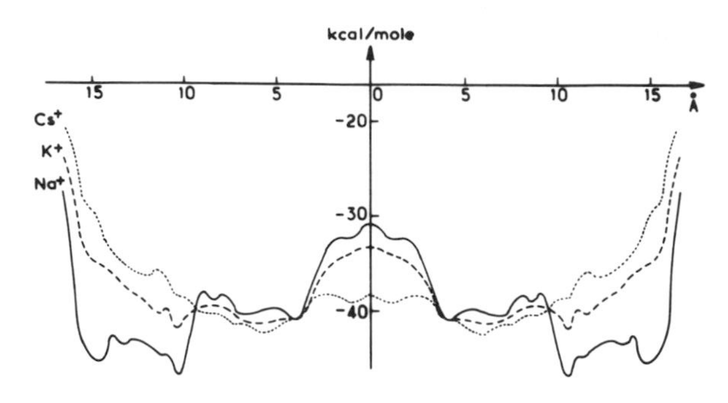

FIGURE 8. Energy profiles for K^+ and Cs^+ compared with that for Na^+ in the gramicidin A channel (tail mobile, no side chains).

being intermediate everywhere. This comes from the fact that, in the first zone, the favorable interaction with the hydroxyl oxygen, although present for the three ions, is less efficient for the larger ones. Beyond 9.5 Å, the effect of the tail disappears and the largest ion is favored because its size allows its favorable interactions with more groups simultaneously. The central barrier is in the order Cs^+ < K^+ < Na^+, in agreement with deductions drawn from experiment (20).

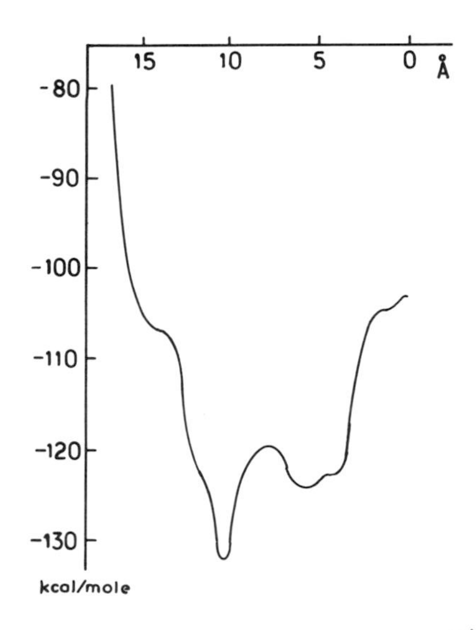

FIGURE 9. The profile calculated for Ca^{2+} in GA (complete structure, mobile tail), (half curve only).

2. Calcium. Divalent cations are not transported by GA (21,22), but it appears that they are capable of "entering the binding site" situated similarly to that of Na^+ (23,24).

As a step towards understanding the reasons for this behavior, calculations were done on the hypothetical energy profile which would be felt by a Ca^{2+} ion in a gramicidin A channel, using the same successive structural approaches as for the other cations (25). This has led to results very reminescent of those obtained for Na^+ concerning the structural effects on the profiles, namely a minimum close to the channel center when the tail is omitted, the appearance of another minimum at 10.5 Å upon inclusion of the fixed tail, a strong deepening of the latter into a minimum minimorum upon labilization of the tail, an overall lowering of the profile by the presence of the side-chains, etc.... The most conspicuous differences observed with respect to Na^+, illustrated in Fig. 9, on the profile obtained with the complete structure and the mobile tail, are i) the very deep negative values of the energies involved (compare with Na^+ in Fig. 5); ii) the steepness and facility of the descent towards the deep energy minimum at 10.5 Å; iii) the high barrier from this minimum to the center (much higher than that for Na^+). To these features must be added a supplementary difference with respect to Na^+ concerning the possibility of entrance of the ion, since all the calculations described so far concern the profiles _in vacuo_. That is for a bare ion, provision must be made for the necessity of dehydrating the ion (if ever partially, see below) upon penetration. An evaluation of the corresponding entrance barrier can be made, at least on a comparative basis, for different ions by substracting from the calculated energy at the beginning of the profile the experimental value of the dehydration energy of the ion. Although desolvation occurs progressively and not completely, it may be expected, especially in view of the fact that the state of hydration of the ions in the channel cannot differ much from one to another (_vide infra_), that the order of the barriers obtained is significant. Such an evaluation gives the relative values of roughly 30, 6, 5, and 4 for Ca^{2+}, Na^+, K^+, and Cs^+, respectively (14,25).

Altogether, the comparison of the results concerning Ca^+ and the alkali ions indicates that i) the difficulty apparently encountered by calcium to enter the channel is obviously related to the very large value of its dehydration energy, which would be only very partly compensated by the interaction energy of the ion with the channel; ii) should the ion penetrate, it would fall very easily into a site of tight-binding located similarly to that of Na^+, but then the large activation energy necessary to overcome the high central barrier would prevent it from being transported.

III. THE EFFECT OF WATER

The study of the effect of water on the profile has been performed using the same point of view as in the previously described computations, namely by following the variations of the optimal global energy of the system GA - water - cation along the progression of the ion in the channel, thus by energy optimization of the position of the ion in successive planes perpendicular to the axis, optimizing simultaneously the positions of all the water molecules in the system (12). This study, carried out for Na^+, has required a preliminary investigation, in the same fashion, of the interaction of Ga with water alone. The computations have been done using the complete channel structure, with the ethanolamine tail provisionally fixed.

A. Water in the Gramicidin A dimer (Tail Fixed)

The interaction of GA with water was considered by adding water molecules to GA in succession, optimizing their position each time, in the presence of all the others until the last molecule added (the sixteenth)did not bind directly to GA but rather to another water molecule, itself bound to the channel mouth. Figure 10 indicates in the optimal final

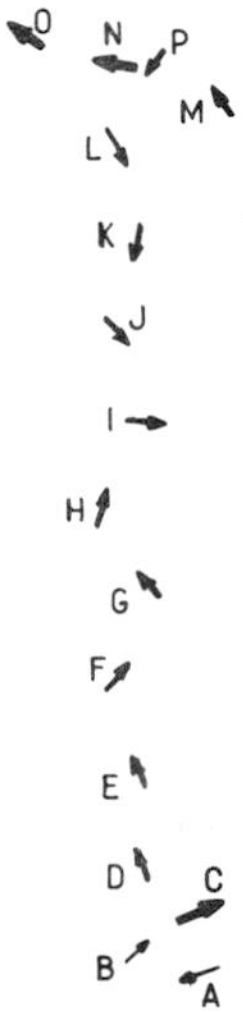

FIGURE 10. The orientations of the water molecules in the absence of ion indicated by the orientation of their dipole moments.

structure the labels of the water molecules and the directions of their dipole moments represented by vectors departing from the oxygen atoms. The disposition of the O and H atoms for each water can be seen in the first diagram of Fig. 11

Under the conditions adopted, 15 water molecules are hydrogen-bonded to carbonyl oxygens or NH groups of GA, each molecule being also hydrogen-bonded to one neighbouring water on each side. The oxygen of the water is generally in the vicinity of a hydrogen of the NH groups of GA, but, except for the waters C and N, no real hydrogen bond is formed involving these groups. One of the waters (P) is only bound to other water molecules (N, M, L) and does not interact directly with GA.

The observed mesh of hydrogen bonds leads to a strong interaction energy $E_{X\ W}$ of the same order of magnitude as that of an ideal water dimer (-6 kcal/mole calculated with the same methodology). The interaction energy E_{GA-W} has the strongest value of -9.4 kcal/mole for the molecules F and J.

Among the 15 waters directly bound to GA, one can define those (A, B, C, D, L, M, N, O) bound to the free carbonyl or the tail oxygens as being "outside" the channel and the others as being "inside" the channel: 7 water molecules are then inside the channel and 8 outside. This result is in fair agreement with the conclusions (n = about 6) of a recent review of the rather controversial experimental evidence (26). On the other hand, molecules D and L are just at the entrances of the channel (with their oxygen atoms respectively at 11.1 and 11 A of the center) and might be considered as "inside" it. This would lead to 9 water molecules "inside" the channel in agreement with the point of view of Dani and Levitt (27) and of Decker and Levitt (28). It is clear that the number of "inside" waters depends strongly on the definitions chosen for the exterior and interior of the channel. Recent molecular dynamics (29) as well as Monte Carlo (13) calculations find 8/9 water molecules within what seems our larger limits. Figures 10 and 11 show very clearly the "file" disposition of the water molecules inside the channel and the end view of Fig. 12 indicates that this file occupies just the space available. It may further be observed (Fig. 11) that in each GA monomer the water dipoles are directed towards the center, corresponding to an alignment of the overall dipole moment with that of the helical oligopeptide monomer. The angle between adjacent vectors (representing two hydrogen-bonded molecules) departs from that of an ideal water dimer by values ranging from -10 to +15 degrees, an effect clearly due to the influence of the GA backbone in which the helical distribution of the attractive centers induces a helical structure within the chain of water

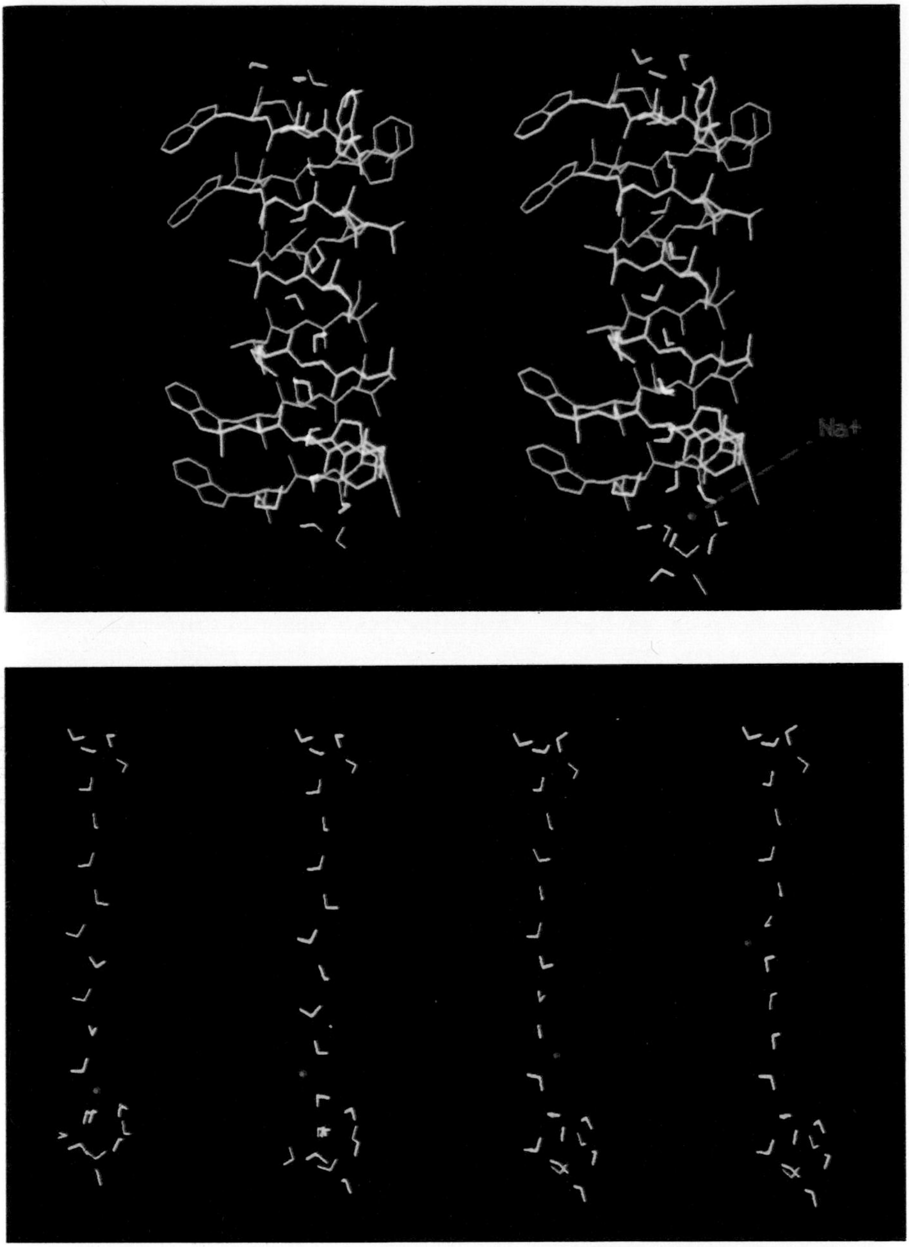

FIGURE 11. The distribution of the water molecules in the system GA-16 waters, and the evolution of the system upon inclusion and progression of an initially hexahydrated Na^+ ion. Ion at 14.5; 11.7; 10.3; 8.4 and 0.3 Å from center respectively.

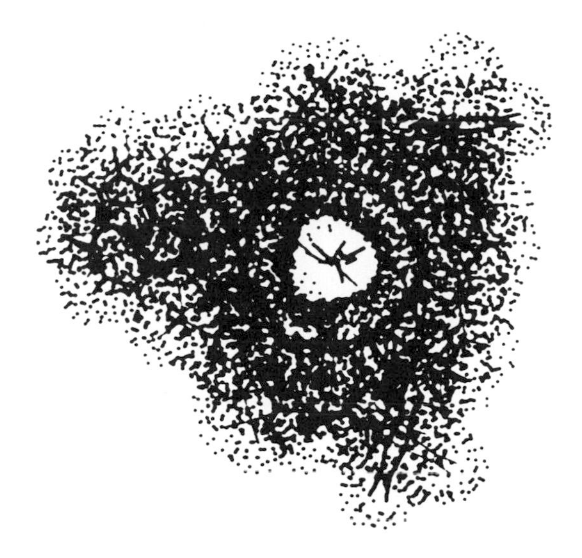

FIGURE 12. End view point of the GA channel filled with the water molecules within the Van der Waals envelope of the dimer.

molecules.

B. The GA-Water-Na System (Tail Fixed)

Starting with the optimal system GA-16 water molecules, a sodium ion carrying six water molecules in an octahedral first hydration shell was introduced at the exterior of the channel at a distance Z>15 Å from the center and the most favorable interaction energy of the entire system was obtained by optimizing simultaneously the position of the ion in the chosen plane perpendicular to the Z axis and the positions of all the 22 water molecules of the global system. The ion was then progressively displaced along the channel, the most favorable interaction energy being obtained by a similar complete optimization. In order to mimick the fact that the presence of the bulk of water at the outside of the channel insures a constant flux of waters, a water molecule coming out of the channel under the push of the ion was allowed to fill the "hole" created in the water following the ion, thereby keeping a constant number of 22 water molecules in the system. The variation of the total interaction energy of the system GA-Na^+-22 waters so obtained is represented in Fig. 13. This may be considered as the energy "profile" of the system including water and thus compared to the corres-

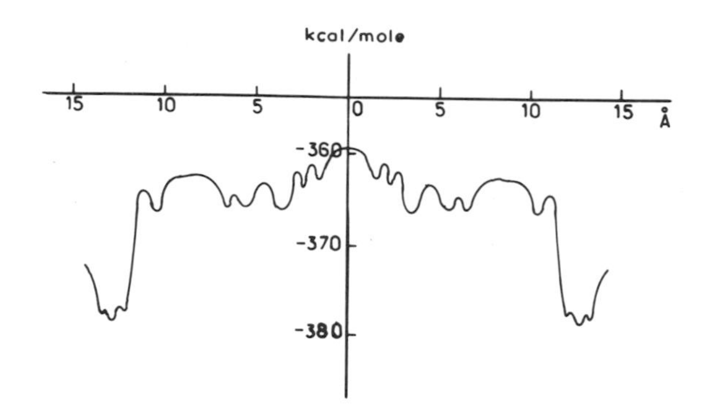

FIGURE 13. Energy profile for Na^+ in GA (tail fixed) in the presence of water.

ponding profiles computed _in vacuo_. The profile of Fig. 13 presents a deep and wide global minimum centered at 12.5 Å from the center. Then, a rather steep energy barrier is observed, followed by a small minimum at 10.5 Å and another lower barrier. Beyond 7 Å, the curve presents a number of local minima and maxima with relatively small energy variations (7 kcal/mole between 10.5 Å and the center). Only 18 kcal/mole separate the deepest energy minimum from the highest value at the center, the first steep barrier amounting to 13 kcal/mole. These are small values compared to the total interaction energies reaching -370 kcal/mole, and their order of magnitude is quite reasonable in comparison to the values deduced from experiments (30).

A detailed analysis of the profile of Fig. 13 (12) has been performed in the light of the underlying components of the energy E_{Na^+-W}, E_{W-W}, E_{Na^+-GA}, and E_{W-GA} and by considering the effect of the ion presence and progression upon the distribution of the water molecules in the system. It was found that the shape of the profile in the first region is dominated by the large and opposite variations of the terms E_{Na^+-W}, E_{W-W}, the ion-water interaction rapidly becoming less and less favorable, while at the same time the water-water interaction becomes more and more negative, compensating, but only partly, the loss in the former energy. The oscillations, followed by a steep rise, in E_{W-Na^+} are due to the progressive loss of water molecules in direct interaction with the ion: near the "entrance" of the channel (14.3 Å), the cation has already lost several water molecules from its original hydration shell but it still interacts strongly with five water molecules, two of them initially "belonging" to the chain of water molecules observed in the absence of the cation. As the ion penetrates the channel, it progressively loses water molecules until finally interacting with only two water molecules inside the channel (see the diagrams of Fig. 11). The unfavorable dehydration process is in part balanced by the simultaneous increase of a favorable interaction between the water. At first, in the octahedral hydration shell, the water molecules prefer to interact strongly with the cation than more weakly with one another but when the cation enters the channel the water molecules which it loses become able to interact favorably with one another by hydrogen bonds.

The two other components of the interaction E_{GA-Na^+} and E_{GA-W} also show complementary variations, translating a competition between the cation and water for interacting with the channel, but the range of these variations (roughly ±10 kcal/mole) is considerably smaller than that of the W-Na^+ and W-W interactions. Overall, the deep minimum in the global energy profile results from the combination of several favorable interactions at the channel mouth, the high ensuing barrier appearing essentially as a desolvation barrier.

In the second region of the profile (from 11 Å to the center) the minimum observed at 10.5 Å corresponds to the minimum found <u>in vacuo</u> and manifests as before the attraction of the hydroxyl oxygen of the (fixed) tail. Further inside, the local minima observed correspond to the interaction of the cation with the different L-carbonyl oxygens, again similarly to the situation <u>in vacuo</u> discussed in section II, the oscillations, observed from 11 to about 3 Å resulting from the interplay of the opposite variations in the Na^+-GA and W-GA interactions, the latter ones "dampening" the variations in the Na^+-GA interactions observed <u>in vacuo</u> and yielding a consecutive flatter overall profile. The central barrier is lowered with respect to that <u>in vacuo</u>, another consequence of the interplay of opposite individual energy variations. The perturbation brought about by the cation in the chain of water molecules is clearly observable in the diagrams of Fig. 11. At first, upon the approach of the ion the file of water molecules present in the GA dimer undergoes a change in orientation, several waters having been rotated so as to turn their oxygen atom towards Na^+. In these new orientations the water molecules still form a hydrogen-bonded chain in the channel, even between the two monomers, as indicated by their interaction energies. Moreover, hydrogen bonds have formed between the water molecule previously bound to free carbonyl oxygens and some of those from the hydration shell of the cation.

At all points of the progression the chain of water molecules inside the channel is maintained, progressively re-adopting its original configuration as the cation approaches the center of the channel.

The results of the energy optimizations upon progression of the ion, while maintaining a constant flux of water molecules, indicate that the cation pushes the chain of water molecules in front of itself and cannot pass waters, except perhaps one close to the center, an observation which requires further explorations [see Etchebest and Pullman (12) for a detailed discussion in connection with the experimentally based conclusions concerning single-file transport (26).

The persisting role of the tail hydroxyl group in the presence of water in the genesis of a minimum at 10.5 Å corresponding to the experimental location of the "site" is intriguing. It appears likely that the introduction of the lability of the tail will also have a deepening effect in this region for the same reasons as *in vacuo* thus emphasizing the significance of the minimum observed at 10.5 Å. Of course, the precise final relative location and depths of the outer and inner minima in the first region of the global curve can be assigned only by explicit computations which are now in progress.

REFERENCES

1. Veatch W. R., Mathies, R., Eisenberg, M., and Stryer, L., J. Mol. Biol. 99:75. (1975)
2. Urry, D. W., in this volume, 1986.
3. Wallace, B., in this volume, 1986.
4. Urry, D. W., Proc. Natl. Acad. Sci. USA 68:672 (1971).
5. Urry, D. W., in "Fifth Jerusalem Symposium of Quantum Chemistry and Biochemistry" (E. D. Gergmann, and B. Pullman, ed.), p. 723. Israel Academy of Sciences, Jerusalem, 1973.
6. Urry, D. W., Venkatachalam, C. M., Prasad, K. U., Bradley, R. J., Parenti-Castelli, G., and Lenaz, G., Int. J. Quant. Chem. Quantum Biology Symposium Nr. 8:385 (1981).
7. Pullman, A., and Etchebest, C., FEBS Lett. 163:199 (1983).
8. Pullman, A., in "Methods in Enzymology: Biomembranes: Protons and Water. Structure and Translocation" (L. Packer, ed.), Academic Press, New York, in press, 1986.
9. Urry, D. W., Prasad, K. U., and Trapane, T. L., Proc. Natl. Acad. Sci, USA 79:390 (1982).
10. Etchebest, C., and Pullman, A., FEBS Lett. 163:199 (1983).
11. Etchebest, C., and Pullman, A., FEBS Lett. 170:19 (1983).
12. Etchebest, C., and Pullman, A., J. Biomol. Struct. Dyn.

3:805 (1986).
13. Kim, K. S., Vercauteren, D. P., Welti, M., Chin, D., and Clementi, E., Biophys. J. 47:327 (1985).
14. Etchebest, C., Ranganathan, S., and Pullman, A., FEBS Lett. 173:301 (1984).
15. Venkatachalam, C. M., and Urry, D. W. J. Comput. Chem. 5:64 (1984).
16. Etchebest, C., and Pullman, A., J. Biomol. Struct. Dyn. 2:859 (1985).
17. Heitz, F., Spach, G., and Trudelle, Y., Biophys. J. 39:87 (1982).
18. Heitz, F., Gavach, G., and Trudelle, Y., Biophys. J. 45: 97 (1984).
19. Mazet, J. L., Andersen, O. Z., and Koeppo, RE II, Biophys. J. 45:273 (1984).
20. Eisenman, G., and Sandblom, J. P., in "Physical Chemistry of Transmembrane ion motions" (G. Spach, ed.), p. 329. Elsevier, Amsterdam, 1983.
21. Hladky, S. B., and Haydon, D. A., Biochim. Biophys. Acta 274:294 (1972).
22. Mayers, V. B., and Haydon, D. A., Biochim. Biophys. Acta 274:313 (1972).
23. Urry, D. W., Trapane, T. L., Walket, J. T., and Prasad, K. U., J. Biol. Chem. 254:6659 (1982).
24. Urry, D. W., Walker, J. T., and Trapane, T. L., J. Membr. Biol. 69:225 (1982).
25. Etchebest, C., Pullman, A., and Ranganathan, S., Biochim. Biophys. Acta 818:23 (1985).
26. Finkelstein, A., and Andersen, D. S., J. Membr. Biol. 59:155 (1981).
27. Dani, J. A., and Levitt, D. G., Biophys. J. 35:385 (1981).
28. Decker, E. R., and Levitt, D. G., Biochim. Biophys, Acta 730:178 (1983).
29. Mackay, D. H. J., Berens, P. H., Wilson, K. R., and Hagler, A. T., Biophys, J. 46:229 (1984).
30. Urry D. W., Trapane, T. L., Prasad, K. U., Jnt. J. Quant. Chem., Quant. Biol. Symp. 9:31 (1982).

STRUCTURE-FUNCTION STUDIES ON LINEAR GRAMICIDINS: SITE-SPECIFIC MODIFICATIONS IN A MEMBRANE CHANNEL

O. S. Andersen[1]

Department of Physiology and Biophysics
Cornell University Medical College
New York, New York, U.S.A

R. E. Koeppe II[2]

Department of Chemistry and Biochemistry
University of Arkansas
Fayetteville, Arkansas, U.S.A.

J. T. Durkin[3]

Department of Biological Cehmistry
Harvard Medical School Boston
Massachusetts, U.S.A.

J.-L. Mazet[4]

Laboratoire de Physiologie Comparee
Universite de Paris-Sud
Orsay Cedex, FRANCE

---

[1]Supported by NIH Grant GM 21342
[2]Supported by NIH Grant GM 34968 and RCDA NS 00648, and NSF Grant NSF-INT-8413704
[3]Supported by NIH Grant AM 07300 to E. R. Blout
[4]Supported by CNRS Grants ATP 83.717 and AI 85.113

I. INTRODUCTION

The linear gramicidins have proven useful to study molecular and physico-chemical aspects of ion movement through membrane-bound channels (for recent reviews, see Ref. 1-3). The gramicidins are well suited for such studies because their primary structures are known, and because there exists a generally accepted channel structure, the β-helical dimer proposed by Urry (4). The structure of valine gramicidin A (5) is:

```
Formyl-L-Val---Gly-L-Ala-D-Leu-L-Ala-
         1      2     3     4     5

-D-Val-L-Val-D-Val-L-Trp-D-Leu-L-Trp-
   6     7     8     9    10    11

-D-Leu-L-Trp-D-Leu-L-Trp-Ethanolamine.
   12    13    14    15
```

The transmembrane channel is formed by two β-helical monomers joined by hydrogen bonds at their formyl-NH_2-termini. The site of dimerization is thus in the middle of the membrane. Support for these general structural features was obtained through investigations on channels formed by gramicidins modified at their NH_2-terminal or COOH-terminal ends (6-10), and spectroscopic studies on native or specifically labeled gramicidins (9,11-16). The channels are permeable to water and impermeable to urea (17). The luminal diameter is thus about 4 Å, consistent with channels formed by $\beta^{6\cdot3}$-helices. The narrow channel lumen constrains ions and H_2O to move in single-file (18,19).

The organization of the peptide backbone of a $\beta^{6\cdot3}$-helical dimer is illustrated in Fig. 1. There is a central cavity lined by the polar groups of the peptide backbone along the axis of the helix. The peptide groups solvate the permeating ions and water molecules, and decrease the energy barrier for ion movement through the channel. The magnitude of the single-channel conductance depends critically upon the existence of this central solvation path (21,22). The hydrophobic side chains project from the exterior surface and permit channel incorporation into the hydrophobic core of a lipid bilayer.

Figure 2 depicts an amplitude histogram of valine gramicidin A single-channel current transitions. The narrowness of the peak reflects the uniform characteristics of the channels: Each channel is similar to any other channel, and each channel is stationary in time, at least on the millisecond time scale of our measurements. The channel structure is thus remarkably uniform among channels and over time. Gramicidin channels are therefore useful prototypes for structure-function studies on transmembrane channels, because one can monitor changes in function relative to a well-defined reference state. The amplitude of the current transitions reflects the efficient catalysis of transmembrane ion movement mediated by the channels: Single-channel current steps of 1.4 pA correspond to the transmembrane movement of about 10^7 ions per second. At high alkali metal cation concentrations and high potentials the currents can be 10 - 30 times greater.

The magnitude of the single-channel conductance is primarily determined by the polar backbone, but the channel conductance varies when the amino acid sequence is altered (23-29). There conductance changes could result from changes in channel structure, or they could result from side chain-dependent modulations in the energy profile for ion translocation through the channels. These problems are general in structure-function studies (e.g., Ref. 28,30-32). We have exploited the fortunate fact that gramicidin channels are symmetrical dimers to develop a criterion for structural invariance among channels formed by different gramicidin ana-

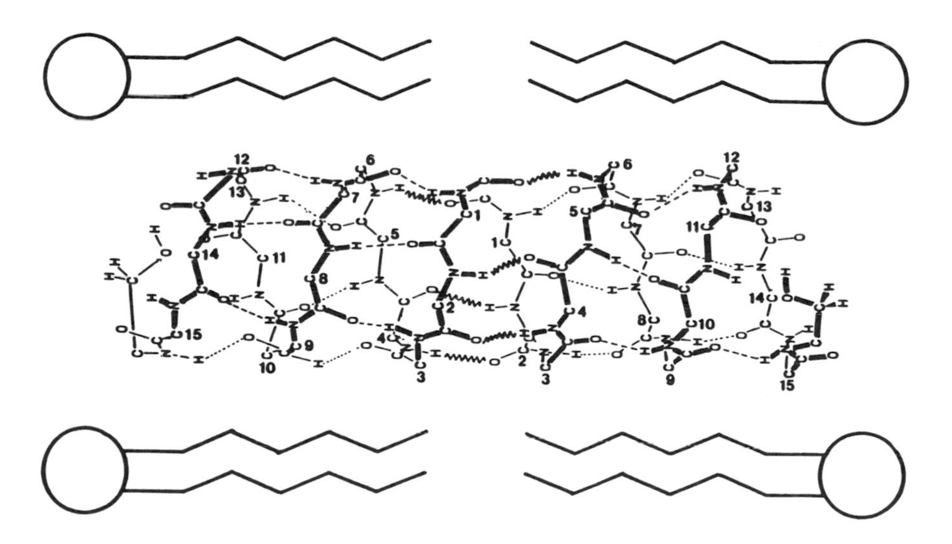

FIGURE 1. Schematic drawing of a $\beta^{6\cdot3}$ helical gramicidin dimer in a lipid bilayer. The side chains, which project from the exterior surface, are removed for clarity. Intramolecular hydrogen bonds are drawn as interrupted lines, intermolecular hydrogen bonds are drawn as zig-zag lines. The numbers refer to the α carbons in the amino acids. Modified from Etchebest and Pullman (20).

logues; and we have used natural and semi-synthetic analogues of valine gramicidin A to probe how side chains at the exterior surface of the channel, 5 - 10 Å from the ion, modulate the channel conductance (27-29).

II. ION TRANSPORT THROUGH GRAMICIDIN CHANNELS

The symmetrical structure of the gramicidin channel, in particular the repeating series of carbonyl oxygens which line the permeation path, leads naturally to a simple model for ion translocation through the channel: A particle (an ion or an H_2O) moves through the channel in a sequence of jumps from one site to another (or to the aqueous phases) with sojourns of different duration at each site. The probability that a particle is at a particular coordination site, and the direction of its next jump, is determined by the energy profiles and the number (and type) of particles at other sites in the channel.

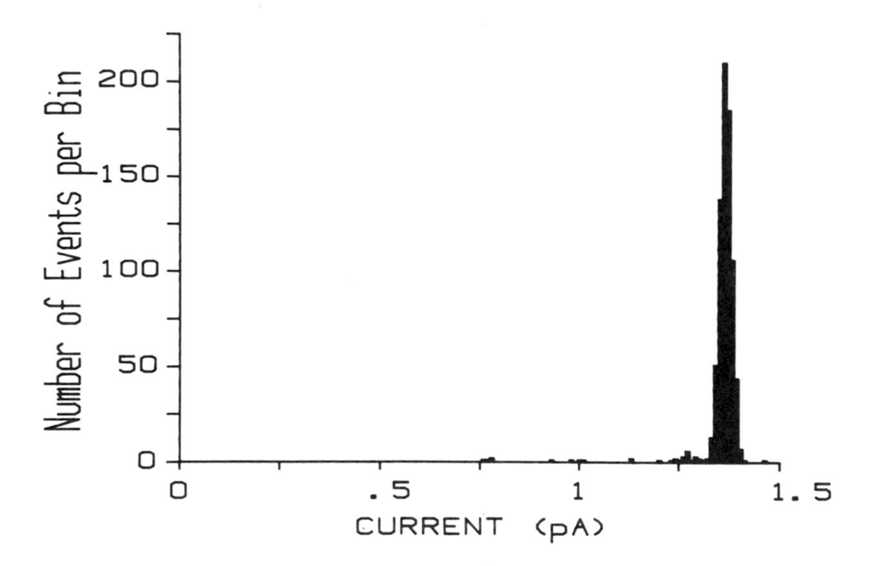

FIGURE 2. Amplitude histogram of single-channel current steps obtained with valine gramicidin A in diphytanoylphosphatidylcholine/n-decane bilayers. The single-channel current was 1.363 ± 0.014 pA (mean ± s.d.). There were 789 channel transitions in the histogram, of which 758 (or 96%) were in the major peak between 1.309 and 1.410 pA. Data from a single experiment in 1.0 M NaCl, 25°C, and 100 mV applied potential.

Equilibrium binding and competitive displacement experiments show that gramicidin channels have a relatively high affinity for small monovalent cations (33-35), and that the channels can be simultaneously occupied by only a small number of ions, most likely one or two. Ion tracer flux studies show no interaction among Na^+ in gramicidin A channels in diphytanoylphosphatidylcholine/n-decane bilayer membranes (36), and the shapes of single-channel current-voltage relations vary as function of $[Na^+]$ (37). These findings suggest that at most one Na^+ occupies a gramicidin channel at attainable $[Na^+]$ - not all the ion binding sites are simultaneously occupied. (Other alkali metal cations [K^+ through Cs^+] exhibit clearcut evidence for multiple ion occupancy, at least two ions can simultaneously bind to the channel (e.g., Ref. 38,39). The Na^+ data can also be fit assuming multiple ion occupancy (39). If this were the case, it would not affect our general conclusions. We will therefore not deal further with the question of multiple ion occupancy.)

The major cation binding sites are located at the COOH-terminal ends of the channel, close to the channel entrances (13,14,40). There are, however, no specialized groups in the channel wall, only the repeating series of carbonyl oxygens. This leaves tow possibilities for the formation of localized cation-binding sites: an end-of-channel effect (41,42), and/or specific interactions with the amino acid side chains at the COOH-termini (20).

The energy profile for a permeating ion is determined by several factors: Favorable interactions between the peptide carbonyl oxygens and the permeating ions (42,43) allow ion binding into the channel, while the electrostatic potential associated with moving an ion through a channel that spans the low dielectric constant bilayer tends to exclude ions from the channel interior (22,44). The superposition of the solvation and electrostatic energies with the local energy barrier at the adjoining formyl-NH_2-termini (42) produces an energy profile with a central barrier, which separates ion binding sites close to the channel entrances. This picture of the channel is consistent with results of ^{13}C NMR studies on gramicidin/lysolecithin bilayer structures, which show that ions bind to the exterior carbonyl oxygens (No.s 11,13, and 15), while carbonyl oxygens in the channel interior are unaffected by ions (14). Ion translocation through the channel interior is fast on the ^{13}C NMR time scale. The relatively long sojourns in the major ion binding sites, taken together with the short duration of the translocation step, implies that it should be possible to approximate this latter transfer by a one-step process.

A priori, one would expect that the central barrier would be a significant barrier for ion movement through the channel, but this does not appear to be the case for channels formed by the parent valine gramicidin A. A comparison of the rate of ion movement with the rate of diffusional H_2O movement indicates that an ion crosses the channel interior at a rate which is comparable to that of a H_2O (37,45). Analysis of the concentration-dependence of the shape of single-channel current-voltage shows further that the central barrier is fairly flat (38). It appears that one or more terms are missing in the present description of the energy profile. One possible omission is the two ordered arrays of H_2O dipoles with their nagative ends pointing towards an ion in the channel interior (46).

Ion movement through a gramicidin channel may be decomposed into five steps (see Fig. 3): First, diffusion through the aqueous phases up to the channel entrance to form the equivalent of an outer-sphere complex; second, association with the channel proper through partial dehydration and solvation by the carbonyl oxygens and the ethanolamine -OH in the channel; third, translocation through the channel interior; fourth, dissociation from the channel through desolvation and rehydration; and fifth, diffusion away from the channel entrance into the other aqueous phase.

Channels formed by valine gramicidin A are so permeable to small monovalent cations that the single-channel currents can become limited by ion movement through the aqueous phase close to the channel entrance (47). This aqueous diffusion or access limitation step can be studied at low permeant ion concentrations and high applied potentials after minimizing, or correcting for, the changes in interfacial ion concentration that occur upon application of a potential difference across a membrane (48). The rate of ion transfer up to the

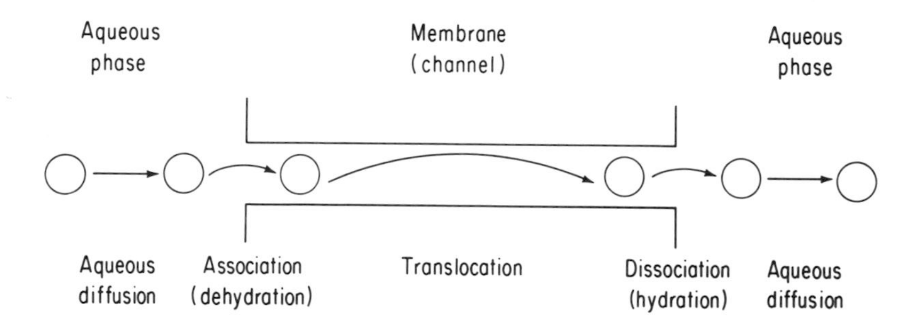

FIGURE 3. Schematic representation of the steps involved in the transfer of an ion through a gramicidin A channel, from one bulk aqueous phase to another. Each depicted step may be composed of many elementary steps. Modified from Andersen and Procopio (37).

channel entrance can be quantified by an access permeability coefficient, p_a. For Na^+ through Cs^+, p_a is proportional to the aqueous diffusion coefficient for the ions (47).

If there are no other voltage-dependent steps involved in ion association with the channel, one should be able to estimate p_a from a simple (macroscopic) description of diffusive ion movement to an absorbing sink (the channel entrance) (e.g., Ref. 47,49). Such estimates are, however, suspect because of the assumptions involved, particularly because water close to the channel entrance may not behave as bulk water (47). Killian and de Kruijff (50) have for example recently found that gramicidin A channels are strongly hydrated with ∿70 H_2O bound per channel entrance region, a qualitatively similar conclusion was reached in Monte Carlo simulations of water structure in and around gramicidin A channels (51).

The uncertainties involved in the measurement or calculation of p_a are serious: It would probably not have been possible to deduce the existence of an aqueous diffusion limitation in the absence of the high-voltage measurements; and present knowledge about the behavior of water close to channel entrances is so limited that we at this time can only expect qualitative agreement between theory and experiment.

The aqueous diffusion limitations are significant, as more than 50% of the resistance in the ion entry steps may reside in the aqueous convergence regions (47). It will thus be difficult to characterize molecular events involved in ion entry, e.g., the role of the ethanolamine moiety (42), because changes in intrinsic (dehydration/solvation) rate constants may be masked by the diffusion step. This should be less of a problem for studies of ion translocation through the channel interior, particularly at high permeant ion concentrations where ion permeation becomes limited by the translocation and dissociation steps.

The data analysis becomes particularly simple if we restrict ourselves to the single ion occupancy approximation, which seems to be appropriate for Na^+ in valine gramicidin A channels in diphytanoylphosphatidylcholine/*n*-decane membranes. The relation between the measured small-signal conductance, g, and permeant ion activity, a, can then be expressed as:

$$g = g_{max} \cdot a/(a + K_g) \qquad [1];$$

g_{max} is the maximal single-channel conductance, whose magnitude is determined by the translocation and dissociation rate constants; and K_g is the activity for half-maximal conductance, which is related to the equilibrium ion affinity of the channel (47). Most of our experiments on channels formed by modified gramicidins are therefore done with Na^+ as the permeating ion.

III. EXPERIMENTAL APPROACH

To study the mechanism(s) by which changes in the amino acid sequence can alter the single-channel conductance, we made use of two naturally occurring variants of valine grami-

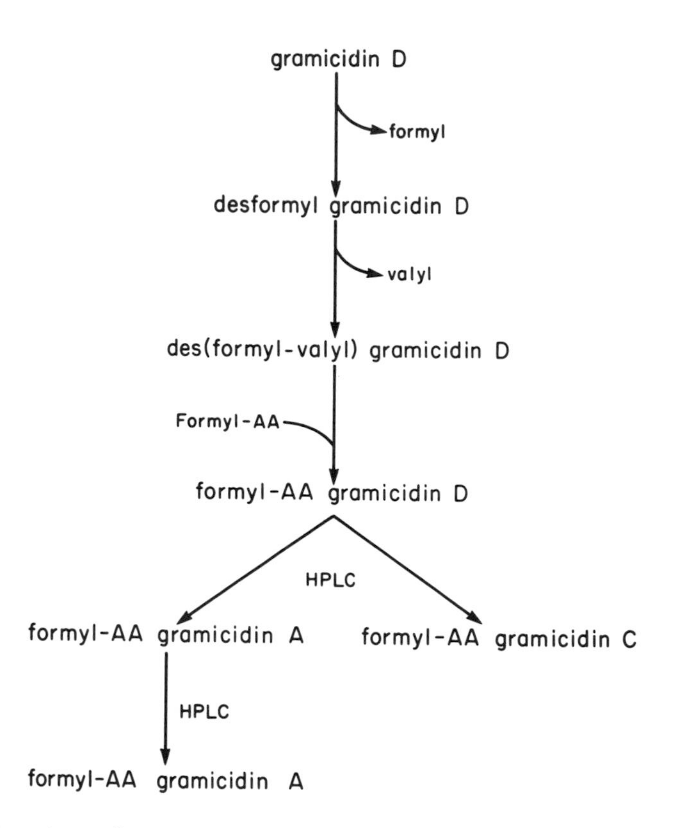

FIGURE 4. Diagram illustrating the major reactions in the replacement of the naturally occurring valine at position # 1 with another amino acid. Gramicidin D, the naturally occurring mixture of valine and isoleucine gramicidin A, B, and C, is the starting material. It is deformylated, the NH_2 -terminal amino acid is removed by Edman degradation, and a formylated amino acid is coupled to the peptide using diphenylphosphorazidate to produce a gramicidin with any desired residue at position # 1 (52). Gramicidin B's are synthesized starting with purified gramicidin B.

cidin A, where tryptophan at position # 11 is replaced by either phenylalanine (gramicidin B) or tyrosine (gramicidin C), and we produced semi-synthetic amino acid substitutions at the formyl-NH_2-terminus. (Substitutions at position # 11 will be identified by A, B, or C; substitutions at position # 1 will be identified by the name of the substituted amino acid.)

Figure 4 shows the semi-synthetic scheme. The products are purified by a two-step HPLC procedure, which also separates the modified peptide into their A and C variants (27, 52). The resulting analogues are sufficiently pure to allow characterization of the modified channels in planar lipid bilayers. The experiments are done using the bilayer punch (53).

One advantage of this chemical modification approach is that we are not limited to the twenty "genetic" amino acids. The use of non-genetic amino acids is necessary to produce systematic variations in side chain bulk and polarity.

IV. MODULATION OF SINGLE-CHANNEL CONDUCTANCES

Gramicidin single-channel conductances vary when the

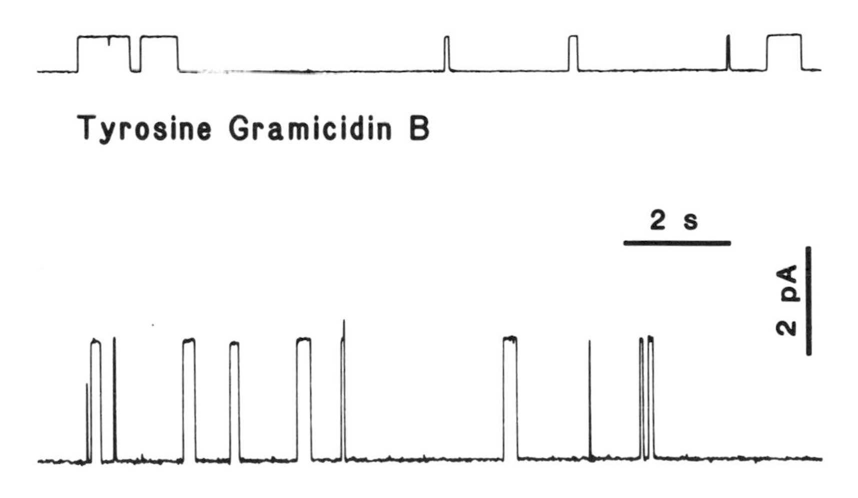

FIGURE 5. Comparison of single-channel events seen with tyrosine gramicidin B and phenylalanine gramicidin C in diphytanoylphosphatidylcholine/n-decane bilayers. Both traces were obtained at the same amplification and chart speed. 1.0 M NaCl, 200 mV, 25°C. Modified from Mazet et al. (27).

amino acid sequence is altered. A particularly dramatic example is shown in Fig. 5 (27). The conductance of channels formed by phenylalanine gramicidin C is more than three-fold greater than the conductance of channels formed by tyrosine gramicidin B, yet these compounds have the same overall amino acid composition. Indeed, they differ only in the shift of a single -OH between position # 1 and position # 11.

The amino acid side chains cover the exterior surface of the channel. The chemical changes we have produced in the peptides, and thus in the channels, are therefore spatially separate from the permeation path. How, then, do the amino acid substitutions alter the single-channel conductances?

A. Analogue Channel Structure

In principle, side chain substitutions could produce changes in the three-dimensional structure of the channels. If this were the case, it would have consequences for the interpretation of any conductance changes seen in the modified channels. It is therefore important to have a (simple) test for structural invariance among channels formed by "mutant" peptides. We have used the fact that valine gramicidin channels are symmetrical dimers to develop a simple functional criterion for structural invariance: If two gramicidin with different amino acid sequences form channels with identical structures, then they must be able to form hybrid channels between the chemically dissimilar monomers. The basic experiment is illustrated in Fig. 6 (28): Both glycine gramicidin C and phenylalanine gramicidin A form channels, but it is not obvious that two peptides with such extreme variations at their formyl-NH_2-termini will form channels with identical structure; to resolve this question we add the two peptides together, and we see the two pure channel types as well as a new channel type with a conductance in between the conductances of the pure channels. This new channel type must be a hybrid channel formed between the different monomers!

That hybrid channels exist shows that the folding of the formyl-NH_2-termini in the two dissimilar monomers can be very similar in the dimer. But the folding in each of the three channel types could still be different. To examine this possibility, we compared the frequency of observing hybrid channels with the frequencies for observing the two pure channel types. If the different monomers form β-helices with identical structures at their joining ends, the hybrid channel frequency, $\underline{f}_h$, should be related to the pure channel frequencies, $\underline{f}_a$ and $\underline{f}_c$ by (27):

$$\underline{f}_h > 2\ (\underline{f}_a\ \underline{f}_c)^{0.5} \qquad [2],$$

where the equality holds if the two orientations of the hybrid channels (a-c and c-a) occur with equal probability. The r.h.s. of Eq. 2 can thus be used to predict $\underline{f}_h$. The ratio $\underline{f}_h$(obs.)/$\underline{f}_h$(pred.) was 1.2 ± 0.4 in 12 experiments with phenylalanine gramicidin A and glycine gramicidin C (28). We conclude that these peptides form channels that have remarkably similar peptide backbone structures.

Similar experiments with gramicidins that have either

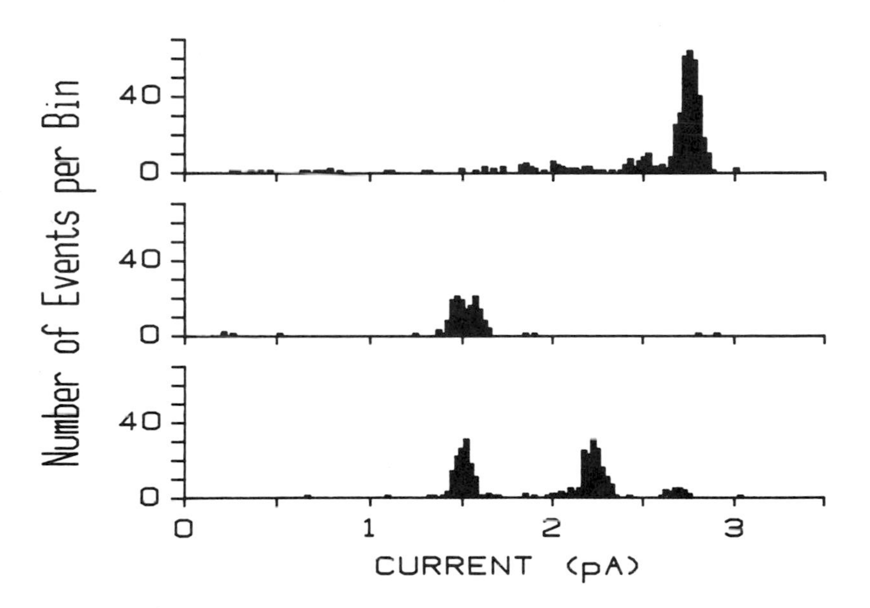

FIGURE 6. Amplitude histograms of single-channel current steps obtained with mixtures of phenylalanine gramicidin A and glycine gramicidin C. Top: Histogram with phenylalanine gramicidin A alone, the predominant single-channel current step was 2.743 ± 0.048 pA (mean ± s.d.). There were 444 channel transitions, of which 319 (or 72%) were in the major peak. Middle: Histogram with glycine gramicidin C alone, the single-channel current was 1.510 ± 0.062 pA. There were 157 channel transitions, of which 145 (or 92%) were in the major peak. Bottom: Histogram with a mixture of phenylalanine gramicidin A and glycine gramicidin C. The new peak that appears between the two peaks that were observed with the pure compounds represent hybrid channels. The average currents for the three peaks were, from left to right: 1.493 ± 0.040 pA [127]; 2.219 ± 0.048 pA [148]; and 2.673 ± 0.043 pA [24]. The numbers in brackets denote the number of channels in each population. There were 331 channel transitions, we can account for 90% of the transitions. 1.0 M NaCl, 200 mV. Modified from Durkin et al. (28).

phenylalanine, tryptophan, or tyrosine at position # 1 and position # 11 showed likewise that the side chain substitutions leave the basic channel structure unchanged even though they produce more than four-fold changes in the single-channel conductance (27). This structural invariance raises the question as to how an amino acid side chain at the exterior surface of the channel, 5 - 10 Å from the permeation path (54), can modulate the energy profile for the permeating ions.

B. Relative Importance of Side Chain Bulk and Polarity

The side chain-induced alterations in single-channel conductances could arise from steric restrictions of (small) conformational motions essential for ion movement (24); from electrostatic interactions between the side chains and the

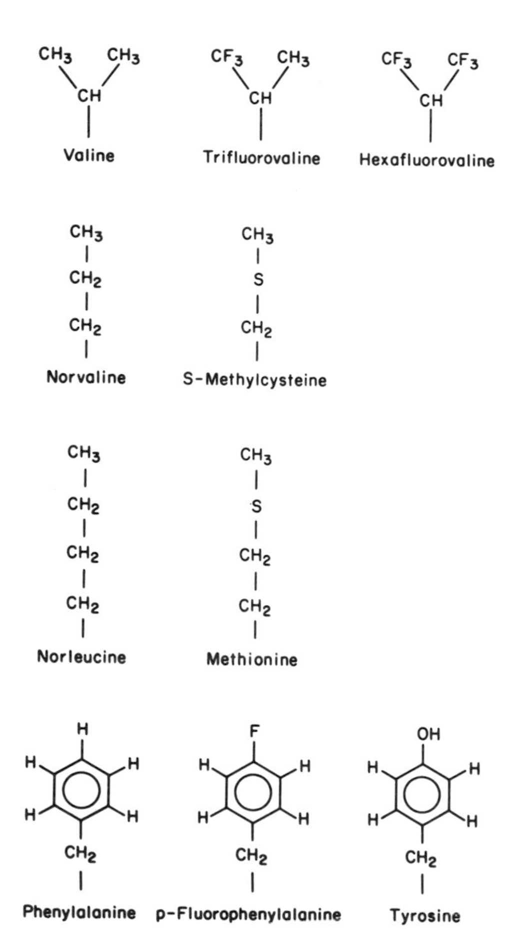

FIGURE 7. Examples of amino acid side chains that were introduced semisynthetically at position # 1 in gramicidin A. The four horizontal rows denote the four families of essentially isosteric side chains. In each family the side chain polarity increases from left to right.

permeating ions, either directly through ion-dipole interactions (24,27,55) or indirectly through inductive shifts in the electron densities of the peptide carbonyl oxygens (56); or from a combination of these factors.

We could distinguish among these possibilities in experiments where the bulk and polarity of the side chain at position # 1 in gramicidin A was systematically varied. Figure 7 shows a few of the side chains we have used. The importance of the bulk of the side chain was examined by comparing the characteristics of channels formed by peptides with one of the four nonpolar side chains at position # 1. The importance of side chain polarity was examined by comparing the charcteristics of channels formed by peptides that had essentially isosteric side chains with different polarity at position # 1.

The side chains vary in length, in branching, and in bulk. But hybrid experiments show that all the analogues form channels with the same structure as channels formed by the parent valine gramicidin A (27,29; Koeppe, Mazet, and Andersen, manuscript in preparation). The analogues thus satisfy a basic requirement in structure-function studies.

These seemingly innocuous amino acid substitutions produced more than ten-fold variations in the single-channel conductances (see Table I). Qualitatively similar variations were observed with Na^+ and with Cs^+ and the permeating ions. Data in 0.1 and 1.0 M NaCl and in 1.0 M CsCl are

TABLE I. Single-channel Conductances in NaCl and CsCl

Amino Acid at position # 1	[NaCl] 0.1 M	[NaCl] 1.0 M	[CsCl] 1.0 M
Valine	5.32	12.4	50.5
Trifluorovaline	1.01	1.93	16.0
Hexafluorovaline	1.06	1.42	18.8
Norvaline	5.66	14.7	
S-Methyl-cysteine	4.16	9.90	
Norleucine	6.34	14.7	49.4
Methionine	3.70	8.28	38.2
Phenylalanine	4.65	10.2	44.8
p-F-Phenylalanine	3.17	5.86	31.2
Tyrosine	2.32	3.94	

Diphytanoylphosphatidylcholine/*n*-decane bilayers
Small-signal conductances in pS, 25°C.

summarized in Table I. Despite the qualitative similarities between the conductance variations observed with Na^+ and with Cs^+, it is easier to interpret the Na^+ data because they to first approximation can be analyzed using a single-ion occupancy model. Single-channel conductances measured at several Na^+ activities can thus be used to fit Eq. 1 and yield estimates for $\underline{g}_{max}$ and $\underline{K}_g$. The major conductance variations result from changes in $\underline{g}_{max}$, which can vary more than ten-fold within a family of analogue channels (see Fig. 8). The variations in $\underline{K}_g$ are less pronounced, $\underline{K}_g$ is in most cases $\sim$ 0.1 M. The exceptions are channels formed by tyrosine or hexafluorovaline gramicidin A, which have two- to four-fold increased Na^+ affinities relative to the other channels (27, 29). Interestingly, these peptides are the only gramicidins with 15 amino acids where we find that the hybrid channels with valine gramicidin A have conductances that are less than the conductances of the pure channels types.

Importantly, the decreases in $\underline{g}_{max}$ are associated with an increased voltage-dependence (steepness) of the single-channel current-voltage characteristics and with a decreased concentration-dependence of their shapes (29; Koeppe, Mazet, and Andersen, manuscript in preparation). We thus conclude that the conductance decreases primarily result from increases in

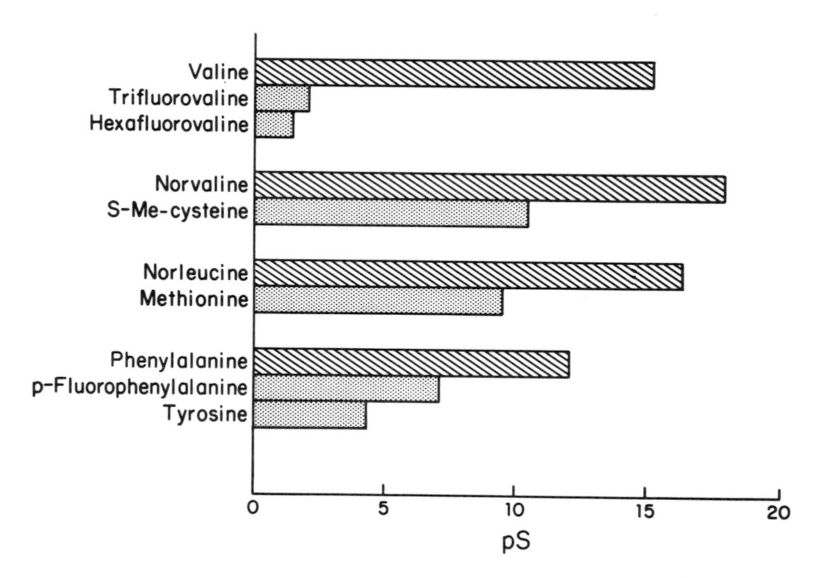

FIGURE 8. Bar diagram summarizing the changes in $\underline{g}_{max}$ for Na^+ among gramicidin A channels that have side chains with different bulk and polarity at position # 1. The four groups of bars indicate the four families of isosteric side chains.

the height of the energy barrier for ion translocation through the channel interior. We will not quantitate the changes in translocation rate constant, because such an analysis would be quite model-dependent and sensitive to the single ion occupancy approximation.

The data in Table I and Fig. 8 show that the single-channel conductances can be modulated by changes in the steric configuration and by changes in the polarity of the side chain at position # 1. The major conductance changes are related to changes in polarity! The dipoles associated with the polar side chains have their positive ends pointing towards the channel lumen (29; Koeppe, Mazet and Andersen, manuscript in preparation), and the dipole monents are sufficiently large that the conductance variations could arise from unfavorable electrostatic interactions between the side chain and an ion traversing the middle of the channel (27).

The changes in polarity of the side chains could also give rise to inductive (through bonds) electron shifts at the formyl-NH_2-termini, such as to decrease, or increase, the negative charge density at these carbonyl oxygens. We have examined this possibility with a series of gramicidins, where the amino acid # 1 was either phenylalanine or phenylalanine derivatives chosen such that the side chains varied in their electron-donating or -withdrawing capabilities (56). The conductance variations could not be accounted for by inductive electron shifts in the molecule. We conclude therefore, in part through exclusion, that the major mechanism by which these side chains affect ion permeability through the channels is through ion-dipole interactions.

Depending on the orientations of the side chains, ion-dipole interactions could be favorable or unfavorable for ions traversing the channel - one cannot conclude that an increased polarity per se will lead to a decreased single-channel conductnace. The indole ring in tryptophan has, for example, a dipole moment ∿ 2.1 D (57), while channels formed by gramicidins with tryptophan at position # 1 have conductances that are comparable to the conductances of channels formed by the corresponding valine gramicidins (cf. data in Tables II and III in Ref. 27). Not only the magnitude of the dipole moment but also its orientation must be important.

The invariance in Na^+ affinity among most analogue channels suggests that the ion-dipole interactions exert their major effects close to the center of the channels, i.e., that the conductance changes arise primarily from an increase in the height of the central energy barrier. Simple model calculations of ion-dipole interactions support this notion of (fairly) local effects (27), as do the much more extensive calculations of Etchebest and Pullman (20).

The affinity changes seen with channels formed by hexafluorovaline and tyrosine gramicidin A are also compatible with ion-dipole interactions between a side chain and an ion in one of the major binding sites, *if* the dipole is strongly oriented with its negative end pointing towards the binding site close to the channel entrances. A single side chain could in this stituation induce a quite asymmetrical modulation of the energy profile experienced by a permeating ion, $\Delta E'(\underline{x})$, with a minimum around one binding site and additionally a maximum towards the other end of the channel ($\underline{x}$ denotes the ion's distance from the center of channel, from left to right). If this were so, we have a simple explanation why these peptides that form channels with an increased Na^+ affinity also form hybrid channels with valine gramicidin A that have conductances *less* than the conductances of the pure channel types (27,29). In a symmetrical channel the total change in the energy profile, $\Delta E(\underline{x})$, is the sum of contributions from the two side chains, $\Delta E'(\underline{x})$, and $\Delta E''(\underline{x})$, where $\Delta E'(\underline{-x}) \simeq E''(\underline{-x})$, and where $\Delta E'(\underline{x})$ and $\Delta E''(\underline{x})$, will have their maxima towards opposite ends of the channel. If the maximum of the symmetrical $\Delta E(\underline{x})$ is less than the maxima of the asymmetrical $\Delta E'(\underline{x})$, or $\Delta E''(\underline{x})$, and the energy barrier in the interior of channels formed by valine gramicidin A is fairly flat (as seems to be the case), then the maximum of the total energy profile may be higher in a hybrid channel than in a symmetrical channel formed by the polar peptides. The rate constant for ion translocation through the hybrid channels could thus be less than for translocation through the pure channels.

V. CONCLUSIONS

Amino acid substitutions in linear gramicidins can have dramatic effects on the conductances and ion selectivities of the "mutant" channels. These changes are important in their own right, as they allow us to study some of the molecular mechanisms by which a peptide channel can interact with permeating ions. These studies on modified gramicidins also have a broader importance as they allow us to use chemically modified gramicidin channels as a prototype for chemical or genetic manipulations of channel function.

The use of genetic amino acids allowed us to state some qualitative consequences of amino acid substitutions and of the importance of the position in the sequence where the substitution is made (27). The general framework in these experiments was similar to that used in molecular genetic studies on protein structure-function relations. Our use of hybrid channel experiments is for example conceptually similar to the use of the double mutations to establish structural equivalence (31). We were, however, in these experiments only able to arrive at fairly qualitative conclusions (e.g., on the importance of the position of the side chain in the channel) and could not draw conclusions about the mechanisms by which the side chains modulate channel function.

A more refined understanding of the possible mechanisms required a distinction between effects due to side chain-induced steric restrictions of conformational motions essential for ion movement through the channel and electrostatic interactions between the side chain and the permeating ions. This separation was possible in experiments where we used semi-synthetic sequence alterations to extend structure-function studies beyond what is feasible with "conventional" molecular biological techniques: i.e., insertions or deletions in the sequence, and point mutations among the twenty amino acids that are coded by mRNA. By systematically and independently varying the steric configuration and the polarity of a side chain we could conclude that the magnitude and orientation of the side chain dipole moment are important determinants of channel function. This conclusion may not be surprising, but it is to our knowledge the first time the functional importance of the side chain dipoles has been demonstrated experimentally in a transmembrane channel.

We note that molecular biological methods cannot easily deal with amino acids other than the 20 that are specified by mRNA. Nevertheless, one can conceive of using _in vitro_ translation of mRNA together with a purified misacylated tRNA to introduce a non-genetic amino acid. The introduction of trifluorovaline, for example, into the core of a protein is therefore not in principle an insurmountable problem.

The interpretation of the functional data was greatly helped by our knowledge of the general channel structure, by our use of families of isosteric side chains, and by our functional criterion for structural invariance among channels formed by the "mutant" peptides. If the functional changes were associated with side chain-dependent alterations in channel structure, it would have been necessary to establish the structure of each analogue channel _de novo_. It would probably also have been difficult to derive mechanistic information from the conductance changes.

If the structure of the channel were unknown, we might even have concluded that residue # 1 constitutes a critical part of the wall of permeation pathway, and attempted to construct a model around this constraint! This would be erroneous, because we know that the side chains are outside the channel, and not in direct physical contact with the permeating ion. Our data show that significant ion-dipole interactions may occur over $\sim$ 10 Å. Simple reasoning, based on an assumption of proximity between the point of structural perturbation and the point of functional perturbation, would thus have led us astray. It follows that one cannot rely on such simple assumptions in the interpretation of data from site-directed mutagenesis experiments in systems far more complex than the gramicidin channel!

The possibility of introducing any amino acid into the sequence permits us to investigate the interactions between channel peptide and permeating ions in molecular detail, because we can produce analogues with graded changes in their physico-chemical characteristics and thus pursue increasingly detailed studies of channel function. Importantly, we have reached the point where the design and interpretation of experiments will depend upon the availability of theoretical calculations: conformational energy calculations (54); energy profile calculations (42,58); and Monte Carlo (51), Brownian dynamics (59), and molecular dynamics simulations (46,60) on the energetics and dynamics of ion, H_2O, and peptide motions. The next few years should therefore witness a fruitful interaction between theory and experiment, where experimental results will stimulate the development of increasingly detailed (and complicated) calculations of channel properties and theoretical calculations will suggest new experiments. The theory should also allow us to decrease the total experimental effort needed to elucidate a specific question, while experimental observations will provide the yardstick against which the theories eventually must be measured.

REFERENCES

1. Andersen, O. S., Ann. Rev. Physiol. 46:531 (1984).
2. Hladky, S. B., and Haydon, D. A., Curr. Top, Membr. Transp. 21:327 (1984).
3. Hinton, J. F., and Koeppe, R. E. II., in "Metal Ions in Biological Systems" (H. Sigel, ed.), vol. 19, p. 173. Marcel Dekker, New York, 1985.
4. Urry, D. W., Proc. Natl. Acad. Sci. USA 68:672 (1971).
5. Sarges, R., and Witkop, B., J. Am. Chem. Soc. 87:2011 (1965).
6. Urry, D. W., Goodall, M. C., Glickson, J. D., and Mayers, D. F., Proc. Natl. Acad. Sci. USA 68:1907 (1971).
7. Bamberg, E., and Janko, K., Biochim. Biophys. Acta 465: 486 (1977).
8. Bamberg, E., Apell, H.-J., and Alpes, H., Proc. Natl. Acad. Sci. USA 74:2402 (1977).
9. Veatch, W., and Stryer, L., J. Mol. Biol. 113:89 (1977).
10. Szabo, G., and Urry, D. W., Science 203:55 (1979).

11. Weinstein, S., Wallace, B. A., Blout, E. R., Morrow, J. S., and Veatch, W. R., Proc. Natl. Acad. Sci. USA 76: 4230 (1979).
12. Weinstein, S., Wallace, B. A., Morrow, J. S., and Veatch, W. R., J. Mol. Biol. 143:1 (1980).
13. Urry, D. W., Walker, J. T., and Trapane, T. L., J. Membr. Biol. 69:225 (1982).
14. Urry, D. W., Trapane, T. L., and Prasad, K. U., Science 221:1064 (1983).
15. Weinstein, S., Durkin, J. T., Veatch, W. R., and Blout, E. R., Biochemistry 24:4374 (1985).
16. Boni, L. T., Connolly, A. J., and Kleinfeld, A. M., Biophys. J., 49:122 (1986).
17. Finkelstein, A., in "Drugs and Transport Processes" (B. A. Callingham, ed.), p. 241. Macmillan, London, 1974.
18. Levitt, D. G., Elias, S. R., and Hautman, J. M., Biochim. Biophys. Acta 512:436 (1978).
19. Rosenberg, P., and Finkelstein, A., J. Gen. Physiol. 72: 327 (1978).
20. Etchebest, C., and Pullman, A., J. Biomol. Struct. Dyn. 2:859 (1985).
21. Parsegian, V. A., Ann. NY Acad. Sci. 264:161 (1975).
22. Jordan, P. C., Biophys. J. 39:157 (1982).
23. Bamberg, E., Noda, K., Gross, E., and Läuger, P., Biochim. Biophys. Acta 419:223 (1976).
24. Morrow, J. S., Veatch, W. R., and Stryer, L., J. Mol. Biol. 132:733 (1979).
25. Heitz, F., Spach, F., and Trudelle, Y., Biophys. J. 40:87 (1982).
26. Prasad, K. U., Trapane, T. L., Busath, D., Szabo, G., and Urry, D. W., Int. J. Peptide Protein Res. 22:341 (1983).
27. Mazet, J.-L., Andersen, O. S., and Koeppe, R. E. II., Biophys. J. 45:263 (1984).
28. Durkin, J. T., Andersen, O. S., Blout, E. R., Heitz, F., Koeppe, R. E. II, and Trudelle, Y., Biophys. J. 49:118 (1986).
29. Russell, E. W. B., Weiss, L. B., Navetta, F. I., Koeppe, R. E. II, and Andersen, O. S., Biophys. J. 49:673 (1986).
30. Villafranca, J. E., Howell, E. E., Voet, D. H., Strobel, M. S., Ogden, R. C., Abelson, J. N., and Kraut, J., Science 222:782 (1983).
31. Carter, P. J., Winter, G., Wilkinson, A. J., and Fersht, A. R., Cell 38:835 (1984).
32. Mishina, M., Tobimatsu, T., Imoto, K., Tanaka, K.-I., Fujita, Y., Fukuda, K., Kurasaki, M., Takahashi, H., Morimoto, Y., Hirose, T., Inayama, S., Takahashi, T., Kuno, M., and Numa, S., Nature 313:364 (1985).
33. Veatch, W. R., and Durkin, J. T., J. Mol. Biol. 143:411 (1980).

34. Hinton, J. F., Koeppe, R. E. II, Shungu, D., Whaley, W. L., Paczkowski, J. A., and Millett, F. S., Biophys. J. 49:571 (1986).
35. Hinton, J. F., Whaley, W. L., Shungu, D., Koeppe, R. E. II, and Millett, F. S., Biophys. J. 50:538 (1986).
36. Procopio, J., and Andersen, O. S., Biophys. J. 25:8a (1979).
37. Andersen, O. S., and Procopio, J., Acta physiol. scand. suppl. 481:27 (1980).
38. Urban, B. W., Hladky, S. B., and Haydon, D. A., Biochim. Biophys. Acta 602:331 (1980).
39. Eisenman, G., and Sandblom, J. P., in "Physical Chemistry of Transmembrane Ion Motions" (G. Spach, ed.), p. 247. Elsevier, Amsterdam, 1983.
40. Andersen, O. S., Barrett, E. W., and Weiss, L. B., Biophys. J. 33:63a (1981).
41. Levitt, D. G., Biophys. J. 22:221 (1978).
42. Etchebest, C., Ranganathan, S., and Pullman, A., FEBS Lett. 173:301 (1984).
43. Renugopalakrishnan, V., and Urry, D. W., Biophys. J. 24: 729 (1978).
44. Levitt, D. G., Biophys. J. 22:209 (1978).
45. Finkelstein, A., and Andersen, O.S., J. Membr. Biol. 59: 155 (1981).
46. Mackay, D. H. J., Berens, P. H., Wilson, K. R., and Hagler, A. T., Biophys. J. 46:229 (1984).
47. Andersen, O. S., Biophys. J. 41:147 (1983).
48. Andersen, O. S., Biophys. J. 41:135 (1983).
49. Läuger, P., Biochim. Biophys. Acta 455:493 (1976).
50. Killian, J. A., and de Kruijff, B., Biochemistry 24:7890 (1985).
51. Fornili, S. L., Vercauteren, D. P., and Clementi, E., Biochim. Biophys. Acta 771:151 (1984).
52. Weiss, L. B., and Koeppe, R. E. II., Int. J. Peptide Protein Res. 26:305 (1985).
53. Andersen, O. S., Biophys. J. 41:119 (1983).
54. Venkatachalam, C M., and Urry, D. W., J. Comput. Chem. 4: 461 (1983).
55. Tredgold, R. H., Hole, P. N., Sproule, R. C., and Elgamal, M., Biochim. Biophys. Acta 471:189 (1977).
56. Koeppe, R. E. II., Andersen, O. S., and Mazet, J.-L., 8th Int. Biophys. Congr. 273 (1984).
57. Weiler-Feilchenfeld, H., Pullman, A., Berthod, H., and Giessner-Prettre, C., J. Mol. Structure 6:297 (1970).
58. Jordan, P. C., J. Membr. Biol. 78:91 (1984).
59. Cooper, K., Jakobsson, E., and Wolynes, P., Prog. Biophys. Mol. Biol. 46:51 (1985).
60. Fischer, W., Brickmann, J., and Läuger, P., Biophys. Chem. 13:105 (1981).

GRAMICIDIN: A MODULATOR OF LIPID STRUCTURE

Ben de Kruijff

Institute of Molecular Biology
University of Utrecht
Utrecht, The Netherlands

J. Antoinette Killian

Department of Biochemistry
University of Utrecht
Utrecht, The Netherlands

I. INTRODUCTION

Gramicidins are polypeptide antibiotics, produced by *Bacillus brevis* strain ATCC 8185. They are linear pentadecapeptides, consisting of alternating D- and L-amino acids. Their N-terminal and C-terminal parts are blocked by a formyl group and ethanolamine group, respectively. Due to the absence of any charged or polar amino acids, these molecules are extremely hydrophobic. The natural mixture consists for 85% of gramicidin A, the structure of which is shown in Fig. 1. The less abundant gramicidin species B and C differ from gramicidin A in the substitution of Trp^{11} by a phenylalanine or tyrosine residue, respectively. In 5-20% of the molecules, the Val^1 is substituted by an isoleucine residue (1).

HCO-L-Val(1)-Gly-L-Ala-D-Leu-L-Ala(5)-D-Val-L-Val-D-Val-

L-Trp-D-Leu(10)-L-Trp-D-Leu-L-Trp-D-Leu-L-Trp(15)-NHCH$_2$-CH$_2$OH

FIGURE 1. Chemical structure of gramicidin A.

The gramicidins are synthesized by multi-enzyme complexes (2) and most likely are, together with the tyrocidine type of peptide antibiotics produced by *B. brevis*, involved in gene regulation during the shift from the vegetative phase to the sporulation phase of growth of the bacteria (3). Gramicidin can interact *in vitro* with the DNA-tyrocidine complex, which results in activation of overall RNA synthesis (4). Furthermore, gramicidin is able to inhibit transcription by binding to the σ-subunit of the RNA polymerase (5,6).

The main interest in gramicidin has arisen from the observation that it can dramatically influence the barrier properties of membranes (7,8). The addition of very small amounts of the peptide to either model or biological membranes results in a loss of barrier function due to the formation of transmembrane channels through which water and small cations can pass the membrane (recent review: 9). In the channel conformation the molecule is completely hydrophobic at the outside and it spans, most likely as a dimer, the membrane. Since this conformation resembles that of the membrane-spanning part of intrinsic membrane proteins, gramicidin has been a popular model for such proteins. In particular the molecule has been used to study various aspects of lipid-protein interactions in model membrane systems. Studies with lamellar lipid systems formed by saturated phosphatidylcholines (PC's) showed that gramicidin can perturb acyl chain packing of the lipid molecules (10-13). Below the transition temperature it decreases chain order. In the liquid-crystalline state for gramicidin-lipid ratios less than 1:15 (molar) the chains are more ordered, at higher concentrations the polypeptide causes chain disordering.

In 1981 (14), it was found that gramicidin can greatly influence lipid polymorphism. Since then many new aspects of the gramicidin-lipid interaction were revealed. It is the aim of this review to summarize these studies, and to discuss the molecular basis of the lipid structure-modulating activity of the polypeptide.

Lipid polymorphism is the ability of lipids to organize in different macroscopic structures depending on the environmental conditions (for recent reviews see 15,16). For membrane lipids under physiological conditions of pH, ionic strength, temperature and hydration, typically 3 types of structures are observed. The minor membrane components, gangliosides and lysophospholipids, organize in excess water in micellar structures (Fig. 2). The more abundant PC's and sphingomyelins are examples of lipids which organize under such conditions in the lamellar phase. Phosphatidylethanolamines (PE's) and monogalactosyldiglycerides are membrane lipids which prefer an organization in the hexagonal H_{II} phase. In this phase, the lipid molecules are oriented with their headgroups towards the aqueous interior of the hexagonally organized tubes of which

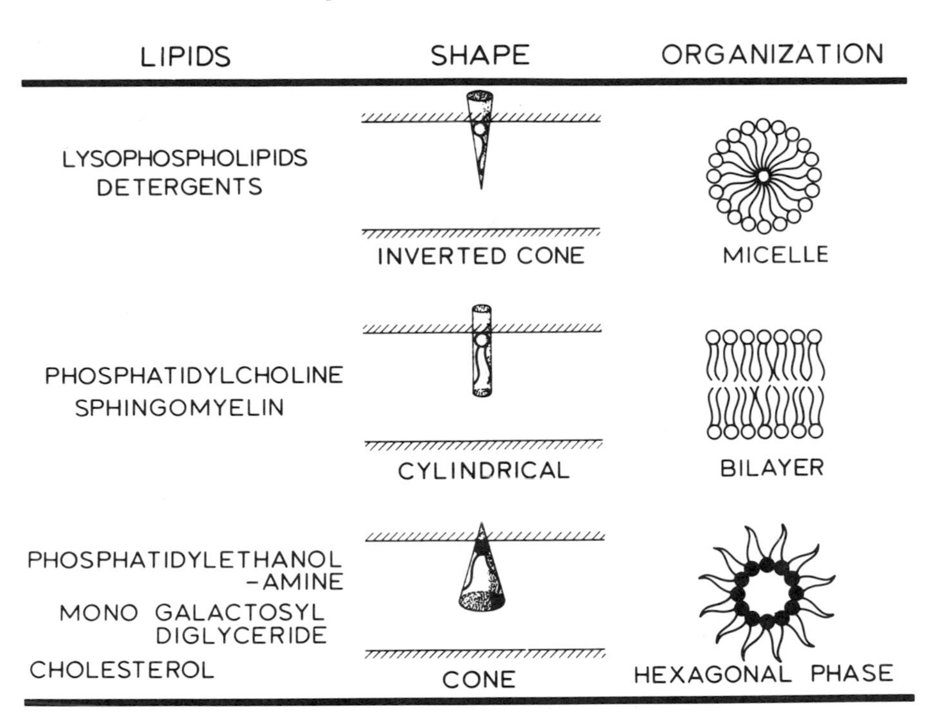

FIGURE 2. Shape-structure concept of lipid polymorphism. The H_{II} phase is schematically represented by a cross-section through one of the tubes of which this phase is composed.

this phase is formed. The most important feature of the hexagonal H_{II} phase is the inverted orientation of the lipid molecules. H_{II} type of lipids most likely are involved in a number of membrane functions such as fusion, transbilayer transport of lipids and polar compounds, protein insertion and translocation (15,16).

Figure 2 illustrates the shape-structure concept, which provides a basis in the understanding of lipid polymorphism. When the headgroup area exceeds that of the hydrocarbon part of the molecule, the lipids have an inverted cone shape, and thus prefer a micellar organization. When the size of the headgroup is in balance with the hydrocarbon part, the molecules on average are cylindrical, and thus are optimally suited to organize in a two-dimensional structure such as a lipid bilayer. When the headgroup is relatively small, the molecules have a cone shape and thus prefer an organization in the hexagonal H_{II} phase. It should be noted that shape in the shape-structure concept of lipid polymorphism is inclusive. It is not only determined by the geometrical space occupied by the molecule, but includes dynamics as well as intra- and intermo-

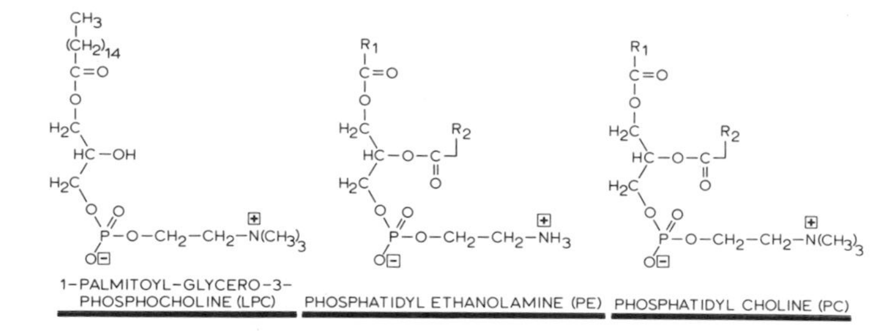

FIGURE 3. Chemical structures of lipids.

lecular interactions and headgroup hydration. The chemical structure of representatives of lipids forming the 3 main phases are shown in Fig. 3. The effect of gramicidin incorporation on the structure of these three lipids is described in the following chapters.

II. GRAMICIDIN-LYSOPHOSPHATIDYLCHOLINE INTERACTIONS

Palmitoyl-lysophosphatidylcholine (lyso-PC) in excess water forms micelles. The ^{31}P NMR spectrum of such a micellar solution is characterized by an isotropic signal (Fig. 4). This is the result of rapid reorientation of the lipids by micelle tumbling and lateral diffusion by which the chemical shift anisotropy of the phosphorus atom is almost completely averaged (17,18). Incorporation of gramicidin by hydration of a mixed lipid-gramicidin film results in the gradual appearance of a broad spectral component with an asymmetrical line shape (19). This line shape is typical for a lamellar organization of the phospholipids in which the chemical shift anisotropy is only partially averaged by fast axial rotation of the lipid molecules (17,18). The growth of this bilayer component goes at the expense of the isotropic NMR signal, such that at a molar ratio of 1 gramicidin to 4 lyso-PC molecules, the isotropic signal has almost completely vanished. Quantification of the spectral component reveals that as a result of the gramicidin-lipid interaction, a lamellar gramicidin-lyso-PC (1:4, molar) complex is formed (19). The bilayer stabilization exerted by gramicidin on lyso-PC was an unexpected observation. Detergents, such as lyso-PC, normally destabilize bilayer structures.

Information on acyl chain order in the lamellar gramicidin-lyso-PC complex can be obtained by ^{2}H NMR using specifically deuterated compounds (20). The quadrupolar splitting ($\Delta\nu_q$) is related to the order parameter of the C-D segment and in case of dipalmitoyl-PC (DPPC) is constant for the first 9 methylene

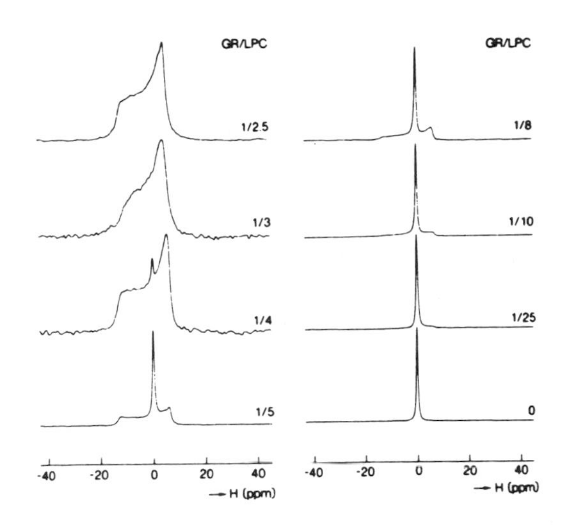

FIGURE 4. 81.0 MHz ^{31}P NMR spectra of aqueous dispersions of gramicidin (GR) and lyso-PC (LPC) at various molar ratios. For details see ref. (19).

groups and decreases towards the terminal part of the acyl chain as is shown in Fig. 5. A very similar order parameter profile is observed for lyso-PC in lamellar structures with either gramicidin or as present in dispersions of lyso-PC with DPPC (molar ratio 1:4) (21). That the quadrupolar splitting of chain labelled lyso-PC is very sensitive towards changes in lipid packing can be observed in Fig. 5 for the lamellar complex formed with cholesterol. The condensing effect of cholesterol (22) is manifested as an overall increase of the quadrupolar splitting which translates into an increase in acyl chain order. The lack of any significant effect of gramicidin on $\Delta\nu_q$ despite the very high peptide content, e.g. 4 acyl chains per gramicidin monomer, suggests that either the acyl chains are unperturbed by the presence of the gramicidin molecule, or indicates that extensive self-association of gramicidin occurs, whereby the contact area between the peptide and the lipid decreases. Formation of structured gramicidin aggregates in mixtures with lyso-PC has previously been suggested on the basis of tryptophan fluorescence measurements and electron microscopy (23). The presence of ordered gramicidin aggregates is also suggested from freeze-fracture electron microscopy. Although the gramicidin-lyso-PC 1:4 molar complex is characterized by smooth fracture faces, in case of a molar ratio of 1:2.5 regularly structured fracture faces are observed (Fig. 6). Previously it was shown that mixtures of cone-shaped lipids

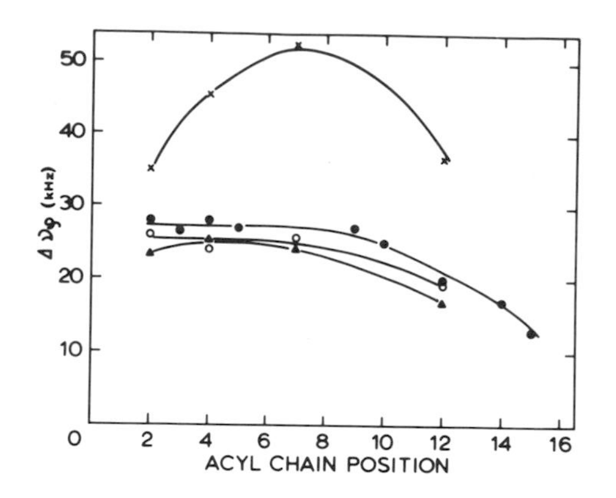

FIGURE 5. Quadrupolar splitting at 45°C of the labelled carbon atom in the acyl chain, in aqueous dispersions of DPPC (●—●), LPC/DPPC (1/4, m/m) (o—o), LPC/gramicidin (4/1, m/m) (▲—▲) and LPC/cholesterol (1/1, m/m) (×—×). Reproduced with permission from ref. (21).

such as PE and inverted cone-shaped lipids such as lyso-PC are organized in a lamellar phase (24) due to the complementarity of the shapes of the molecules. This implies that the gramicidin monomer is cone shaped. The location of the 4 bulky tryptophan residues at the C-terminal part of the molecule suggests that this part will be at the wider end of the cone and will be located in the interior of the lipid bilayer. However, this is in contradiction to what has been postulated for the structure

FIGURE 6. Freeze-fracture micrograph of an aqueous dispersion of lyso-PC/gramicidin (2.5/1, m/m), quenched from room temperature. Magnification 100,000 x.

of the gramicidin dimer in mixtures with lyso-PC. In the channel conformation, gramicidin monomers are believed to be present as $\beta^{6.3}$ helices, joined via hydrogen bonding by their N-terminal parts (7). This will then result in a location of the tryptophan-rich C-terminal part at the membrane water interface. This is opposite to the location inferred from the effect of gramicidin on structure of the lyso-PC. Recent energy calculations(25), however, revealed that due to the hydrophobicity of the tryptophans, a location of the monomer with the tryptophan-rich C-terminal part in the hydrocarbon layer would be more favorable.

Gramicidin-lyso-PC interactions

1. Gramicidin upon co-dispersion with lyso-PC induces a micelle to bilayer transition in lyso-PC, resulting in an 1:4 molar complex.
2. Gramicidin is probably self-associated in this complex.
3. For the $\beta^{6.3}$ structure of gramicidin, the shape concept and conformational analysis predict a location of the N-terminal part at the bilayer/water interface.

III. GRAMICIDIN-PHOSPHATIDYLETHANOLAMINE INTERACTIONS

Aqueous dispersions of PE's can undergo in the lamellar phase a gel to liquid-crystalline phase transtition and in addition a lamellar liquid-crystalline to hexagonal H_{II} transition (15). This is shown in the thermogram of dielaidoylphosphatidylethanolamine (DEPE) dispersed in buffer (Fig. 7). The gel to liquid-crystalline phase transition occurs around 35°C. The bilayer to hexagonal transition occurs around 60°C and has a much smaller transition enthalpy due to the fluid character of both the lamellar and hexagonal H_{II} phase.

Incorporation of even 1 gramicidin per 5 DEPE molecules has virtually no effect on the gel to liquid-crystalline phase transition (Fig. 7 and 8) (26). However, in case of dielaidoylphosphatidylcholine (DEPC) incorporation of gramicidin results in a decrease in transition enthalpy up to a molar ratio of gramicidin to lipid of about 0.05 (26). At a higher peptide concentration, no further decrease is ovserved. These results

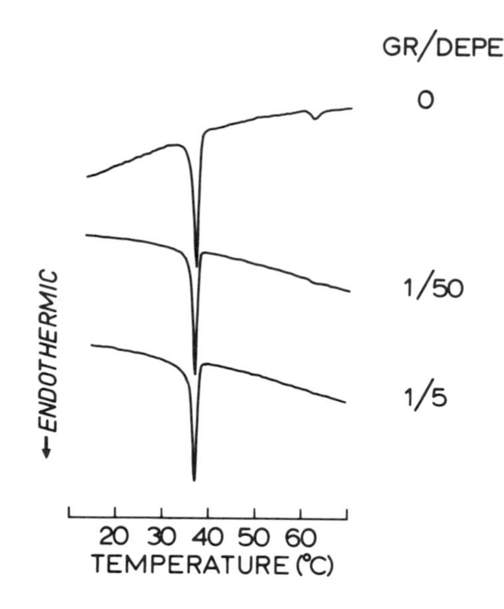

FIGURE 7. DSC-heating curves of aqueous dispersions of mixtures of DEPE and gramicidin (GR) with the indicated molar composition. Reproduced with permission from ref. (26).

indicate that gramicidin has a limited solubility in the gel state of PE and PC. For DEPE apparently virtually no gramicidin-lipid interactions occur in the gel state, suggesting extensive segregation between the peptide and the lipid. In case of DEPC {and also DPPC (26,27)}, apparently the peptide can dissolve in the bilayer up to a molar ratio of 1 gramicidin per 20 PC molecules.

The effect of gramicidin incorporation on the lamellar to hexagonal transition in the DEPE is shown in Fig. 9. The presence of only one gramicidin molecule per 100 DEPE molecules causes already a significant downward shift in transition temperature. The presence of higher concentrations of gramicidin even lowers the bilayer to hexagonal transition temperature till that of the gel to liquid-crystalline phase transition.

^{31}P Chemical shift anisotropy measurements showed that in case of co-existing liquid-crystalline and hexagonal H_{II} phases the gramicidin molecules are preferentially located in the hexagonal H_{II} phase and cause a decrease in headgroup order (26). The possible mechanism of the gramicidin-induced H_{II} phase formation in PE systems is shown in the upper part of Fig. 10. In the lamellar gel state, gramicidin forms aggregates at all concentrations of the peptide. Upon raising the temperature till the gel → liquid-crystalline phase transition temperature T_c,

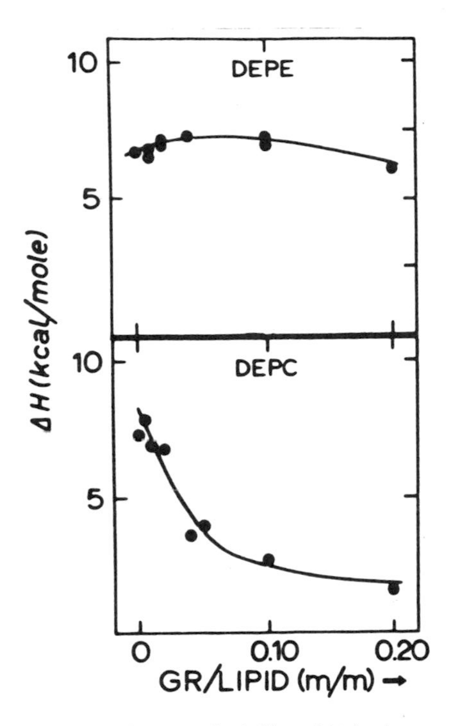

FIGURE 8. Effect of gramicidin (GR) incorporation on the enthalpy of the gel → liquid-crystalline phase transition in aqueous dispersions of DEPE and DEPC. Details see ref. (26).

phase separation occurs between a fluid bilayer of pure DEPE (route 1) and one which is very rich in gramicidin (route 2). Upon further increasing the temperature, more PE will organize in this gramicidin-containing H_{II} phase. The remaining part of the lipids, the extent of which is dependent on the gramicidin concentration, will undergo a normal bilayer to hexagonal H_{II} phase transition (route 3).

Gramicidin-PE interactions

1. Gramicidin strongly promotes H_{II} phase formation.
2. Gramicidin self-associates in the gel phase and possibly in the H_{II} phase.

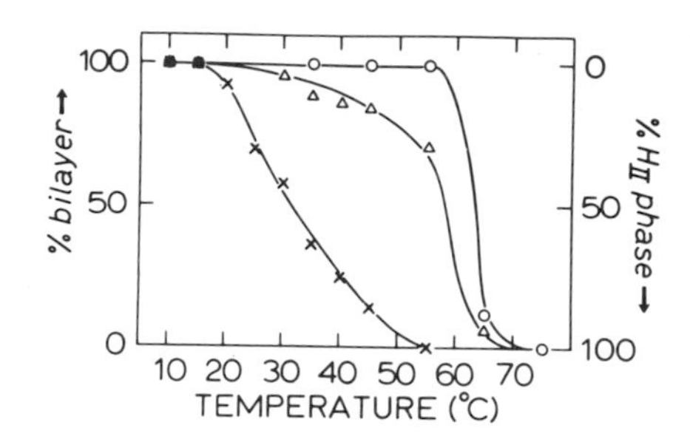

FIGURE 9. Percentages of bilayer and H_{II} component in the ^{31}P NMR spectra of dispersions of DEPE (o—o) and gramicidin/DEPE mixtures of molar ratios of 1/100 (Δ—Δ) and 1/10 (×—×) as a function of temperature. Details see ref. (26).

IV. GRAMICIDIN-PHOSPHATIDYLCHOLINE INTERACTIONS

Naturally occurring PC's are typically bilayer-forming lipids. As compared to PE, the presence of the quaternary ammonium group increases headgroup hydration and decreases intermolecular interactions, both resulting in an increase of effective size of the headgroup.

The gramicidin-induced bilayer destabilization observed in studies with PE's is so strong that even in case of PC's an H_{II} phase can be induced under the appropriate conditions (14,28). The ^{31}P NMR spectra of a series of dispersions of saturated liquid-crystalline PC's containing gramicidin in a 10:1 molar ratio are shown in Fig.11. For acyl chain lengths up to 16 carbon atoms, incorporation of gramicidin results in the appearance of a new spectral component with a ^{31}P NMR line shape typical of the hexagonal H_{II} phase (28). Due to fast lateral diffusion of the lipid molecules around the tubes in the hexagonal H_{II} phase, additional averaging of the chemical shift anisotropy occurs, resulting in a reversal in anisotropy of the spectrum and a reduction in line width (17,18). As a result, the dominant spectral component for lipids in the hexagonal H_{II} phase occurs at around -7 ppm. The same behavior is observed for a series of unsaturated PC's (Fig. 12). In the absence of the peptide, all species are organized in a lamellar phase with their characteristic ^{31}P NMR spectrum. However, the incorporation of 1 gramicidin per 10 PC molecules also now results in hexagonal H_{II} phase formation when the acyl chain length exceeds 16 carbon atoms (28). H_{II} phase formation is maximal for the longest chain tested ($22:1_c$-PC). The difference in amount

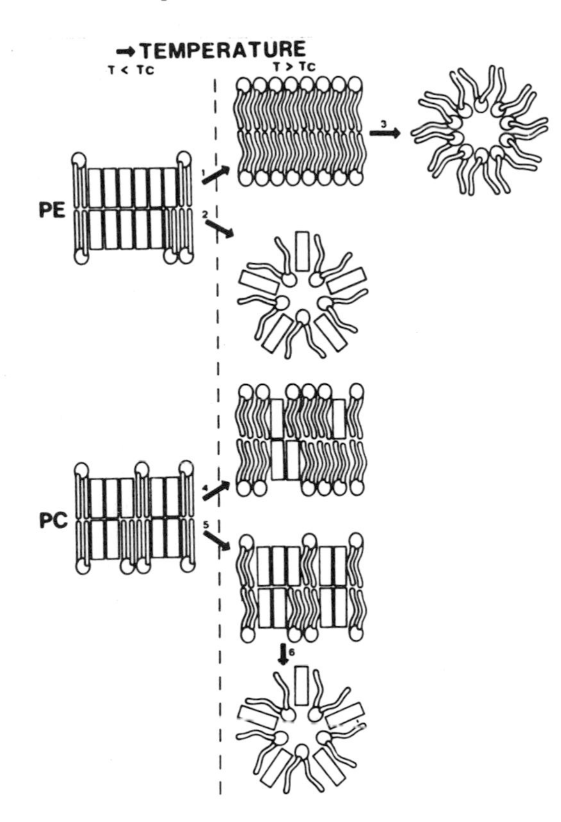

FIGURE 10. Schematic representation of the possible mechanism of gramicidin-induced H_{II} phase formation in PE and PC systems. Reproduced with permission from ref. (26). See text and this reference for details.

of hexagonal H_{II} phase induced by gramicidin in DEPC ($18{:}1_t$) and dioleoyl-PC ($18{:}1_c$, DOPC) suggests that in addition to the chain length, also the nature of unsaturation plays an important role in the gramicidin-induced H_{II} phase formation.

From both the thermodynamical and structural data, the following picture of gramicidin-induced H_{II} phase formation in PC model membranes emerges (lower part of Fig. 10, compare with the results on PE-gramicidin mixtures). In PC systems aggregation of the peptide in the lamellar gel state only occurs at high gramicidin concentrations, whereby probably smaller aggregates are formed as in PE systems. Upon increasing the temperature at low peptide concentrations gramicidin will be present in the fluid bilayer as mono- or oligomers (route 4). At higher concentrations the aggregate size increases (route 5). These large aggregates are able to induce H_{II} formation (route 6)

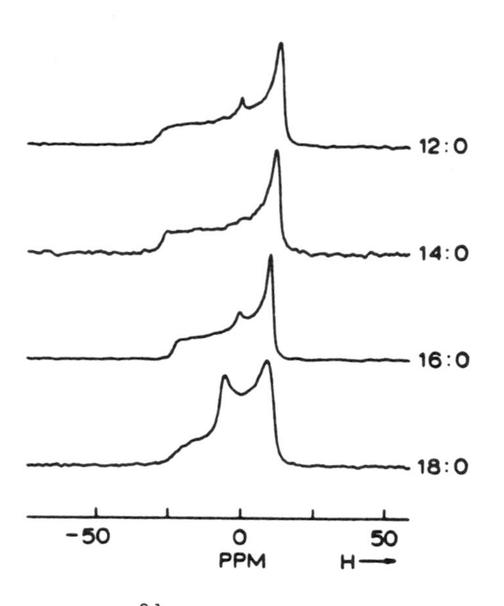

FIGURE 11. 81.0 MHz ^{31}P NMR spectra of aqueous dispersions of mixtures of gramicidin with disaturated liquid-crystalline PC species in an 1:10 molar ratio. Reproduced with permission from ref. (28).

when the chain length exceeds 16 carbon atoms.

In the experiments shown in Figs. 11 and 12, samples were prepared by hydration of a mixed PC-gramicidin film. However, also the addition of gramicidin as an ethanolic solution to preformed DOPC liposomes results in H_{II} phase formation, demonstrating that the spontaneous insertion of the peptide in a bilayer can trigger a bilayer $\rightarrow$ H_{II} transition (29).

DOPC was chosen to investigate in more detail the molecular aspects of the gramicidin-induced H_{II} phase formation (30). Since H_{II} phase-preferring lipids typically have a low headgroup hydration (31), it can be expected that hydration might play an important role in the gramicidin-induced H_{II} phase formation. Investigations by ^{31}P NMR and small-angle X-ray diffraction of mixed DOPC-gramicidin films as a function of the ^2H_2O content revealed a number of interesting observations (30). Figure 13 illustrates this with the ^{31}P NMR spectra of DOPC-gramicidin films in a molar ratio of 10:1, containing increasing amounts of ^2H_2O. Whereas in the absence of the peptide, DOPC at all stages of hydration adopts the typical ^{31}P NMR spectrum indicative of a bilayer organization, the presence of gramicidin has a number of different effects. For values of N (which is the number of ^2H_2O molecules per PC) lower than 5,

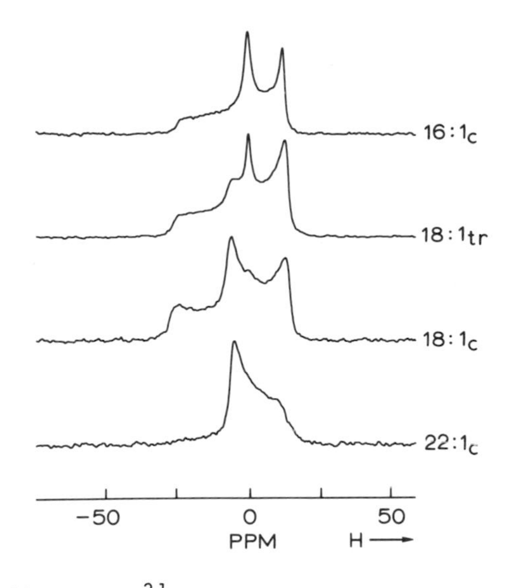

FIGURE 12. 81.0 MHz ^{31}P NMR spectra of aqueous dispersions of mixtures of gramicidin with di-unsaturated PC species in an 1:10 molar ratio. Reproduced with permission from ref. (28).

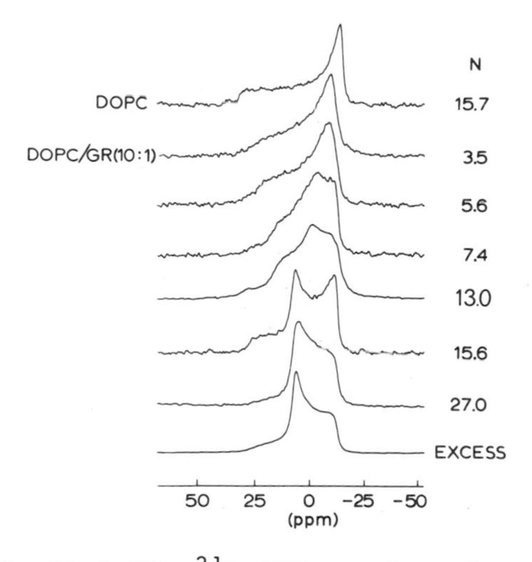

FIGURE 13. 81.0 MHz ^{31}P NMR spectra of aqueous dispersions of DOPC and DOPC/gramicidin (10:1, m/m) samples containing different amounts of $^{2}H_{2}O$. Reproduced with permission from ref. (30).

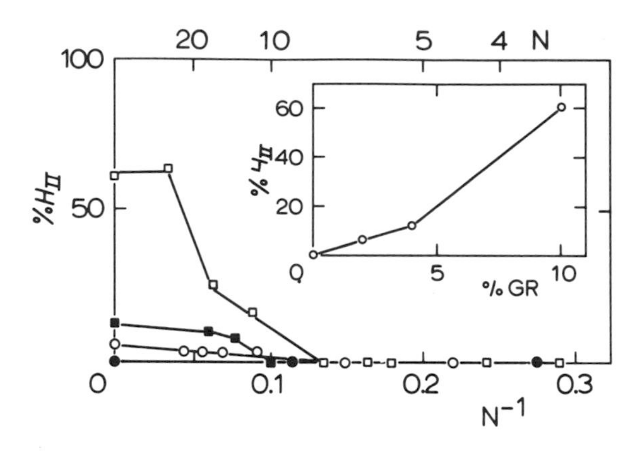

FIGURE 14. H_{II} phase formation in DOPC/gramicidin systems of varying ^2H_2O content. The insert shows the relation between the amount of H_{II} phase versus gramicidin concentration (mol%) in excess ^2H_2O. Molar ratios gramicidin/DOPC 0 (●—●), 1:50 (o—o), 1:25 (■—■) and 1:10 (□—□). Reproduced with permission from ref. (30).

typical bilayer ^{31}P NMR spectra are observed. For intermediate ^2H_2O contents (N = 5-15) a second bilayer spectral component appears, which has a marked reduced chemical shift anisotropy. The hexagonal H_{II} phase is induced for N > 15. *Thus, the presence of water is essential for gramicidin-induced hexagonal H_{II} phase formation in PC systems.* This is opposite to what one would expect from the phase behavior of pure lipid systems and suggests that a specific hydrated conformation of gramicidin is essential for H_{II} phase formation.

Very similar behavior was also observed for other gramicidin concentrations and the data are summarized in Fig. 14. The fraction of DOPC in the hexagonal phase decreases with decreasing gramicidin and ^2H_2O content. The molecular efficiency of the H_{II} phase formation by gramicidin is rather low: 2.5 mol DOPC per gramicidin monomer in excess water for DOPC-gramicidin ratios up to 25:1 (insert in Fig. 14).

The new lamellar phase observed at intermediate ^2H_2O content was also detected by X-ray diffraction and was characterized by a smaller repeat distance. In this intermediate phase headgroup order appears to be greatly decreased, as indicated by the large reduction in chemical shift anisotropy (30). Similar behavior was noted when the samples were investigated by ^2H NMR using $N|-C^2H_3|_3$-labelled DOPC (Fig. 15). Now the samples were hydrated with deuterium-depleted H_2O. Increasing the water content results in a marked reduction in the quadrupolar split-

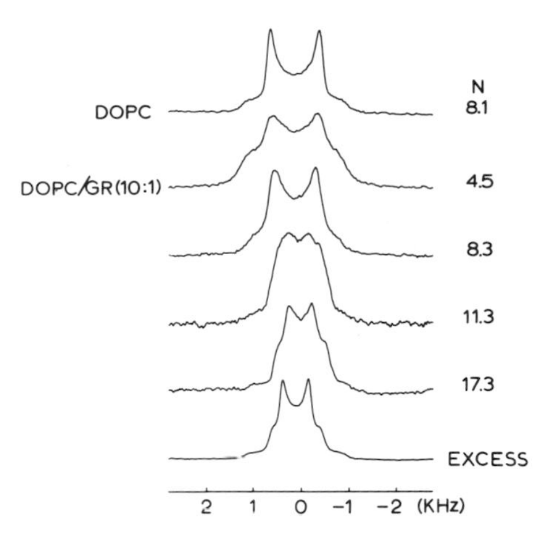

FIGURE 15. 30.7 MHz ^2H NMR spectra of $|N\text{-}C^2H_3|$-DOPC and $|N\text{-}C^2H_3|$-DOPC/gramicidin (10:1, m/m) samples at different H_2O contents. Reproduced with permission from ref. (30).

ting especially at intermediate water concentrations (for N = 8-15). Hexagonal H_{II} phase formation occurring at higher water content manifests itself in these NMR spectra as the appearance of a second set of peaks with a two-fold reduced quadrupolar splitting.

Information on the behavior of water in these systems was obtained by ^2H NMR using samples which have been hydrated with ^2H_2O. A set of representative spectra is shown in Fig. 16. For pure DOPC from the lowest amount of ^2H_2O tested up to N = 21, a single quadrupolar splitting is observed which value decreases in a biphasic way with increasing ^2H_2O content (Figs. 10 and 16 of ref. 30). Above N = 21 an isotropic component becomes visible which intensity increases with further increase of N. This is due to phase separation between the maximally swollen DOPC/^2H_2O phase and excess ^2H_2O. These latter molecules cannot exchange any more with the ^2H_2O molecules present in the other pools, at least on the time scale of the ^2H NMR measurement.

Gramicidin incorporation has a dramatic influence on the property of the ^2H_2O molecules as revealed by ^2H NMR. Two effects are most noticeable. At the lowest stage of hydration and for N = 3-5, the spectra are broadened, but a single quadrupolar splitting is observed with a value which decreases linearly with increasing gramicidin concentration up to 4 mol% of the peptide. Above N = 5, multi-component spectra are observed demonstrating phase separation (30). One component has the quad-

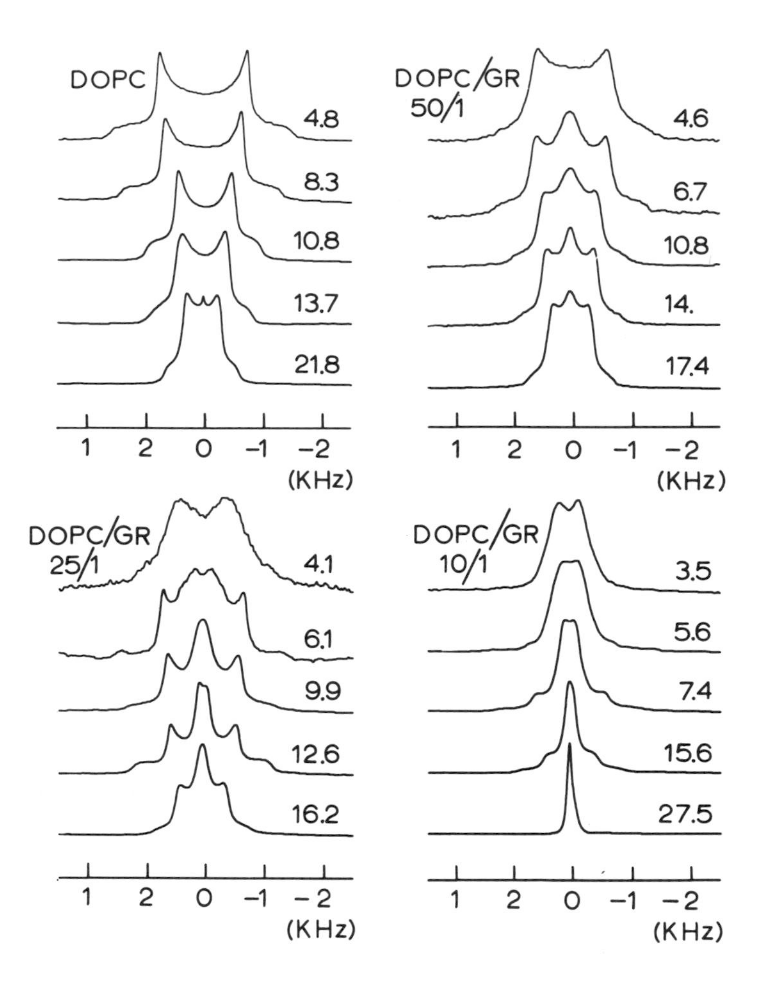

FIGURE 16. 30.7 MHz ^2H NMR spectra of DOPC/gramicidin/^2H_2O samples. Reproduced with permission from ref. (30).

rupolar splitting very similar to that of the gramicidin-free bilayer. The other component has a broad isotropic line shape or has a very small value of the quadrupolar splitting. The intensity of these components increases with increasing gramicidin concentration at any given stage of hydration. Analysis of these data revealed that for N = 5, gramicidin becomes preferentially hydrated over DOPC and takes up approximately 140 ^2H_2O molecules per gramicidin monomer. Thus, the gramicidin molecule dehydrates the PC headgroup.

These results are summarized in the hydration model shown in Fig. 17. This scheme depicts the situation for a DOPC/grami-

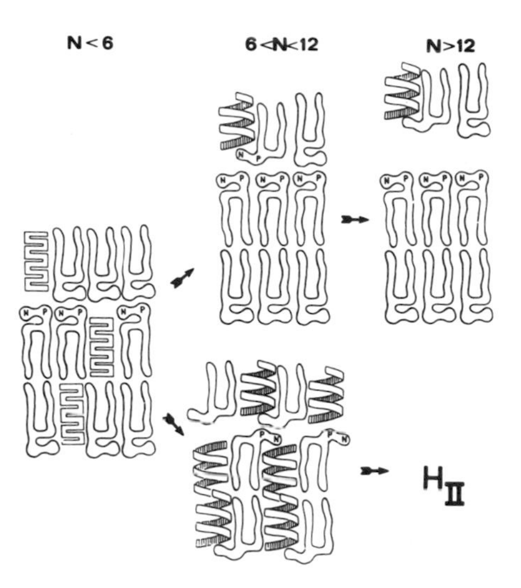

FIGURE 17. Schematic representation of the effect of hydration on the phase properties of DOPC/gramicidin mixtures. Reproduced with permission from ref. (30). See text and this reference for details.

cidin ratio of 25:1. For N < 6, one lamellar phase exists with similar structural properties as the pure DOPC system. The gramicidin molecules, represented by the zig-zags, are thought to be randomly oriented in the hydrophobic part of the bilayer in a condensed and anhydrous conformation. Increasing water concentration will cause preferential hydration of the gramicidin molecule. Above N = 6, phase separation occurs resulting in a lamellar phase which is highly enriched in the hydrated gramicidin molecule, shown here in the $\beta^{6.3}$ head-to-head dimer channel conformation. This bilayer is relatively thin and the headgroups of the lipid molecules have obtained increased motional freedom. The other lamellar phase behaves like the pure DOPC lamellar system but contains some hydrated gramicidin. Above N = 12, the gramicidin-rich lamellar phase converts to a gramicidin-rich hexagonal H_{II} phase possibly as the result of an interbilayer fusion event, involving linear arrays of gramicidin molecules. The gramicidin molecules in the H_{II} phase most likely are oriented with their axis perpendicular to the aqueous tubes present in the lipid cylinders as there is rapid exchange between the gramicidin-associated water and the water within these tubes.

Gramicidin-PC interactions

1. Gramicidin in a hydrated conformation induces an H_{II} phase for acyl chain lengths in excess of 16 carbon atoms.
2. In the H_{II} phase gramicidin is highly self-associated. The H_{II} phase is rich in both gramicidin and water. Gramicidin:PC:^2H_2O = 1:1:0.9 (w/w/w).

V. GRAMICIDIN-NEGATIVELY CHARGED LIPID INTERACTIONS

The induction of the H_{II} phase by gramicidin in bilayer-forming lipid systems is not restricted to PC (32). Under conditions that the negative charge on the anionic phospholipids cardiolipin, phosphatidylserine (PS) and phosphatidylglycerol (PG) is expressed (100 mM NaCl, pH 7.4, for instance), these lipids are typically organized in bilayers (33), which often are widely separated due to strong electrostatic interbilayer repulsive forces (34). Incorporation of gramicidin in these lipid systems also results in H_{II} phase formation as is evidenced for DOPG in Fig. 18. The freeze-fracture micrograph shows the characteristic striated fracture patterns observed for lipids organized in the H_{II} phase. Quantification of the extent of H_{II} phase formation by ^{31}P NMR revealed that the presence of a negative charge on the headgroup virtually has no effect on H_{II} phase formation (Table I).

FIGURE 18. Freeze-fracture micrograph of an aqueous dispersion (100 mM NaCl, 10 mM Tris, pH 7.4) of gramicidin and DOPG in a 1:10 molar ratio. Final magnification 100,000 x. See ref. (32) for details.

TABLE I. H_{II} phase formation in aqueous dispersions (100 mM NaCl, 10 mM Tris, pH 7.4) of different lipids in the absence and presence of 10 mol% gramicidin*.

Lipid	% of lipid in H_{II} phase	
	- gramicidin	+ gramicidin
DOPC	0	44
DOPS	0	48
DOPG	0	46

*For details see ref. (32).

Since a location of negatively charged headgroups around the narrow aqueous cylinders present in the H_{II} phase must be energetically unfavorable, due to strong intermolecular repulsive forces, these observations suggest that the H_{II} phase-promoting ability of gramicidin must be very strong. Recently, gramicidin-induced H_{II} phase formation was also observed in total lipid extracts of rat liver microsomes and mitochondria (G. van Duijn, A. Rietveld and B. de Kruijff, unpublished observations), demonstrating that this effect is not restricted to single component synthetic lipid systems.

Gramicidin-negatively charged lipid interactions

Gramicidin also promotes an H_{II} phase in dispersions of negatively charged lipids.

VI. RELATIONSHIP BETWEEN CHEMICAL STRUCTURE OF GRAMICIDIN AND ITS LIPID STRUCTURE-MODULATING ACTIVITY

The lipid structure-modulating activity of gramicidin, most likely is the result of the specific chemical and spatial structure of the molecule. This can be inferred from the results described in Chapter IV, which indicate that a hydrated conformation of gramicidin is essential, but also is suggested from the observation that other hydrophobic polypeptides such as the 23 amino acid long membrane-spanning segment of the red cell membrane glycoprotein glycophorin, in contrast to gramicidin,

GR	HCO-L-Val[1]...(-L - Trp-D-Leu)$_3$-L - Trp[15]-$NHCH_2CH_2OH$
DESF-GR (+)	L-Val[1]..................-L - Trp[15]-$NHCH_2CH_2OH$
N-Suc-GR (-)	$HOOCCH_2CH_2CO$-L-Val[1]..................-L - Trp[15]-$NHCH_2CH_2OH$
O-Suc-GR (-)	HCO-L-Val[1]..................-L - Trp[15]-$NHCH_2CH_2OOCCH_2CH_2COOH$
FTRP-GR	HCO-L-Val[1]...(-L-FTrp-D-Leu)$_3$-L-FTrp[15]-$NHCH_2CH_2OH$

FIGURE 19. Chemical structures of gramicidin A and some derivatives.

acts as a bilayer stabilizer in H_{II} phase-preferring lipid systems (35,36). In an attempt to unravel the structural requirements in the gramicidin molecule for its potency to induce an H_{II} phase in DOPC systems, the derivatives of gramicidin shown in Fig. 19 were synthesized (32). By removing the formyl group (yielding desformyl-gramicidin, DESF-GR) and by subsequent coupling of a succinate group (yielding N-succinyl-gramicidin, N-Suc-GR), N-terminal modified gramicidin analogs were prepared which, when incorporated in model membrane systems, can be expected to behave like positively and negatively charged molecules, respectively. The coupling of succinate to the ethanolamine group at the C-terminal end of the molecule (yielding O-succinyl-gramicidin: O-Suc-GR), also will result in a negatively charged molecule but now with an opposite direction of polarity.

Incorporation of the N- and C-terminal modified gramicidin analogs in DOPC systems in a 1:10 molar ratio in all cases results in H_{II} phase formation to an extent which is of the same

TABLE II. Effect of gramicidin and C- and N-terminal modified analogs on lipid structure in aqueous dispersions.

	% H_{II} in DOPC*	Structure** with lyso-PC	Decrease (°C) in bilayer → H_{II} transition temperature in DEPE***
No peptide	0	micelle	0
Gramicidin	44	bilayer	10
DESF-GR	33	bilayer	13
N-Suc-GR	40	bilayer	13
O-Suc-GR	31	bilayer	17

*DOPC/peptide: 10/1 (molar), **lyso-PC/peptide: 4/1 (molar), ***DEPE/peptide: 50/1 (molar).

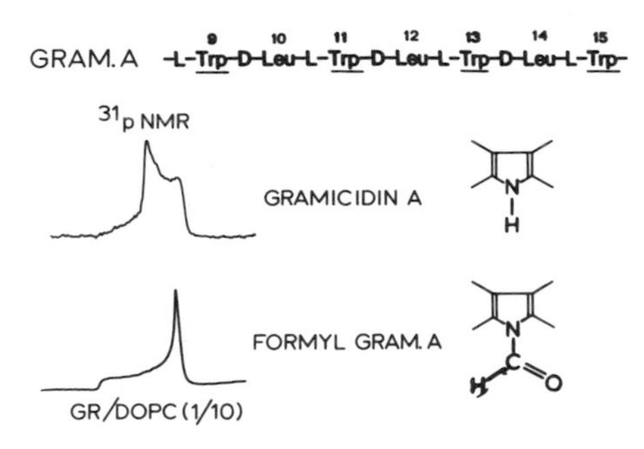

FIGURE 20. Effect of formylation of the tryptophans of gramicidin on the structure of peptide/DOPC (1:10) model membranes as detected by 81.0 MHz ^{31}P NMR spectroscopy. See ref. (37) for details.

magnitude as that observed for the unmodified molecule (Table II). Moreover, also the two other characteristic lipid structure-modulating activities of the peptide, e.g. bilayer formation in mixtures with lyso-PC and a downward shift in bilayer to hexagonal transition temperature in DEPE were observed for these analogs (32). These data clearly demonstrate that the presence of ionizable groups at the C- and N-terminal part of the gramicidin molecule, which can be expected to be charged under the experimental conditions used, do not interfere with the specific interactions of the peptide with the lipid.

Most interestingly, the tryptophan residues in the gramicidin molecule appear to be very important for the gramicidin-lipid interaction (37). Replacement of the indol proton of all 4 tryptophans by a formyl group (Fig. 19, FTRP-GR), results in a complete loss of H_{II} phase formation in DOPC model membranes (Fig. 20). Deformylation of the tryptophans results in the reappearance of the H_{II} phase-inducing ability of gramicidin (37).

Structural requirements of gramicidin for lipid structure modulation

1. The presence of ionizable groups at either the C- or N-terminal part of gramicidin does not result in significant changes in lipid structure-modulating activity of the peptide.
2. Formylation of Trp^9, Trp^{11}, Trp^{13} and Trp^{15} results in complete and reversible loss of H_{II} phase formation in DOPC systems.

VII. SUMMARIZING DISCUSSION, CONCLUDING REMARKS AND FUTURE PROSPECTS

Besides the long-known, well-studied channel-forming properties of gramicidin (review 9) the more recent observations on gramicidin-RNA polymerase interaction (5,6) and the lipid structure-modulating activity of the peptide reviewed in this paper, make gramicidin an even more intriguing multi-functional compound. Before we will try to relate these various properties of gramicidin, we will first discuss the molecular basis of the lipid structure-modulating effect of the peptide. We consider this effect to be a very specific property of the gramicidin molecule. This is because of: (1) the sharply defined stoichiometry in the interaction with lyso-PC, (2) the very strong H_{II}-promoting ability in bilayer structure-preferring lipids, the extent of which is almost independent on temperature (28) and the nature of the headgroup, and (3) the requirement for a specific hydrated conformation to induce this H_{II} phase in DOPC model systems.

Since the mode of organization of lipids in aqueous dispersions is determined by the balance of several attractive and repulsive forces, the strength of which is dependent on many different conditions, it can be expected that also in case of the gramicidin-induced changes in lipid polymorphism, various factors will be involved. We will subsequently consider: (a) the shape of the molecule, (b) its chain and headgroup disordering ability, (c) its dehydrating capacity, and (d) its tendency to self-associate into different structures. It should be realized at forehand that the relative importance of any particular factor could be different for the various lipid structure-modulating activities of the peptide.

In the $\beta^{6.3}$ helical conformation space-filling models reveal a pronounced cone shape of the molecule, which is due to the location of the 4 bulky tryptophan residues at the C-terminal part of the molecule. The shape-structure concept of lipid polymorphism as applied to gramicidin in this conformation can explain most of the lipid structure-modulating effects of gramicidin, but two aspects deserve special attention. First, it is in contradiction with the channel conformation (N-N terminal dimer), but see the arguments at the end of Chapter II. Second, and more severely, is the notion that the N- and O-succinyl monomers, which most likely are charged and therefore will have an opposite orientation at the membrane-water interface, affect lipid structure in a similar way as the native gramicidin molecule. However, it should be realized that these discrepancies might be only apparent, because the three-dimensional structure of gramicidin and these analogs in the different lipid systems is virtually unknown.

Gramicidin causes a disordering of the acyl chains when incorporated in high amounts in bilayers of saturated PC's (10-13). Furthermore, the peptide causes a decrease in headgroup order, especially in the lamellar phase of DOPC at low water content (30), and in the H_{II} phase of PE systems (26), which is possibly caused by a location of the headgroup in part over the gramicidin molecule. Both observations are consistent with the suggestion that the shape of the complex of gramicidin and its surrounding lipids will be more cone-like than the sum of their individual shapes, thus favoring H_{II} phase formation. The mismatch in length between the gramicidin dimer (30 Å in the lipid-associated form; 38) and the thickness of the hydrophobic part of the bilayer {30 Å; DPPC (39), thus longer for those PC species in which the H_{II} phase is induced} could result in dimple formation and thus contribute to H_{II} phase formation. However, it should be realized that due to geometrical reasons longer chained lipids are better packed into an H_{II} phase than shorter ones. It is clear that these factors will be less important for gramicidin-induced bilayer formation in lyso-PC systems.

The lipid headgroup dehydrating capacity of gramicidin observed at low water contents also most likely will contribute to H_{II} phase formation. However, it is not clear whether such dehydration effect also occurs in the presence of excess water. Consistent with this dehydration idea is the visual observation that dispersion of mixed gramicidin-lipid films (including those of negatively charged lipids) is more difficult and swelling less extensively as in case of peptide-free systems.

In several lipid systems it has now been observed that gramicidin has a pronounced tendency to self-associate. In case of mixtures with lyso-PC, a two-dimensional array-wise organization of gramicidin hexamers was postulated (23). Such an organization could well be involved, also in H_{II} phase formation. Close approximation of bilayers at the site of parallel organized long arrays of such structures could result in line fusion and subsequent H_{II} phase formation. Alternatively, it could be that gramicidin molecules in lateral self-association have themselves a strong tendency to be organized in tubular structures such as those found in the H_{II} phase. The notion that the tube diameter of the H_{II} phase induced by gramicidin is virtually independent on acyl chain length and nature of the headgroup supports this suggestion. The idea that peptides can form the backbone of a tubular structure such as found in the H_{II} phase is biologically intriguing in that such stable tubes are found for instance in case of the tight junction (40,41). It will be clear that information on the structure of gramicidin in the various lipid systems is essential to answer some of the questions raised above.

The 4 tryptophan residues of gramicidin play an important role in the lipid structure-modulating effect of the peptide.

In case of the lyso-PC-gramicidin lamellar system, it was suggested that by stacking interactions the tryptophan indol rings are involved in the formation of ordered gramicidin aggregates (23). We would like to propose that also for H_{II} phase formation intermolecular indol ring stacking interactions are essential.

If we now return to the two other well-studied properties of gramicidin: channel formation and inhibition of RNA polymerase, then it is intriguing to note that also for these effects the tryptophans have been shown to be absolutely essential (42, 43). We hypothesize that tryptophan-tryptophan interactions occur which may determine the conformation of the peptide and are functionally essential in mediating gramicidin-gramicidin or gramicidin-protein interactions. Recently, the importance of aromatic-aromatic interactions for the structure of proteins was emphasized (44). To what extent other structural characteristics of the molecule, such as the alternation of D- and L-amino acids in the sequence which gives the molecule the ability to organize in β-type helices is important for the various effects of the peptide, remains to be seen.

For the membrane scientist the challenging prospect is now emerging that due to the relative simplicity of the molecule, the modern methods of chemistry and purification of hydrophobic peptides and the availability of powerful biophysical and computational methods to monitor both lipid and peptide structure and dynamics, it will become possible to obtain for the first time a detailed molecular picture of the lipid structure-modulating activity of gramicidin.

ACKNOWLEDGMENTS

We would like to thank A.J. Verkleij, J. Leunissen-Bijvelt, C.J.A. van Echteld, R. van Stigt, J. de Gier, F. Borle, J. Seelig, R. Brasseur, V. Cabiaux, J.M. Ruysschaert, C.W. van den Berg, H. Tournois, S. Keur, A.J. Slotboom, G.J.M. van Scharrenburg and W.J. Timmermans for their expert contributions to parts of the research described in this paper.

REFERENCES

1. Sarges, R., and Witkop, B., Biochemistry 4:2491 (1965).
2. Lipmann, F., Science 173:875 (1971).
3. Pschorn, W., Paulus, H., Hansen, J., and Ristow, H., Eur. J. Biochem. 129:403 (1982).

4. Ristow, H., Biochim. Biophys. Acta 477:177 (1977).
5. Sarkar, N., Langley, D., and Paulus, H., Proc. Natl. Acad. Sci. USA 74:1478 (1977).
6. Sarkar, N., Langley, D., and Paulus, H., Biochemistry 18: 4536 (1979).
7. Urry, D. W., Goodall, M. C., Glickson, J. D., and Mayers, D. F., Proc. Natl. Acad. Sci. USA 68:1907 (1971).
8. Hladky, S. B., and Haydon, D. A., Nature 255:451 (1970).
9. Anderson, O. S., Ann. Rev. Physiol. 46: 531 (1984).
10. Rice, D., and Oldfield, E., Biochemistry 18:3272 (1979).
11. Rajan, S., Kang, S.-Y., Gutowski, H. S., and Olafield, E., J. Biol. Chem. 256:1160 (1981).
12. Chapman, D., Cornell, B. A., Eliasz, A. W., and Perry, A., J. Mol. Biol. 113:517 (1977).
13. Lee, D. C., Durrani, A. A., and Chapman, D., Biochim. Biophys. Acta 769:49 (1984).
14. van Echteld, C. J. A., van Stigt, R., de Kruijff, B., Leunissen-Bijvelt, J., Verkleij, A. J., and de Gier, J., Biochim. Biophys. Acta 648:287 (1981).
15. de Kruijff, B., Cullis, P. R., Verkleij, A. J., Hope, M. J., van Echteld, C. J. A., and Taraschi, T. F., in "The enzymes of Biological Membranes" (A. N. Martonosi, ed.), 2nd edition, Vol. I, p. 131. Plenum Press, New York, 1985.
16. de Kruijff, B., Cullis, P. R., Verkleij, A. J., Hope, M. J., van Echteld, C. J. A., Taraschi, T. F., van Hoogevest, P., Killian, J. A., Rietveld, A., and van der Steen, A. T. M., in "Progress in Protein-Lipid Interactions" (A. Watts and J. J. H. H. M. de Pont, ed.), Vol. 1, p. 89. Elsevier Sic. Publ., Amsterdam, 1985.
17. Cullis, P. R., and de Kruijff, B., Biochim. Biophys. Acta 599:399 (1979).
18. Seelig, J., Biochim. Biophys. Acta 515: 105 (1978).
19. Killian, J. A., de Kruijff, B., van Echteld, C. J. A., Verkleij, A. J., Leunissen-Bijvelt, J., and de Gier, J., Biochim. Biophys. Acta 728:141 (1983).
20. Seelig, J., and Seelig, A., Q. Rev. Biophys. 13:19 (1980).
21. Killian, J. A., Borle, F., de Kruijff, B., and Seelig, J., Biochim. Biophys. Acta in press (1986).
22. Demel, R. A., and de Kruijff, B., Biochim. Biophys. Acta 457:109 (1976).
23. Spisni, A., Pasquali-Ronchetti, I., Casali, E., Lindner, L., Cavatora, P., Masotti, L., and Urry, D. W., Biochim. Biophys. Acta 732:58 (1983).
24. Madden, T. D., and Cullis, P. R., Biochim. Biophys. Acta 684:149 (1982).
25. Brasseur, R., Chabiaux, V., Killian, J. A., de Kruijff, B., and Ruysschaert, J. M., Biochim. Biophys. Acta, Submitted (1986).
26. Killian, J. A., and de Kruijff, B., Biochemistry, in press

(1986).
27. Chapman, D., Cornell, B. A., Eliasz, A. W., and Perry, A., J. Mol. Biol. 113:517 (1977).
28. van Echteld, C. J. A., de Kruijff, B., Verkleij, A. J., Leunissen-Bijvelt, J., and de Gier, J., Biochim. Biophys. Acta 692:126 (1982).
29. Killian, J. A., Leunissen-Bijvelt, J., Verkleij, A. J., and de Kruijff, B., Biochim. Biophys. Acta 812: 21 (1985).
30. Killian, J. A., and de Kruijff, B., Biochemistry, in press (1986).
31. Luzzati, V., in "Biological Membranes" (D. Chapman, ed.), p. 71. Academic Press, New York, 1968.
32. Killian, J. A., van den Berg, C. W., Tournois, H., Keur, S., Slotboom, A. J., van Scharrenburg, G. J. M., and de Kruijff, B., Biochim. Biophys. Acta, submitted (1986).
33. Cullis, P. R., de Kruijff, B., Hope, M. J., Verkleij, A. J., Nayar, R., Farren, S. B., Tilcock, C., Madden, T. D., and Bally, M. B., in "Membrane Fluidity" (R. C. Aloia, ed.), Vol. 2, p. 40. Academic Press, New York, 1982.
34. Hauser, H., Biochim. Biophys. Acta 772:37 (1984).
35. Taraschi, T. F., de Kruijff, B., Verkleij, A. J., and van Echteld, C. J. A., Biochim. Biophys. Acta 685:153 (1982).
36. Taraschi, T. F., van der Steen, A. T. M., de Kruijff, B., Tellier, C., and Verkleij, A. J., Biochemistry 21:5756 (1982).
37. Killian, J. A., Timmermans, W. J., Keur, S., and de Kruijff, B., Biochim. Biophs. Acta 820:154 (1985).
38. Wallace, B. A., Veatch, W. R., and Blout, E. R., Biochemistry 20:5754 (1981).
39. Büldt, G., Gally, H. U., Seelig, A., Seelig, J., and Zaccai, G., Nature 271:182 (1978).
40. Kachar, B., and Reese, T. S., Nature 296:464 (1982).
41. Pinto da Silva, P., and Kachar, B., Cell 28:441 (1982).
42. Busath, D. D., and Waldbillig, R. C., Biochim. Biophys. Acta 736:28 (1983).
43. Paulus, H., Sarkar, N., Mukherjee, P. K., Langley, D., Ivanov, V. T., Shepel, E. N., and Veatch, W., Biochemistry 18:4532 (1979).
44. Burley, S. K., and Petsko, G. A., Science 229:23 (1985).

Index

H

I

V

W

X